ICONS OF INVENTION

ICONS OF INVENTION

The Makers of the Modern World from Gutenberg to Gates

VOLUME 2

John W. Klooster

Greenwood Icons

Greenwood Press
An Imprint of ABC-CLIO, LLC

Santa Barbara, California • Denver, Colorado • Oxford, England

Library of Congress Cataloging-in-Publication Data

Klooster, John W.
Icons of invention : the makers of the modern world from Gutenberg to Gates / John W. Klooster.
p. cm. — (Greenwood icons)
Includes bibliographical references and index.
ISBN 978-0-313-34743-6 (set : hard copy : alk. paper) — ISBN 978-0-313-34745-0 (v. 1 : hard copy : alk. paper) — ISBN 978-0-313-34747-4 (v. 2 : hard copy : alk. paper) — ISBN 978-0-313-34744-3 (set : ebook) — ISBN 978-0-313-34746-7 (v. 1 : ebook) — ISBN 978-0-313-34748-1 (v. 2 : ebook)
1. Inventors—Biography. 2. Inventions. I. Title.
T39.K56 2009
609.2′2—dc22 2009009580
[B]

13 12 11 10 09 1 2 3 4 5

This book is also available on the World Wide Web as an eBook.
Visit www.abc-clio.com for details.

ABC-CLIO, LLC
130 Cremona Drive, P.O. Box 1911
Santa Barbara, California 93116-1911

This book is printed on acid-free paper ∞

Manufactured in the United States of America

This book is dedicated to my wife. She sacrificed uncounted hours that enabled me to research, develop, and write this book.

Contents

List of Photos ix

Series Foreword xi

Acknowledgments xiii

Introduction xv

Volume 1

Printing and Johannes Gutenberg 1

The Steam Engine and James Watt 25

Mass Transportation: Fitch, Fulton, Stevens, and Trevithick 53

Food Preservation and Nicolas François Appert 99

The Reaper and Cyrus McCormick 113

Telegraphy: Morse and Marconi 147

The Sewing Machine: Hunt, Howe, and Singer 171

The Telephone and Alexander Graham Bell 187

Internal Combustion Engine: Otto and Diesel 219

Electric Power: Edison, Westinghouse, Tesla, and Stanley 249

Vaccination and Louis Pasteur 315

Volume 2

Affordable Automobiles and Henry Ford 331

The Airplane and the Wright Brothers 371

Radio: Fessenden, de Forest, and Armstrong 391

Penicillin and Alexander Fleming 421

Television and Philo Farnsworth 433

Jet Engines: Hans von Ohain and Frank Whittle 447

Nylon and Wallace Carothers 461

Nuclear Reactor and Enrico Fermi 483

Transistors: John Bardeen, William Shockley, and Walter Brattain 515

Integrated Circuits, Microprocessors, and Computers: Jack Kilby and Steve Wozniak 527

Satellite Communications and John R. Pierce 557

Software and Bill Gates 569

The Internet and the World Wide Web: Kleinrock, Baran, Kahn, Cerf, and Berners-Lee 585

Epilogue 613

Bibliography 619

Index 627

List of Photos

A page of the Gutenberg Bible opened to the beginning of the Gospel of Luke (page 1). Courtesy of Library of Congress Rare Book and Special Collections.

James Watt's steam engine (page 25). Courtesy of Library of Congress Prints and Photographs Division, LC-USZ62-110376.

Richard Trevithick's first passenger-carrying common road locomotive (page 53). Courtesy of Library of Congress, LC-USZ62-110377.

R&R Ham Can (circa 1880) shows how early cans were hand-soldered top and bottom (page 99). Courtesy of Library of Congress.

Cyrus McCormick (page 113). Courtesy of Library of Congress Prints and Photographs Division, LC-USZ62-27710.

Samuel F. B. Morse (page 147). Courtesy of Library of Congress Prints and Photographs Division, LC-DIG-cwpbh-00852.

Singer sewing machine advertisement card, distributed at World Columbian Exposition, Chicago, 1893, showing two people from Ceylon with two Singer sewing machines (page 171). Courtesy of Library of Congress Prints and Photographs Division, LC-USZC4-2764.

The first telephone (page 187). AP Photo.

A diagram of the four-stroke gasoline-powered internal-combustion engine designed by Nikolaus Otto (page 219). Courtesy of Library of Congress, LC-USZ62-110412.

Thomas Edison in his laboratory (page 249). J. Walter Thompson/AP Photo.

Louis Pasteur (page 315). Courtesy of Library of Congress Prints and Photographs Division, LC-USZ62-3497.

Henry Ford at the tiller of the first vehicle, the Quadricycle, in the United States, November 1927 (page 331). AP Photo.

First flight of the Wright Brothers: 10:35 A.M., December 17, 1903, Kitty Hawk, North Carolina (page 371). Courtesy of Library of Congress Prints and Photographs Division, LC-W86-35.

Pittsburgh radio station KDKA broadcasts the first radio news report when it broadcast Associated Press election returns in 1920 (page 391). AP Photo.

Dr. Selman Abraham Waksman, left, discoverer of streptomycin and neomycin, and Sir Alexander Fleming, right, discoverer of Penicillin, in 1949 (page 421). Rutgers University/AP Photo.

Philo Farnsworth (page 433). AP Photo.

This 1939 image shows the world's first jet plane, the Heinkel He 178 (page 447). HEINKEL/AP Photo.

Wallace Carothers demonstrating a piece of his synthetic rubber in 1931 (page 461). AP Photo.

Enrico Fermi in a November 25, 1946 photograph at the University of Chicago (page 483). AP Photo.

John Bardeen, left, William Shockley, center, and Walter Brattain, right, working at Bell Laboratories in Murray Hill, New Jersey, in 1956 (page 515). LSI Corp., National Canal Museum/AP Photo.

Jack S. Kilby (page 527). AP Photo.

The Telstar communications satellite, designed by Bell Telephone Laboratories for relaying telephone calls, data messages, and television signals, in 1962 (page 557). AP Photo.

Bill Gates (page 569). AP Photo.

Leonard Kleinrock stands next to the computer that made the first-ever connection in 1969 to what was to become the Internet (page 585). AP Photo.

Series Foreword

Worshipped and cursed. Loved and loathed. Obsessed about the world over. What does it take to become an icon? Regardless of subject, culture, or era, the requisite qualifications are the same: (1) challenge the status quo, (2) influence millions, and (3) impact history.

Using these criteria, Greenwood Press introduces a new reference format and approach to popular culture. Spanning a wide range of subjects, volumes in the Greenwood Icons series provide students and general readers a port of entry into the most fascinating and influential topics of the day. Every two-volume title offers an in-depth look at approximately 24 iconic figures, each of which captures the essence of a broad subject. These icons typically embody a group of values, elicit strong reactions, reflect the essence of a particular time and place, and link different traditions and periods. Among those featured are artists and activists, superheroes and spies, inventors and athletes—the legends and mythmakers of entire generations. Yet icons can also come from unexpected places: as the heroine who transcends the pages of a novel or as the revolutionary idea that shatters our previously held beliefs. Whether people, places, or things, such icons serve as a bridge between the past and the present, the canonical and the contemporary. By focusing on icons central to popular culture, this series encourages students to appreciate cultural diversity and critically analyze issues of enduring significance.

Most importantly, these books are as entertaining as they are provocative. Is Disneyland a more influential icon of the American West than Las Vegas? How do ghosts and ghouls reflect our collective psyche? Is Barry Bonds an inspiring or deplorable icon of baseball? Designed to foster debate, the series serves as a unique resource that is ideal for paper writing or report purposes. Insightful, in-depth entries provide far more information than conventional reference articles but are less intimidating and more accessible than a book-length biography. The most revered and reviled icons of

American and world history are brought to life with related sidebars, timelines, fact boxes, and quotations. Authoritative entries are accompanied by bibliographies, making these titles an ideal starting point for further research. Spanning a wide range of popular topics, including business, literature, civil rights, politics, music, and more, books in the series provide fresh insights for the student and popular reader into the power and influence of icons, a topic of as vital interest today as in any previous era.

Acknowledgments

The help and services generously provided by my friend and computer consultant and expert Paul A. Loeffel are acknowledged with great appreciation.

The attention, care, and comments provided by my Greenwood editor, Kevin Downing, are acknowledged with deep appreciation.

Introduction

Without a doubt, civilization today would be very much different but for the many iconic inventions that not only grace our lives but indeed make life as we know it possible. What were the inventors who gave us these inventions like?

Millions of inventions have been achieved just in the modern era of our civilization, but only a few handfuls of these are iconic, that is, so significant that they are widely utilized by and known to the public and have even entered the pages of history. They have truly contributed to the advance of civilization.

These iconic inventions are relatively easy to identify, and sometimes, seemingly, are taken for granted. But in fact, before their embodiments could be made, sold and used, each was invented by an inventive entity. What were these inventors? Who were they? How did they ever manage to create and make embodiments of their inventions that became important? Why did they proceed?

Perhaps answers to such questions and others about some of these inventors would be useful information now. Perhaps such answers would encourage individuals and would promote environments conducive to the creation of new and further iconic inventions.

The present work endeavors to consider and review a group of iconic inventions and their inventors of iconic inventions that occurred over approximately the formation and development time of our modern era, this time commonly being taken as about 600 years and extending from about the early 1400s to the early 2000s. Thus, here, twenty-four inventions of iconic status and proportions are chosen and considered, and the inventors responsible for them are briefly considered, with particular attention being given to the respective time periods and inventor environments existing when each iconic invention was being formulated, created, and achieved. Each inventor's relevant background circumstances are reviewed. In the process, usually a substantial amount of relevant technology is described.

WHAT IS AN INVENTION?

For the present purposes, an *invention* is a novel technological creation originating from experimentation, mental fabrication, and insight that results in, or is materialized as, a new entity. The new entity can perhaps be in various forms or represented by one or more structures, steps, or methodology. It can be, for example, a mechanism; a component of a contrivance; a shaped or molded body; an electrical, electronic, or electromagnetic device; software and part of a program usable in a computer's operation; or the like. Characteristically, an invention involves an idea that is embodied or associated (or associable) with physical or material means whereby the idea is reduced to a functional reality as an actual physical embodiment.

For the present purposes, an invention is usually, as so reduced, a material, a structure, or a process (method). An invention, as conceived and reduced to practice, is perceivable, has a physical existence, and is more than a disembodied idea or an abstract sight, sound, odor, or impalpable entity. Thus, for present strict purposes, an invention, even if reduced to a written or electronic form, is not a mere or bare abstraction, or an unassociated idea, such as a mathematical theory, artistic idea, nonphysical representation, or composition. While an invention is a new physical thing, the line in law between patentable and copyrightable subject matter can be indefinite or indistinct.

Many inventions exist now, have existed, and will exist. A degree of novelty exists but the extent of departure from relevant prior art (i.e., what has gone before), can vary for an individual invention. Usually, inventions utilize or involve prior science and technology and are not themselves apart from pure science or natural law. In practice, most inventions are associated with definite goods or services. As science and technology have accumulated and advanced, the number, variety, and/or complexity of inventions can appear to have tended to increase.

ICONIC INVENTIONS

Some inventions are more significant than others. Some can be regarded as iconic. Many, perhaps countless, inventions have occurred, but only a few are iconic. An invention that is iconic, like all inventions, is distinct, has novelty, and characteristically incorporates a technological advance or change over the relevant prior art.

Sometimes an invention, after its creation, never becomes actually embodied, undergoes development, or becomes commercialized, and it has no embodiments that are manufactured, distributed, and sold. An iconic invention, though, characteristically has undergone all these things.

An *iconic invention* characteristically and importantly is an invention that has enjoyed wide utilization by people in civilization. It can be considered to be an important invention.

For simple, present purposes, an invention is or becomes iconic in status not because of the judgment or opinion of knowledgeable scientific, technological, economic, or legal experts but rather because the invention, as embodied, has been sold and used extensively by members of the public. Civilization can be considered to receive benefit from its wide utilization.

Many inventions ratable as iconic are known. Products incorporating an invention need not themselves be wholly new. The study of history indicates that an iconic invention need not be a wholly new innovation or development: it need not be a so-called pioneering invention, in a legal sense or otherwise, or need not be a so-called breakthrough invention in a practical sense or otherwise, relative to whatever previously was available. Historically, many an iconic invention seems to involve not only technological novelty (and the novelty may be modest relative to a prior art or scientific standpoint) but also extensive acceptance, selection, and utilization by members of the public. History indicates various inventions that were novel at their respective times of origin and that also achieved wide public utilization after their introduction. An invention can become iconic and important from a historical standpoint mainly because of its widespread usage. From history, it seems that to become an iconic invention, an invention need not only involve an innovation and novelty, but also be widely utilized by the public. One facet of this matter is that sometimes a widely utilized invention is a development, improvement, variation, or application of some prior or underlying invention, and care must be taken to sort out and distinguish the actual invention that is iconic.

In fact, perhaps even without actual awareness that invention is involved, an innovation that becomes iconic may do so more because of its utilization, or apparent capacity to be utilized (as from its sales) or to produce benefit, as perceived or recognized by members of the using public, than because of its novelty. Thus, briefly, an iconic invention is one that has gone into widespread usage and, knowingly or otherwise, has won a sort of popularity contest. It can be considered to have somehow satisfied a substantial need in civilization, sometimes even when the need was not previously known. Purchase and use by the public of an invention that becomes iconic must be extended or enduring. Mere fad, nonusage, or usage that is soon discontinued because the invention is found to be undesirable, troublesome, burdensome, problematic, without redeeming benefit, or replaceable by something deemed better does not make an invention iconic and potentially could repeal an achieved iconic status, perhaps when a better invention comes along.

An iconic invention that has achieved wide general usage may presumably have an effect upon the development of civilization and may reach into pages of the book of history. Extensive usage suggests an impact upon civilization, but that impact may not be simply measurable or quantifiable. Its effect may be mainly identified only qualitatively. However, for reasons of simplicity, the main criterion here used for determining the presence or existence of an iconic invention is just apparent extent of usage. An invention may be iconic as "born," but its importance and iconic status are not identified or determined or established until significant usage has taken place.

It seems that, commonly, members of the public become aware of the advent of an invention that becomes, or could become, iconic from one of two primary sources: (1) witnessing (including experiencing or using) the invention, perhaps sometimes not directly but as utilized by others; and/or (2) promotion (including advertising), direct or indirect, perhaps by a manufacturer, promoter, or other entity, of the invention or a product incorporating or utilizing the invention. For example, perhaps a promoter has a motivation, such as sales revenues, to sell invention embodiments. Perhaps a customer (in the public domain) has an incentive to acquire the invention or product incorporating the invention from realized or perceived benefit achieved from its use. Usage and acceptance are important. Perhaps the offering price is acceptable but customers fail to receive a benefit or a good result, directly or indirectly, from an invention or the use thereof; then promotion or sales would ultimately result in few or no total further sales or a discontinued use, and the invention would not become iconic.

Iconic inventions are characteristically and usually more beneficial than detrimental to many individuals who are customers or users. Inventions that have achieved iconic status and extensive utilization seem to have contributed to the development, formulation, and sustaining of our civilization. Without the iconic inventions, it seems fair to observe that little progress or advance in civilization beyond the basic and the primitive would probably have occurred.

For most individuals, it is usually sufficient and easy simply to use an embodiment of an invention of interest, whether or not the invention is iconic or merely appreciated, without particular consideration of things such as origin, history, extent of usage, inventorship, or the like. Sometimes, though, such a consideration can be important. For example, background information can be of value concerning possible usage, or how, when, and why an innovation, especially an iconic invention, occurred or might have occurred. Also, such information can suggest conditions, circumstances, environments, equivalents, and alternatives that could encourage usage or application in some contemporary situation.

For some, biographical information about the inventors of iconic inventions is interesting and even intriguing. Such information that includes some details about the origins, lives, and environmental circumstances of the inventors who achieved iconic inventions can suggest factors about how those inventions came to be. Inventor-related circumstances about iconic inventions can be useful in various non-use contexts.

Space and time limits mean that only relatively few iconic inventions could be considered here to a meaningful extent. Selections had to be made.

Criteria had to be considered. History proved to be an important consideration. In cases where the inventor made more than one seemingly qualified (on the basis of extent of usage) iconic invention, only the most significant one of his iconic inventions was selected for in-depth consideration here.

Although sometimes similarities in factors or circumstances can be discerned between selected individual iconic inventions, the reader will find that on the whole there are vast differences in inventor-related circumstances. Sometimes each selected iconic invention could be regarded by some readers more as a representative of a category than as a single individual iconic invention, but this is an unintended circumstance. Statistics based on a single circumstance may not be valid, reliable, or meaningful. Each iconic invention can seem to fit so well into the structure or history of civilization that it may seem to be hard for a reader to imagine alternatives, such as what life would be like now if that invention had not arrived. Actually, there is a school of thought that proposes that if a particular iconic invention had not been invented and developed by a particular inventor, then it would have been invented by another. It is argued that society needed most of the inventions and was in effect waiting for each one, whether or not the need or purpose was realized beforehand. Another school proposes that history suggests that perceived or possible desirability of some (unknown, unrealized, or unconceptualized) invention does *not* mean that such an invention will materialize in response.

For each selected iconic invention, information is here provided about the following:

- The originating circumstances
- The inventors and their enablers
- The origin of the invention
- The early development
- The introductory marketing
- The subsequent activities of the inventors and their enablers

As might be expected, it turns out that some iconic inventions seem to have had an easier arrival, development, and manufacture, or an easier progression to truly iconic status than others. Each iconic invention does seem to have a distinguishable, unique, unexpected, unpredictable, even fascinating individual history, although certainly each iconic invention does have elements that appear to be common to others.

History suggests that an iconic invention characteristically appears to have few or perhaps even no true prior equivalents in structure or function. However, as the facts suggest, the concept of equivalents, as judged or perceived by the using public that has brought an invention into extensive usage and iconic status, can be different from a concept of equivalents as perceived by a technical, economic, or legal analysis. An iconic invention can appear to involve various auxiliary factors, such as marketing, contemporary social standards, applications, and price, among others, that can affect perception of "equivalent." It seems to be easier to identify generally an iconic invention than to identify other specific things about it, such as exactly why it reached iconic status or level.

The elapsed time needed for an invention to reach iconic status can vary. Some iconic inventions after manufacture and sale appear to decline (on a long-time basis) in use over elapsed time. Apparently, one basic reason for such a decline is that an iconic invention over time becomes outmoded or even not fully needed by advancing circumstances. The circumstances can be various, such as, for instance, availability of the following:

- Newer, perhaps more economical alternatives
- Alternatives that are easier to obtain and use but that may not reach iconic status
- Significant improvements or changes in the original invention or embodiments thereof or in the use environment
- Arrival of a different and better invention that becomes iconic

Sometimes portions of an originally iconic invention persist and survive and become part or components of the utility, structure, or combination of a later, improved embodiment. For example, the steam locomotive has been substantially supplanted by the diesel locomotive, yet the basic idea of a rail-supported, wheeled, heavy source of power for moving and transporting cars carrying passengers and freight over distances remains in place and in use. The diesel engine and the so-called diesel locomotive that incorporates that engine are now iconic inventions, but to be considered in context, the original iconic inventions of the high-pressure steam engine and the steam locomotive that incorporated it cannot properly be neglected. The original iconic inventions may be of diminished contemporary importance, but at least from a historical standpoint, they still need to be considered with the later iconic inventions. However, space and time considerations here do, unfortunately, prevent full review and development of many peripheral matters, such as industry growth and change, development of improvement inventions, and general contemporary history, although these matters can perhaps be considered, in the main, to be not relevant to present consideration of origins and circumstances associated with selected iconic inventions and their respective inventors.

Some might argue that the place of some relatively recent iconic inventions in civilization is not yet established fully, and that it takes time, even in this fast-moving age, for an invention to reach and mature into iconic status. However, each iconic invention selected and considered here satisfies the criterion of extensive public use, as indicated above.

For the present purposes, the twenty-four iconic inventions that have been selected and considered here have been individually arranged into their chapters roughly according to the approximate date of first origin (as distinct from other dates, such as date of first commercialization or date of achieving iconic status). Sometimes, in addition to a main text, a chapter includes one or more sidebars about some related matter. Each sidebar presents specialized information of interest relative to, but apart from, material indicated in the main text.

In preparing this work, study of a large amount of background information was deemed necessary. Because of the scope and variety of the individual subject matter and issues presented, taken with the limitations of time and available source material, no original research into the selected iconic inventions, their inventors, or related matters has been possible. It has been necessary to rely on the relevant factual disclosures in prior works, such as those involving biographies of inventors. Every reasonable effort, though, has been made to have all factual information presented here for each iconic invention be accurate; supported by prior publications, both printed and electronic (as in the case of Internet publications, for instance); objective; and inclusive of best available and latest scholarship. Sometimes publications by researchers and historians made near the dates of the invention and respective inventor appeared to be more accurate and complete than similar, later publications. Considerations of space and of limitations on depth and breadth of treatment influenced the scope and selection of material appearing in each chapter.

The material included in each chapter has been selected in the hope of being of interest particularly to contemporary young adult readers, who seem to have potentially the greatest interest in, and also perhaps the greatest desire and need for, reference information about the origins of iconic inventions in the developing world. Ponderous technical descriptions about a subject invention are avoided or minimized; such information may already be available elsewhere, or it may already be known (or of limited interest) to the average young adult. The information presented in this work treads lightly on matters of tiresome structural detail, technical construction, operation, function, and exact patent coverage, particularly regarding outmoded historic structures. The text aims to proceed generally briskly, readably, and objectively through the environments in which the inventors lived and worked while achieving what became their iconic innovations.

As indicated above, the individual chapter texts are concerned primarily with the origins, the inventor(s), and the early development of particular iconic inventions. Each chapter has its own main story, but, for convenient reader study purposes, a similar format, where possible, has been attempted for each chapter for possible comparison purposes. Apart from historical context, each chapter sheds light upon how and why an invention that turned out to be iconic (from the standpoint of the judgment of civilization) happened to be achieved and pursued, but space does not permit a full development of complete pathway details for each invention.

SCOPE

From the standpoint of recorded history, the iconic inventions present in our civilization tend to be relatively recent and can be considered to begin about in the fifteenth century. Some general background information is relevant.

The iconic inventions in our civilization originated in various countries. Historians suggest that our modern culture commenced at about the close of the Middle Ages, mainly in Europe. The existing scientific knowledge, technology, and certain inventions, such as simple hand tools (which seem to be common to all stages of a civilization), plus certain of the knowledge and cultural traditions derived from earlier civilizations and societies, were absorbed and utilized. As our civilization developed, new iconic inventions appeared, at first relatively slowly, and were utilized.

It is usual to wonder what effect the technological developments of East Asia and the Islamic world had, particularly in the early developmental stages of this modern civilization, which had a predominantly Christian heritage. From at least the second millennium BC, the civilization of China dominated the Far East, but this civilization was in many ways different from that in Europe and even that in the Near East. Chinese technologists were familiar with basic tools and mechanics, and they certainly achieved inventions. In various instances, they achieved devices before counterparts appeared in the West. Bronze trigger mechanisms for crossbows, for example, were recognized as part of personal weapons more than 1400 years before they appeared in the West. Until the beginning of the Industrial Revolution in the West (usually dated as extending from the mid-eighteenth to mid-nineteenth centuries), it seems that the Chinese led the world in many inventions and developments. Perhaps due mainly to the nature of the Chinese and European societies, and neglecting things like time delays in travel and in communication, these Eastern inventions and developments were at first not known or appreciated in the West. Western civilization's early development apparently occurred largely independently of Eastern civilization. Owing to such considerations as environmental differences and cultural distinctions, technological inventions and developments that occurred in Muslim-dominated lands appear to have had little impact upon developments in Western Europe.

In modern Western civilization, there is a broad and deep interest in modern inventions and inventors. This has certainly not always been so. In most early civilizations, many inventions were made, but we usually do not know now who the actual inventors were. Those inventions variously include, for example, simple hand tools; the screw; the lathe; the pulley; military body armor (of bronze and iron); wheels and associated simple, cart-like, two- or four-wheeled vehicles; crossbows; toothed gears; simple force pumps; and hand-operated agriculture tools. Their actual inventors, experts suggest, were anonymous, perhaps because these inventions were considered relatively unimportant compared with other contemporary matters; perhaps because of class, education, or social distinctions involving the actual inventors; or perhaps because the inventions were considered to be work products of government functionaries or servants of nobility who did just what they were supposed to do and were without status, recognition, or reward. Sometimes we know the name of an object that was built incorporating an invention, and sometimes we know the name of the professional administrator of a large project involving many persons. In Roman times, for example, engineers seem to have enjoyed a higher status than did engineers in, for example, ancient Sumerian, Egyptian, or Greek times. A book on Rome's aqueducts, for instance, was written by Sextus Julius Frontinus, an administrator. When Frontinus names a man responsible for the building of an aqueduct, that man was the commissioner in charge of several hundred men, including surveyors, plumbers, masons, and engineers, but no staff members or their functions are individually identified, such as the person who designed a particular part of an aqueduct.

Sometimes, in our current civilization, structures, such as building exteriors, are modeled at least in part upon some remarkable Greek and Roman prior structure built of stone and masonry. The various remarkable Roman public works, including the roads and the highway system, the bridges, the aqueducts and water supply system, the sewers (including the Cloaca Maxima), and the various fortifications, were engineering, construction, and structural achievements to note and inspire for all time, but generally they were achievements that proved to have little practical consequence or meaningful physical value in modern civilization, or growth, particularly as our civilization has developed. Those earlier structural achievements undoubtedly involved brilliant originators, performers, managers, engineers, and inventors, albeit commonly unknown, even by name, today. They were constructed using simple and sometimes ingenious hand tools and construction apparatus, including large manually powered and manually operated cranes. Much inexpensive manual labor, mostly ruthlessly driven slaves, was involved. Actual construction was slow and time consuming.

Today, apart from taste and aesthetics, civilization and construction have changed and are profoundly different. In construction, the ancient is replaced by modern apparatus, modern engineering techniques, different tools, and different construction materials, though masonry, stone, and brick survive. Workers are free and paid today and have various employment benefits. The contemporary pressures to complete construction economically and in a relatively brief time frame are probably without prior historical parallel. In modern times, the greater importance of educational, economic, and wealth factors, plus new inventions and machines, has effectively strengthened the hands and potential of "common" people.

After the fall of the Roman empire, during the subsequent dark Middle Ages, some technology became unused, dormant, and unknown. Although historians have shown that some innovation did occur in the Middle Ages, there was little advance in technology or civilization. The compass and gunpowder, though known, were primitive. Developments in mechanical timekeeping and measuring technology occurred largely by inventors now unknown. Mechanical clocks appeared in Europe by the end of the thirteenth century, commonly with a descending weight-driven verge escapement; the verge including the spindle of a balance wheel with a vertical escapement, the escapement including the part of a clock that controls the speed and regularity of the balance wheel or pendulum. A little later, many European cities had clocks of great size and intricacy constructed by craftsmen and experts. Sometimes these great clocks indicated not only time but also various celestial motions, such as those of the known planets, as worked out by astronomers with crude instruments. Sometimes these great clocks were elevated on structures for exterior view by local inhabitants and had various moving automata that delighted spectators when the hour was struck. Clocks small enough for a person to carry eventually appeared, along with specialized experts having the capacity to make and maintain them. Pocket watches and even wristwatches arrived. The capacity to precisely measure the passage of time met basic needs and is believed to have stimulated activities and developments in civilization. Some early clock- and watch-making tools have come down to us, but sometimes their functions are today unknown. Innovations occurred incrementally, but the innovators are largely unknown.

It was not until about the middle of the fifteenth century in Europe that an iconic invention arrived in the developing Western civilization that can be identified, somewhat detailed, and described and for which the inventor can be identified Johannes: Gutenberg's letterpress printing technology with movable type. This advance made possible publications and the associated advances in learning and information distribution on a significant scale. Enough is known about Gutenberg and his inventions to permit their consideration in the opening chapter of the present work and to regard Gutenberg's work as a collective, complete iconic invention.

In the five centuries between Gutenberg's time and the present, other innovators achieved various technological advances in printing and in other fields besides printing. Some of these advances were significant and were themselves iconic inventions.

Between Gutenberg's printing advance and James Watt's steam engine, much time elapsed (perhaps about 325 years). A study of history over the interim suggests that inventions were made, but it appears that none satisfied the criterion of achieving extensive public usage. Watt's earliest steam engine invention, an improved condenser-associated, low-pressure (so-called atmospheric) steam engine, although perhaps relatively modest from a technological standpoint, nevertheless had very great consequences in the Industrial Revolution that extended beyond the engine itself and the immediate uses to which it was put. Watt's work is accordingly considered an iconic invention. His steam engine led to new improvements in factory machines and in transportation, including steamboats and steam locomotives. To provide a better understanding of Watt's important environment, his close associates are briefly reviewed in accompanying sidebars.

Study suggests that, particularly in early iconic inventions, the surrounding circumstances tend to be unique and to have been significant in stimulating invention and enabling it to occur. Sometimes, perhaps because of contemporary or local circumstances, iconic inventions took place more irregularly or intermittently than at later times, such as the twentieth century.

The occurrence of an iconic invention could stimulate a period of development and additional inventions. For present selection and consideration purposes relative to the number and variety of iconic inventions, little attention has been given to certain fields and products, including, for example, hand tools, particular buildings, building construction techniques, architectural creations, bridges, ships, and military weapons and equipment (the latter tend to become outmoded and outclassed, technologically and practically, relatively soon).

Patents and disclosures of inventions seem to have been important for advance since early times. It seems that occasionally, though rarely, an inventive genius was perhaps his or her own worst enemy, mainly because the creations did not surface or become part of the advancing civilization and so perhaps were later independently produced by others. Perhaps the most famous example of such a self-extinguishing inventor is Leonardo da Vinci, who, in addition to his memorable works in art, created various sketched illustrations in notebooks depicting structures of incredible originality relative to his time but that were never revealed during his time and for long afterward. Seldom did he fabricate or even test components of any of his designs. His unpublished notebooks upon his death were left to others; they remained unpublished and apparently unappreciated still longer, until it seems that many of his creations had been subsequently and independently invented by others, who perhaps had lesser abilities. His sketches and conceptual inventions, unfortunately, had little major use or impact upon the development of technology or civilization and cannot properly be classifiable as iconic. Today, his genius and his sketches are considered to reflect the work of a prophet more than that of an inventor of objects of practical value or usage that affect civilization.

Iconic inventions are considered to be apart from theory, science, or scientists. Scientists discover science, scientific facts, and natural laws. Inventions involve inventors, who can also be scientists. Iconic inventions involve inventors who achieve embodiments of inventions that are usually significant and that are physically realized, embodied, and, importantly, utilized in civilization. An invention involves the creation of both an idea and an entity (embodiment) incorporating at least a part of the idea and generally having a physical existence. The entity is made from a fabrication or assembly of components or elements. To become an iconic invention, an incorporating embodiment of an invention must have a capacity to be replicated and be used in civilization. For example, Nicolaus Copernicus, often considered the father of modern astronomy, formulated the theory that in our local area of the universe, the sun is the center about which the planets, including the earth, revolve. His theory contradicted both that which Ptolemy had propounded over 1,300 years previously and that which the established church orthodoxy in Rome had propounded, that the sun and planets revolved around the earth, but Copernicus deduced, from his own studies of astronomy using crude instruments, a table of planetary positions and certain other data, and his own mathematical calculations, that in fact the planets must orbit about the sun. His book was finally published in 1543 with a misleading preface that was not corrected until 1609 by Johannes Kepler. It took about a century for his theory to be accepted, and subsequent astronomers, such as Kepler, Galileo Galilei, and Christian Huygens, built upon his work. But his theory and observations, together with his conclusion, did not result in an iconic invention.

Huygens, though, apart from astronomy, did arguably achieve an invention that could have become iconic under different circumstances. He produced the first accurate mechanical clock in 1637. He discovered that the pendulum clock, as previously proposed by Galileo, inherently involved an error; he provided a theory and a mechanism for overcoming and avoiding this error and achieved an accurate mechanical clock. His innovation built upon the earlier achievements by predecessors. However, much time elapsed, even beyond his lifetime, before Huygens's invention produced any change in clock or watch manufacture. Practical improvements in clock and watch manufacture arguably had a greater impact upon civilization over the same time period.

ICONIC INVENTIONS IN EUROPE

After Watt's low-pressure steam engine, Richard Trevithick invented a high-pressure steam engine and associated it with wheel-based vehicles, including the important first steam locomotive in 1804. The locomotive was not destined to become an iconic invention until after George Stevenson developed an improved steam locomotive in about 1825 and railroading began in England. George Stevenson and his son can be considered to be enablers of the locomotive invention by Trevithick.

The needs of a modern civilization influence and promote invention. Napoleon in France recognized the need to supply food to his armed forces, but safe, nutritious preserved foods, particularly in the quantities needed, were not available, so he offered a prize for preserved foods. Nicolas Appert took up the challenge and, after research and development extending over years, achieved his iconic canned food invention about 1804. Though at first he used glass containers, he later used metal. The British were quick to pick up on and develop this technology. Wars and the growth of civilization created environments fostering subsequent development in food preservation, with many subsequent improvement inventions and advances by others.

ICONIC INVENTIONS IN THE UNITED STATES

Various iconic inventions have been achieved in the United States. The early history is briefly reviewed.

When the "New World"—including North, Central, and South America—began to be colonized by European sources in the seventeenth century, technology from Europe gradually became utilized there. The thirteen colonies in North America that eventually would become the United States were largely under English influence until the Revolutionary War, and a carefully formulated and practiced British mercantile policy toward the colonies inhibited and minimized local innovations and development. In the colonies, little new scientific knowledge and technological development originated or accumulated, and invention was a rare occurrence until about the Revolutionary War, although before the war, the ingenious Ben Franklin invented a heating stove and a lightning rod, and later he invented bifocal glasses. Some development in ships and shipbuilding occurred, especially development of the schooner, and during the war, David Bushnell of Saybrook, Connecticut, invented the submarine. The famous American traitor, Benjamin Thompson, also known to history as Count Rumford, made some discoveries and inventions in Europe during and after the war.

Immediately after the Revolutionary War, in the new United States, commerce, particularly international commerce, was at first practically in a condition of ruin, but mainly from New England "Yankee" entrepreneurs and barter, commerce developed. The political conditions at first were disruptive and included undertones of despair and even rebellion. The individual states became divided and struggled against one another. The Articles of Confederation, produced during the war, were useless, but public-spirited men worked out a new governing instrument, the federal Constitution. The people strove to make things improve, and they did. By about 40 years after the war, the American mood was optimistic, the nation was firmly established, and commerce had become substantial.

Contributing to the apparent growing success of the new republic in this national development period that followed the Revolutionary War were the pioneering and influential technological contributions of a relatively few men, such as the following:

High-pressure steam engines (1773–1802) (Oliver Evans [1755–1819])
Steamboat (1785) (John Fitch [1743–1798])
First successful fabric factory (1790) (Samuel Slater [1768–1835])
Steam engines, steamship, locomotive (1792–1825) (John Stevens [1749–1838])
Cotton gin (1793), factory mass production (1798) (Eli Whitney [1765–1825])
Navigation (1799–1810) (Nathaniel Bowditch [1773–1838])
Steamship (1807) (Robert Fulton [1765–1815])
Erie Canal (1817–1825) (various)

Innovation and development were undertaken individually, and in this early period, the risks and costs of research, development, and manufacture of innovations were generally more than could be taken by poor individuals. Little governmental funding was available, but some other incentives, such as state patent grants, were available. Little manufacture, sale, or distribution of innovative new products occurred.

Prior to the passage of any nationally effective patent statute by Congress, many of the colonies and states granted patents providing for specific exclusionary rights as part of their inherent sovereignty. Any state patent grant was limited to the territory of the state. Sometimes a state patent was granted not to support invention or invention development or a return on investment but for some other purpose, such as to provide a monopoly to stimulate and promote introduction or development of a known technology, a trade, or a business in that state.

The granting of state patents was an important incentive. From their beginning, patents on inventions granted to inventors were intended to provide an economic incentive to inventors and also to investors, manufacturers, and the like to risk capital investment and development in inventions. Inventions and patents, though, have always been associated with high risk. Practical or commercial failure of an invention or of the subject matter of an individual patent is probably more common than success.

Various state and colony patents were awarded before 1793. A 1641 patent grant by Massachusetts is apparently the first made by any of the colonies for an invention, but it seems to be uncertain whether the method presented in this patent was a new invention rather than a prior method being patented for the purpose of establishing a new industry in the colony. If the method was old (prior), then a subsequent 1646 Massachusetts patent grant of 14 years to Joseph Jenks for a scythe is the first patent for an invention granted to an inventor in America.

State patents were granted to John Fitch for a new steam-powered vessel in New York, New Jersey, Pennsylvania, and Delaware. James Rumsey contested the claim of Fitch to this invention, apparently, in New York and Pennsylvania. Among the records is an endorsement of Rumsey's boat by George Washington.

The first federal patent statute was introduced in the first session of Congress in 1789 and was enacted into law on April 10, 1790. Apparently, the statute was drafted by John Stevens. It provided for a three-man Patent Board comprising the Secretary of State (then, Thomas Jefferson), the Secretary of War, and the Attorney General, who were empowered to grant a U.S. patent for a term of 14 years if "they shall deem the invention or discovery sufficiently useful and important." Congress recognized previously granted state patent rights, and this federal patent statute operated to supersede the granting of additional state patents. The courts also recognized patent rights previously granted by the states.

The first U.S. patent was granted on July 31, 1790, to Samuel Hopkins for "Making Pot and Pearl Ashes"; the second on August 6, 1790, to Joseph S. Sampson for "Making Candles"; and the third on December 18, 1790, to Oliver Evans for "Manufacturing Flour and Meal."

Jefferson was constantly aware of the burden of responsibility in granting patents, but he took great pride in his capacity as a member of the Patent Board. Jefferson himself made various inventions, but he himself never sought or obtained patents on his own inventions or on those of others. To prevent others from patenting his inventions, he sought to describe each "anonymously in the public papers, in order to forestall ... its use by some interloping patentee." He favored the granting of a patent for invention to an inventor for a limited period but quibbled over the period of the grant.

However, the Act of 1790 had a short life because it soon proved to be impractical for cabinet members to act as patent examiners. A new patent law was enacted in 1793 that was a registration system, comparable to a conventional copyright registration system for authors, under which an inventor could get a patent without examination for novelty or inventiveness by simply submitting a patent application disclosing the invention's specifications and identifying what was claimed as the invention together with a government filing fee and an affidavit stating that the inventor did "verily believe" that he or she was the "true inventor." The federal courts could later decide whether or not the invention actually existed and, if so, determine infringement and enforcement. This patent registration law resulted in confusion and much litigation.

Subsequently, on July 4, 1836, Congress passed a new Patent Act that established a Patent Office, headed by a Commissioner of Patents, with its own complete staff and requiring examination of patent applications for determining whether or not an invention was presented. (With amendments and changes, this type of patent system has continued.)

One of the most famous state patents ever to be issued was that granted March 19, 1787, for 14 years by the New York state legislature to John Fitch for the sole and exclusive right and privilege of making and using commercially practical boats propelled by "fire or steam" on the waters of New York state. As also reviewed in a later chapter, on the basis that Fitch failed to exercise such right, that state patent was repealed by the legislature on March 27, 1798, and conditionally regranted to Chancellor Robert R. Livingston for the next 20 years, measured from that date. The condition was that the patent would not be effective unless Livingston succeeded in producing and operating such a commercial steamboat on the Hudson by October 5, 1805.

Livingston had partnered with Stevens, but no steamboat was produced within the initially specified conditional period. However, at the request of Livingston (and his agents), the New York legislature on April 5, 1803, extended the conditional period of his state patent for two years, to October 5, 1807, with the 20-year monopoly to become effective from the 1803 extension date in the names of Livingston and Robert Fulton.

Thereafter, Livingston partnered with Robert Fulton, and with a purchased and imported Watt steam engine and a steam-powered vessel designed by Fulton and built in a New York shipyard with Livingston's money, the patent's condition was satisfied in October 1807. The patent transportation monopoly to Livingston and Fulton became effective, and subsequently the New York legislature amended the patent grant, giving five years' additional time for each new vessel built and placed in service under the monopoly.

In 1811, with the New York state patent in force, one James Van Ingen had illegally (under the patent) operated steamboats in the Hudson, and Livingston and Fulton filed a bill in equity for an injunction. The lower court denied the injunction, but the New York appellate court overruled and ordered a perpetual injunction (*Livingston and Fulton v. Van Ingen*, 9 Johns. 506 [1812]).

The Stevens interests pursued a challenge to the New York state patent transportation monopoly in the federal courts. Subsequently, but not until 1824, in *Gibbons v. Ogden* (9 Wheaton 1, 240, 22 US 1, 6 L Ed 23), the United States Supreme Court held, in an opinion by Chief Justice John Marshall, that the injunction previously granted was erroneous and that the New York state patent was invalid, on the grounds that the New York patent interfered with the superior constitutional right of Congress to regulate commerce. Navigation on the Hudson was thus open to anyone.

By about 1830, in the United States, a change in society was discernable. Education had improved, some manufacturing was occurring, increased availability of raw materials was apparent, and wealth in urban centers was increasing. A strong work ethic existed. Religion was an important part of life. Technological innovators had better incentives and were better appreciated by the public. The public was receptive to developments, improvements, and technological changes.

By about 1830, more information about science and more scientific knowledge had generally developed in various fields. An inventor using, at first, his or her own assets could conceive and achieve a new and usually simple device embodiment with relatively little work and could develop it using generally available tools and equipment. An innovative device could be made, marketed, and sold to members of the public who were willing and able to try it out. Usually, equivalent or substitutable products commonly either were not at hand or were not as attractive; perhaps they were more expensive or not as promising. Money, certainly, was needed and was critical, but an inventor's initial monetary lack was not an insurmountable block, and the initial money needed was generally not a great amount (relative to the times). This combination of environmental circumstances served to encourage and promote innovation by individuals. The operating factors favoring individual innovation were good but not so good as to overwhelm other factors.

Thus began an extraordinary period of perhaps 70 years when many substantial and iconic inventions were made in the United States. The period can be called the Golden Age of American invention, although certainly many other iconic inventions were achieved before and since in the United States, Europe, and elsewhere. The Golden Age inventions and their respective inventors include the following:

Reaper (1831), Cyrus McCormick
Basic electrical devices (1831–1834), Joseph Henry
Telegraph (1837), Samuel F.B. Morse
Vulcanized rubber (1839), Charles Goodyear
High-speed printing press (1846), Richard Hoe
Practical sewing machine (1845, 1849), Elias Howe, Isaac M. Singer
Ether anesthesia (1846), William T.G. Morton
Steel preparation (1856), William Kelly
Railroad sleeper (1858), George M. Pullman
Typewriter (1867), Christopher L. Sholes
Railroad brake (1868), George Westinghouse
Celluloid (1869), John W. Hyatt
New plant varieties (1871–1911), Luther Burbank
Telephone (1875), Alexander G. Bell
Practical electric light and direct current power (1789), Thomas A. Edison
Practical alternating current electric power (1886), Nikola Tesla, William Stanley, Jr., George Westinghouse
Aluminum by electrolysis (1886), Charles M. Hall
Cast lines of type (1886), Ottmar Mergenthaler
Camera with film (1888), George Eastman
Carborundum (1891), Edward G. Acheson
Affordable automobile (1902), Henry Ford
Powered, controlled aircraft (1903), Wilbur and Orville Wright

In the twentieth and early twenty-first centuries, many other inventions occurred. Only a few of these iconic inventions are reviewed in the present work.

CONCLUSION

The present work, which includes only twenty-four iconic inventions and their respective inventors, can be regarded as unfinished. Without question, in the future more wonderful iconic inventions will be created, made, and extensively used.

The United States Patent and Trademark Office (USPTO), as just one of the national patent offices in the world, has now issued well over 7 million patents, and more are issued every week, as shown in the *USPTO Official Gazette* (available on the Internet).

It is hoped that the present work will stimulate interest in and development of more iconic inventions.

FURTHER RESOURCES

Ernest Bainbridge Lipscomb III. *Lipscomb's Walker on Patents.* 3rd ed. Vol. 1. Rochester, NY: Lawyers Co-Operative Publishing Co., 1984. (See also similar text in *Walker on Patents*, Deller's Edition. Vol. 1, 1937.)

L. Sprague de Camp. *The Heroic Age of American Invention.* Garden City, NY: Doubleday & Company, Inc., 1961.

AP Photo

Affordable Automobiles and Henry Ford

Although Henry Ford (1863–1947) has tended to be regarded mainly as a successful automobile manufacturer, study of his endeavors, particularly over the period of about 1895–1910, reveals him also to be a significant innovator whose important innovations resulted in practical, sturdy, affordable automobiles that enjoyed immense public acceptance and promoted the motor age. Ford added his own distinct and original significant contributions to those of innovative and capable designers and engineers with whom he teamed and whom he managed. Over time, Ford became immensely wealthy and gained control of his company. After success with autos, and in spite of his developing personal faults and eccentricities, he pursued other projects. It was with his automobiles, though, that he changed and advanced civilized life.

BACKGROUND

Automobile History: Ford's Time and Place

To get a perspective on Ford and his cars and to better understand his actual automobile innovations and developments & a brief look into prior history is worthwhile.

Over previous centuries, various competent innovators dreamed of achieving a self-powered, driver-controlled, wheeled vehicle. However, it was not until about the 1860s, when better machine tools and shop equipment, improved raw materials (especially metals and metal treatments), more knowledge, and better worker training, know-how, and expertise became available, that such dreams could become real and actual vehicle manufacture became possible.

A major factor in automobile development was the availability of internal combustion, Otto-type engines, the invention of which, as well as their early development and incorporation into vehicles, is considered in the earlier chapter on Nikolaus Otto. Available steam engines and electric motors were not as advantageous. For vehicles, steam engines and electric motors had more problems than internal combustion engines and were ultimately eliminated from automobiles. The electric starter for automobiles with internal combustion engines, which was introduced in 1912 by Cadillac (C. F. Kettering), hastened the demise of short-range, low-speed electrics and steamers.

The development of the modern automobile can be traced from the work of Karl Benz and Gottlieb Daimler in Germany, but their early machines had various defects. No one in the United States apparently knew details of their work at least until about 1887. Practicality and commercial success of Daimler's and of Benz's cars was apparently demonstrated and achieved first in France. The Parisian thoroughfares and the primary outward-extending roads were well paved and smooth. New vehicles enjoyed evaluation and

Early American Automobile Development

In 1871, Dr. J. M. Carhard built a steam carriage in Racine, Wisconsin, and another in 1875 in Green Bay, Wisconsin, in response to an offer of $10,000 made 2 years earlier by the state. The latter carriage was apparently invented by E. P. Cowles in 1874. Some claim that John William Lambert of Ohio City, Ohio, created a gasoline-powered automobile in 1891. None of these machines was developed.

George Brayton, an Englishman living in Boston, developed first a gas-fueled engine and then in 1873 a petroleum-fueled engine, that may have antedated Lenoir's work. Brayton's engine, was a two-cycle, single-acting type like the Otto and Langen first engine. Though the piston moved upward rapidly, it descended slowly, giving a characteristic irregular action. (The Otto four-cycle engine characteristically had a regular, uniform action.)

In the early 1890s, various people in the United States were endeavoring to produce horseless carriages besides Henry Ford. Charles Duryea and his twin brother Frank (machinists from Springfield, Massachusetts) publicly demonstrated a vehicle of their design and construction on September 20, 1893. Operating through the Duryea Motor Wagon Company, in 1896 the brothers built thirteen vehicles and apparently achieved the first commercial sale of a manufactured car in the nation. However, both Frank and Charles left their company in 1898, and their car production ceased.

Hiram P. Maxim, working for the Pope Manufacturing Company in Hartford, Connecticut, reportedly had a four-wheeled, steam-powered carriage operating haltingly in 1895.

Elwood Hayes, trained at Worcester Polytechnic and at Johns Hopkins, teamed with Edgar Apperson, and they completed a car in 1894. The Haynes-Apperson Company by 1899 had evidently sold some cars.

In 1896, Alexander Winton, in Cleveland, achieved a car that carried passengers. The Winton Motor Carriage Company was formed in 1897 and produced a car that was sold in 1898.

The supposed superiority of gasoline-powered vehicles over steam and electric power was perceived to be demonstrated by vehicle races in Europe and the United States. A famous race in Chicago on Thanksgiving Day (November 28 of 1895) sponsored by the *Chicago Times-Herald*, extended through snow and slush over a 52.4-mile course and was won by the Duryeas' American-built car, racing against various imported cars.

Colonel A. A. Pope in Hartford, Connecticut, by 1904 was making the Pope-Tribute runabout with a 6-horsepower front-end engine, which sold at $650; the Pope-Hartford with a 10-horsepower engine and with a tonneau (detachable rear seat), selling at $1200; the Pope-Robinson, a gasoline-driven car; the Pope-Toledo; and the Pope-Waverly, an electric car.

(*continued*)

The first automobile show was held in New York at Madison Square Garden in February 1899 and was held in association with a bicycle exhibit. The first independent automotive show was subsequently held at this location on November 3–10, 1900; thirty-four models were displayed, comprising nineteen gasoline models, seven steam, six electric, and two that combined gasoline and electric power. Among the gasoline cars were the Winton, the Duryea, the Haynes-Apperson, three Pope products (the Pope-Hartford, Pope-Toledo, and Pope-Robinson), the Autocar, the Knox, the Packard, the Gasmobile, the Meteor, the Dorris, the Upton, and the Holyoke, plus foreign products. Also exhibited were the already well-known steamers Stanley, Locomobile, and Victor. The electrics included the Waverly, National, Baker, Riker, and Columbia.

By 1900, various wealthy patrons in Newport, Connecticut, were enthusiastic car owners, who in effect made the automobile fashionable. During the early years, sales of automobiles originated mainly in urban areas.

usage not possible with most contemporary American or European roads. Competitors appeared.

The French firm of Panhard & Levassor, in 1891, was the first to mount the gasoline engine in the front of a vehicle, and in 1898, Louis Renault was the first to replace the chain drive (which evidently had been copied from the bicycle) with a drive shaft and apparently a differential means to transfer front engine power to rear wheels.

Before the advent of Ford's Model A in 1903, there were various American endeavors. For those interested, these are briefly reviewed in the accompanying sidebar. Until 1899, the American automobile industry had been profoundly influenced by the bicycle industry, but this influence was then waning. The available machine tools and the known factory systems and procedures from the velocipede industry aided early automobile manufacture. Automobile manufacturers often, for parts, turned to foundries and manufacturers that had previously made bicycles or bicycle parts. Assembly of a car by progressive steps adapted procedures previously used in bicycle manufacture. Car production also borrowed manufacturing procedures, including standardized, interchangeable parts, previously used by Eli Whitney in rifles, by Cyrus McCormick in farm machines, by Isaac Singer in sewing machines, by the Remington Arms in firearms, and by various inexpensive clock makers in New England.

The relatively large number of American car developers appeared relatively suddenly and mushroomed. By 1900, there were perhaps at least fifty-seven American manufacturers making motorcars, but many of the cars then produced apparently can be classified as prototypes or experimental models. The producers were widely scattered geographically. Of the about four thousand cars that some have estimated were manufactured that year in America, more than three-fourths were steam and electric. Apparently, several hundred foreign cars were imported to America in 1900, but Europeans were then starting to purchase American cars. At first, automobiles were very expensive and enjoyed a limited market, but between about 1904 and 1908, apparently about 241 companies joined the American car industry.

Many have wondered why so many Americans began to work on automobiles at about the same time, especially starting about 1894. Apparently, many Americans worked on autos without any definite negative consequence from the fact that others were similarly working. The availability of adequate tools and technology, the favorable economic circumstances, and the anticipated large potential market were stimulants. The railroads were good, but not good enough for family and business utilization. The bicycle created a market that the railroads could not satisfy, but it depended on leg power and moving loads presented a problem. A mechanically-powered vehicle of adequate size was simply needed. The public generally had a strong perceived need, and particularly in cities, the demand for automobiles was great.

The period roughly from about 1890 to 1910 was one of activity and development in the automobile industry marked by a wide variety of ideas, experiments, developments, expenditures, production, and promotion. The birth and death rates of motorcar manufacturers were high in this period. In these years, the automobile business was a pioneer industry, without standards but with speculators and speculations, yet it was growing rapidly. Unusual circumstances existed.

Henry Ford was not the first American to make an affordable American auto. Ransom E. Olds (a machinist experimenter in Lansing, Michigan) achieved and tested a prototype steam-driven carriage, which was reported in *Scientific American* on May 21, 1892. By 1896, Olds had achieved and developed a gasoline-powered car that would become the Oldsmobile. He founded a strong, well-financed company; hired Charles B. King, who was a sound engineer; developed the curved dash Olds, which had a single-cylinder engine; and produced 425 cars in 1901, 2,500 in 1902, and 3,100 in 1903. His roadster sold for $650, was being marketed by twenty-six agencies, and was in substantial demand. A highly successful business in affordable automobiles seemed to be developing. In 1904, with Roy D. Chapin aboard as sales manager, sales rose to 4,000, and in 1905 to 6500. However, the principal financial backer, F. L. Smith, desired to create a business for his two young sons, and they insisted that more expensive models be made and produced, thinking that profits would thus be increased. Olds was forced out, and, against Chapin's opposition, high-priced models were made and pushed. Though the low-priced roadster continued to be sold and to develop revenues, the high-priced, heavier cars were pushed, but they produced only accumulating losses. Cash reserves became exhausted, and the company sank in debt. Chapin and three other executives left, and the Olds Company was almost wrecked. This mess was probably a benefit to Ford and his company in selling their early models beginning in 1903.

In 1908, the Oldsmobile Company was merged by William C. Durant into the new General Motors Company that also then included the Cadillac and Oakland companies. By 1910, Durant, with 14,000 employees, was producing one-fifth of all U.S.-produced automobiles, and in its first year General Motors had net profits exceeding $9 million.

Other moderately priced cars available in the early 1900s included the Auburn by the Auburn Company of Auburn, Indiana, and the Century steamer from the Century Motor Vehicle Company of Syracuse, New York.

Ford's Prospects

When Henry Ford was first developing his gasoline engines and autos in the 1890s, the automobile industry existed but was still relatively undeveloped. Technologies regarding engines, vehicles, and their components were still primitive but were well suited for creation, development, and manufacture.

Had Ford and his activities relating to automobiles come along later, then he would have faced a substantially different environment and competitive situation, but had they come earlier, then he would have faced another and different set of problems having to do with the relatively undeveloped state of the needed relevant technologies and of the public mind.

Study of Ford suggests that he was not a likely candidate for success, riches, and a place in history. As a group, his talents and training basically can mostly be regarded as limited, but by 1902, they were focused and well suited for his automobile endeavors, and he was motivated partly from prior failures to achieve. The world into which he stepped as an automobile designer, creator, developer, manager, and manufacturer was really very diverse, challenging, and competitive. Even though his early vehicle embodiments and components were not individually wholly pioneering or even substantial innovations compared with others, he evidently initially made an important and correct business judgment that he could design and produce new, improved, useful, durable, low-cost, and marketable vehicle embodiments that would sell readily. However, the times and the circumstances were not a likely combination in which one might expect that he and his innovations would flourish. He certainly had very good competitors, but he succeeded against them all.

HENRY FORD

Early Years

Ford's early life, including his training, seems to have been cumulative and to have fitted him well for working with automobiles.

Boyhood

Henry was born July 30, 1863, as the second child of his pioneer parents, William Ford and his wife Mary Litogot, on their 90-acre farm in Wayne County near Dearborn, Michigan. Their first child had not lived. The Civil War was raging. William was an Irish immigrant, but his ancestors included English Protestants. He was a farmer and a carpenter, was adept with tools, and had some mechanical aptitude. Mary was apparently Dutch or Belgian Flemish by descent. She was a good mother and homemaker, and she evidently instilled into her children an appreciation for ethics, quality, and education.

In winter, when work on the farms diminished, from 1871 to 1879, Henry attended one-room public schools. He gained little skill in reading and writing from the classroom McGuffey Readers but displayed some proficiency in arithmetic. He was mischievous, had mechanical aptitude, could memorize literature passages, and had leadership ability.

He liked projects. In one early project, he led a group of classmates in a waterwheel construction effort in a ditch that they dammed. The wheel developed sufficient power to operate a grinder, but the project ended abruptly when one evening the dam they had constructed caused the ditch to overflow onto an adjacent potato patch. In another project, he and his companions constructed a "steam turbine." Though the machine ran at perhaps 3,000 rpm, it developed little power. However, the boiler exploded, causing injuries; one piece was blown through his lip, as Henry later acknowledged. Also, a fire resulted in the nearby school that partially consumed the building. William Ford repaired the damage to the school and told Henry that such play could be dangerous. In another project, Henry led his school mates in constructing a forge equipped with a blower that was used experimentally for melting and recasting lead, brass, and glass.

William recognized Henry's mechanical ability and gave him opportunities on the farm to exercise it. Henry was always tinkering, but he took no pleasure at all in the endless farm work.

In March 1876, when Henry was approaching 13, his mother, Mary, died. In July of that year, Henry chanced to see a self-propelled road vehicle powered by an onboard steam engine. When the vehicle was stationary, the steam engine was used to power threshing or wood sawing. He was impressed. Later that same year, Henry was given a pocket watch, probably as a birthday present, and he soon had repeatedly taken the watch apart and reassembled it. During the next 3 years, Henry continued his schooling, taught himself watch repairing, and continued to investigate steam engines. He decided to become a mechanic.

Detroit Training

At age 16, on December 1, 1879, Henry left the farm and walked the 8 miles to Detroit. At that time, Detroit was already about a century older than Chicago, covered about 17 square miles, and had a population of more than 116,000 people. It was a busy manufacturing and commercial center, with about ten railroads and hundreds of manufacturing businesses.

Fired from his first job after only six days, Henry was next employed by the James Flower & Brothers' Machine Shop, where the products were mainly valves and fire hydrants. In August 1880, he changed to the Detroit Dry Dock Company; there, he worked as an apprentice in the engine shop, where various engines were available. There, beginners, like Henry, were paid $2.50 per week for a sixty-hour work week. To make up the difference between his weekly pay and his weekly room rent of $3.50 and other costs of living, including his horse-drawn streetcar fare and food, he got a job working nights at the McGill Jewelry Store near his lodgings, repairing clocks and watches for $2 or $3 a week, working six hours a night, six nights a week. Henry had plenty of energy and needed little sleep.

He completed his apprenticeship in 1882, left the jewelry business, returned to his father's farm, and began working on a freelance basis installing, operating, and maintaining steam engines.

He got a job with the Westinghouse Engine Company in 1882 and worked for perhaps a year mainly in the Dearborn area as an engine expert on about twenty steam traction engines.

For an unknown time about 1884, Henry attended the Goldsmith, Bryant, and Stratton Business University in Detroit, which was a school for business training.

Henry seems to have seen his first gasoline-powered engine in about 1885. It was a stationary, early Otto-type gasoline engine that had been built by H. K. Shanck, a Dayton carriage mechanic, who was trying to develop a gasoline engine for driving a tricycle. Henry saw the engine being used at a soda bottling plant.

Clara

Apparently in late 1884 or early 1885, at age 21, Henry met popular and attractive 18-year-old Clara Jane Bryant. He demonstrated to Clara his proficiency with watches and steam engines. Clara was the oldest of ten children in farmer Melvin Bryant's family. She became convinced that Henry would accomplish something. By February 1886, they had reached an understanding, and in 1888 they were married.

William Ford, as a wedding present, gave his son Henry 40 acres of timber land on condition that he farm the land. At first, the couple lived in a little house on the land. Henry cut the timber and sawed it into lumber, which he sold to various companies in Detroit, but he did not farm the land. Henry had a small but apparently well-equipped machine shop attached to the house. Sometimes he took on steam engine repair jobs that required him to travel.

As the standing timber approached exhaustion, Henry searched for suitable employment in Detroit. Sometime in August or September 1891, Henry described to and sketched for Clara on the back of a sheet of music for their household organ his concept of a self-propelled carriage that used a gasoline-driven engine. He explained to Clara that he needed to know more about electricity and that he wanted to work in a plant where he could get the knowledge and experience he felt he needed. He related that he had inquired at the Edison Illuminating Company (the EIC) in Detroit and that they would employ him as an engineer. Clara, always supportive of Henry, was disappointed, even depressed, yet she told Henry that he had better take the job and they would move to Detroit. Henry, though, had apparently accepted the position before he talked with Clara.

Living in Detroit

On September 25, 1891, Henry and Clara moved to rented quarters in a duplex house in Detroit, and Henry began receiving a salary of about $40 per month as an EIC night engineer. Henry said later that this was more money than the farm was bringing in.

On November 6, 1893, their only child, Edsel, was born, and also in November 1893, after he repaired steam engines at the company's main power plant, he was transferred there to a day job and his salary was increased, reportedly to $75 per month. His supervisor, Alexander Dow, later president of EIC, described Henry as "very resourceful." Three weeks after the birth of Edsel, Henry was made chief engineer, reportedly at $100 per month. Continuing their pattern over about 1892–1915, as Ford's income increased, the Fords moved into progressively larger and more costly residences.

Once machines at a power station are properly set up, they largely take care of themselves with routine maintenance. Henry displayed the capacity to run a trouble-free station and gained some freedom and leisure time for himself. He taught a course for machinists at the Detroit Young Men's Christian Association (YMCA). He set up an "experimental room" in spare space at the company and apparently also set up a small machine shop in a basement room outside but next to the Edison plant.

In this period at the EIC, Henry had a good job and was an ambitious, energetic, engaging individual with a flair for management. He had considerable mechanical aptitude, was experienced in machines, especially steam engines, and had current knowledge of the state of the art in tools, including machine (i.e., powered) tools. Apparently he reviewed technical publications that were subscribed to and received by his employer. At age 30, though, Henry was a rather unsettled individual who spent nights and spare time tinkering with tools and joking with the boys. In fact, Henry appears to have spent most of his twenties and thirties with undirected tinkering with mechanical things including steam engines, watches, farm machines, and electrical generators. During this period, Henry may have developed his capacity for leadership and for inducing associates to pursue projects and developments that he envisioned.

For years, apparently, he had dreamed about building and developing a horseless carriage, but he knew that in no way would he be the first to invent and produce a motorcar. He also understood that so far, the efforts of others had still been relatively primitive, leaving much room for improvement, but he may not have realized that, actually, he was only one of probably hundreds of individuals in the United States and Europe who had, since about 1885, been thinking about and endeavoring to construct small power plants and vehicles driven therewith. Like most automotive pioneers, he took what he could from prior and contemporary developments but had to adapt, adjust, innovate, improve, improvise, and construct. Factory-made, standardized engines, chassis, carriages, drive trains, and other vehicular components were not yet generally available, just, occasionally, some parts and components. For a newly created or improvised multipart assembly, most of the component parts had to be individually fabricated. Even a simple engine-driven vehicle had many components.

To many, including Ford, the relatively new, gasoline-fueled, internal combustion Otto-type engine appeared to be more practical for a horseless carriage than a steam engine. Comparatively, a gasoline engine seemed to be reliable, easy, and safe in use; required no bulky boiler and associated furnace for heating water; and seemed to be light in weight relative to power delivered. Liquid fuel could be compactly and apparently safely carried near, and drawn upon by, the engine.

Ford could see no future in cars driven by electricity. He reasoned that a road car driven by an onboard electric motor could not run independently of power delivered through an extended arm moving along over an overhead trolley even, if a car contained "a large amount of motive machinery in proportion to the power exerted" (according to Ford in *My Life and Work*). Also, particularly with available motor technology, starting and accelerating an electric motor-driven car produced a big power drain, acceleration seemed slow, cruising speed was limited, and the needed on-board batteries, if used in place of a trolley and overhead wires, were bulky and heavy.

Ford's First Engine

The exact date is debated, but most probably on Christmas Eve of 1895, Henry completed construction of a small, Otto-type, single-cylinder, four-cycle, experimental, gasoline-fueled internal combustion engine with a crude ignition system. He carried the thing home and into the kitchen of their rented duplex near the EIC plant, where Clara was preparing a family Christmas dinner, which her parents were to attend. Henry's son, Edsel, then about six weeks old, was asleep in an adjacent room.

Henry clamped his dangerous little engine to the kitchen sink. Clara, under Henry's direction, splashed gasoline into a metal cup that served as a crude carburetor, while Henry spun the engine's flywheel. The house current available at the sink powered his crude spark plug, and the kitchen light flickered as a spark was delivered. The engine coughed and then began to operate noisily, with flames exiting the exhaust valve and with the sink shaking. It worked. After several minutes, Henry waved Clara away and allowed the engine to die. Henry was now ready to design and build a bigger and better engine.

The Quadricycle

Probably the first vehicle driven with a gasoline engine that Ford actually saw operating was a "testing wagon" that could travel up to about 7 or 8 miles per hour. He saw it at a public demonstration that took place at night on Detroit streets on March 6, 7, and 8, 1896. The wagon was powered by a four-cycle, four-cylinder, Otto-type engine that delivered about 3 horsepower and that was equipped with water-jacketed cylinders, electric ignition, and oil lubrication. The engine had been designed and built by Charles B. King, a friend of Henry's, and the wagon had been built by Emerson & Fisher Company of Cincinnati and modified by King. King had set up a company in Detroit under his name. Ford closely followed this demonstration. However, King, evidently lacking capital, and perhaps personal confidence in a complete auto he might design, did not develop his vehicles further and did not attempt to enter into their production. Charles King operated a mechanical design company in Detroit, and Henry had probably met Charles in 1893 when Henry, in his capacity as chief engineer for Edison, had machine work done by King's design company or by another company with which King was associated. King had studied engineering at Cornell University for 2 years and had previously invented a successful pneumatic hammer and a brake beam for railroad cars. It is possible that King did not have the ambition or confidence needed to start up his own automobile firm.

It was standard practice for American and European automobile developers to carry out runs with their prototypes either at night, when humans, horses, and authorities were few, or on open spaces far from traffic.

By March 1896, Henry was well along in the construction of a gasoline-driven Otto-type four-cycle engine and associated carriage of his own design. He had solicited Charles King and Oliver Barthel to help. Barthel, then 17 years old, was a German-speaking American who had been taught at the YMCA by Ford and who worked with King and expedited work for King by German-speaking, machine-operating employees.

King had entered, then withdrawn from, the historic Chicago *Times-Herald* 1895 motor carriage race, but he had served as an umpire. Ford's friendship with King and King's enthusiastic account of this race and of the accomplishments of others probably strengthened Ford's determination to design and build a motorcar of his own.

In the January 1896 issue of *The American Machinist,* an article on a gasoline-fueled, Otto-type engine, with detailed drawings appeared. That engine, known as the Kane Pennington engine, was discussed by Henry with King and Barthel. Even though King did not like this engine, Henry used some information in this article as a guide in working on his own improved engine, one he thought would work for powering movement of a carriage.

With other power station employees George Cato, Edward S. "Spider" Huff, and James W. Bishop, Henry had a group of mechanically inclined friends who had been talking about engines and ignition systems in Henry's private workspaces at the EIC. For the serious construction of a gasoline engine and carriage for an actual motorcar, Henry shifted the center of their activities to the brick storage shed at the rear of Henry's rented home, which became known as the Bagley Avenue shop.

The second Otto-type, four-cycle engine Henry designed had two cylinders, each with a 2.5-inch bore and 6-inch stroke. The engine developed 3 to 4 horsepower. Somewhat later, construction commenced on a carriage at the same location. For both constructions, Henry had help from his friends.

Ford undoubtedly learned from King and observed what King was building. Ford discussed transmissions with King, particularly after Ford had seen the Duryea horseless carriage at the Detroit Horse Show in April 1896, where King demonstrated his own adapted carriage. Ford evidently switched from his initial belt transmission to one using belt and chain. On May 27, 1896, King obtained a 10-foot drive chain for Henry from the Indianapolis Chain Company. Ford also incorporated into his own vehicle a compensating gear as a differential device to permit appropriate power to be selectively applied to each of the rear wheels during corner turning.

Jim Bishop and Ford worked together after work, usually till 11 or 12 p.m. every night, and had help from King, Barthel, Cato and "Spider" Huff. Following a pattern displayed in his boyhood, Ford directed the design and construction with much of the work being done by the others. No detailed records were kept, and it is not possible today to determine exactly who innovated or constructed what components or vehicular improvements for this auto. Henry called this vehicle the "*quadricycle*".

Materials and a few parts were obtained from various Detroit firms; other parts were fabricated specially. The entire machine when complete weighed about 500 pounds. It somewhat resembled a buggy, ran on four bicycle tires, had a neutral gear and two forward speeds, one for up to 10 miles per hour, the other for up to 20 miles per hour, but no reverse. At first, no brake was included, but later a foot brake was included that was applied with mechanical links to wheels when the engine transmission was manually put in neutral.

Ford used his own earned money for parts and materials, but Clara recalled later that the costs ate up all the family's funds above living expenses. The neighbor next door in the duplex was a friendly man of advanced years who became interested in the project. Helpfully, he removed a dividing partition in the back shed, giving Ford more room. Clara made nightly visits to the shed but refused to tell even her cousins who sometimes visited the family what was happening in the shed.

Subsequent accounts indicate that Ford, particularly with Bishop, kept working progressively longer hours as the vehicle neared completion. During the final 48 hours of construction, he hardly slept at all. On June 4, 1896, at about 1:30 a.m., culminating what was apparently more than a year of work, the car was ready for a trial run, but it was too big to go through the alley door. Clara emerged from the house with an umbrella, for it was raining. Henry seized an ax and broke down part of the brick side wall adjacent to the door to allow the car to emerge. Jim Bishop on his bicycle was ready to ride ahead to warn drivers of horse-drawn vehicles and to be on hand in case of accident. Henry put the clutch lever in neutral, spun the flywheel, started the engine, climbed aboard, grabbed the steering rod, put the quadricycle in low gear, and started away out of the shed, over the gravel in the alley, and into the street beyond. It worked. The car soon stopped when an actuating spring for an "igniter" failed, but a new spring was soon installed, and the car was restarted and driven back to the Bagley Avenue shop.

Later, before Ford could get bricklayers to restore the shed wall, the owner of the duplex appeared to collect the June rent. He became angry and excited when he saw the damaged wall, but upon learning that the car ran, he reportedly shed his anger and suggested that the doorway be enlarged to permit the car to exit and enter. It is supposed that the resulting garage door may have been the first built for a self-propelled vehicle in the United States.

Following this trial run, the quadricycle was substantially gradually rebuilt after a trained carriage maker (a "carriage blacksmith"), David M. Bell, fortuitously applied for work at the EIC and was hired. Ford also hired Bell to work for him to make many changes and improvements in the quadricycle, including replacement of wood parts with metal and strengthening the vehicle.

Henry took many excursions in the quadricycle, and each excursion seems to have led him to improvements. Steering was improved. An engine overheating problem was cured with cylinder jacketing, a cooling tank, and perhaps a water pump. Ford made trips to Greenfield and Dearborn in the quadricycle. Evidently the engine and driver compartments remained open. The car was a success and always attracted curious onlookers. He apparently drove the car about 2,000 miles.

In addition to acquiring experience, Henry developed into his first car qualities such as reliability, responsiveness, simplicity, lightness, features, good appearance, and low cost; these qualities were to contribute to his subsequent success in car manufacturing.

Business Failures and Racers

The Detroit Automobile Company

In August 1896, Ford, now age 33, was sent as a delegate to the Seventeenth Annual Convention of the Association of Edison Illuminating Companies at Manhattan Beach, Long Island, New York, where Henry was introduced to the renowned and famous Thomas Alva Edison, then still in his 40s. Edison was interested in the fact that Ford had built a working horseless carriage, and he asked Ford questions pertaining to the ignition and piston action. Ford always remembered that Edison concluded, "You have it—the self-contained unit carrying its own fuel with it! Keep at it!" He brought his fist down to emphasize his point.

Ford's dreams were further inspired and augmented by this meeting with Edison, Henry, upon his return, immediately began to build a second motorcar. He soon sold the quadricycle for $200. Developing and building the second car was expensive and he needed the money. Through the intervention of his friend William C. Maybury, then the mayor of Detroit, Ford was able to obtain apparently on loan a lathe and other equipment, probably from the Detroit Motor Company. Bell, Bishop, Cato, and Huff assisted Ford. Financial backing was contributed especially by Maybury, and also by Ellery I. Garfield, E. A. Leonard, and Benjamin R. Hoyt, all of whom were prominent members of the Detroit establishment. A written contract with Ford provided that in return for financial help, all patents taken out on this vehicle were to belong jointly to these investors. If manufacturing was undertaken, "Ford's employment on a fair compensation" would be undertaken. At about this time, Ford apparently had one patent, related to a carburetor.

Ford's second vehicle was completed in June or July 1899. An illustration in the *Detroit Journal* of July 29 shows that the car was a two-seater of greater size, weight, and durability than the quadricycle, with all machine components enclosed, but with the driver and passengers in the open.

A new investor, William H. Murphy, a lumber merchant, entered the scene after a demonstration run in the second car with Ford, and, on August 5, 1899, articles of incorporation for the Detroit Automobile Company were filed, which established, apparently, the first firm in Detroit to make motorcars.

Alex Dow, the head of the Edison Illuminating Company and Ford's boss, was interested in an "electric carriage," but Ford could see no future in that. In effect, Dow caused Ford to choose between his EIC job and his automobile. In August 1899, Ford resigned after 8 years with the EIC to become superintendent of the Detroit Automobile Company at a salary of $150 per month and a small amount of stock. He had not invested any money for his shares. Although Ford invited his EIC friends to join him, only Spider Huff did.

Ford had the new company's first vehicle, a delivery wagon, ready in January 1900. In a typical car factory of this period, production was normally about three or four vehicles a year, and each vehicle was essentially hand-made at a work station. A large amount of hand work on each new vehicle then made was required, including motor assembly and mounting, frame or chassis assembly, and body assembly and mounting, with alteration, repair, or refabrication of parts that did not fit (not uncommon), and with associated much drilling, filing, riveting, and bolting. Many components had to be connected separately and adjusted for each vehicle, including the muffler, tail pipes, brake rods, wheels, tires, levers, dashboard, windshield, horns, fenders, steps, and so forth.

In November 1900, though, the Detroit Automobile Company went out of business. Various reasons have been postulated, but the actual details do not seem to be known. Henry was now without employment.

Racing

Ford then began pursing racing cars. Racing proved a vehicle and provided publicity. Murphy provided capital, and Ford got Barthel, Huff, and Ed Verlinden to work on a racer at a shop at Cass Avenue and Amsterdam Street. Barthel reportedly made great contributions to the design of the racer and its engine. The *Detroit Free Press* reported that the machine cost "something over $5,000" for materials alone. The car was ready for the 10-mile race at the local Gross Pointe racetrack on October 10, 1901.

During construction of this racing car, Ford's own training, conceptions, and operating characteristics relative to vehicles became apparent. He had general design concepts that he usually conveyed orally, and he had the ability to manage and direct, but evidently he could prepare blueprints or read ones prepared by others only with difficulty. He required others to perform construction whose results he judged. In the words of Barthel, "Mr. Ford was a 'cut and try engineer.'" Barthel asserted that he, not Ford, performed most of the vehicle design detail work. Ford preferred to work with models prepared by associates. Ford seems to have had general ideas and innovative proposals but to have achieved individually relatively few specific or detailed innovations. Records involving details of work at this time do not seem to have been made. This pattern seems to have prevailed throughout Ford's work with vehicles. Particularly under such circumstances, it can be regarded as remarkable that Ford, often with nonspecific concepts, was able to communicate his ideas; to work with associates; to pursue, lead, and direct the design of cars and components; and to achieve production and manufacture of excellent cars.

Only two competitors started the race: Ford himself driving his racer, rated at 26 horsepower, and Winton from Cleveland in his racer, rated at 70 horsepower. After 3 miles had been covered, Winton was a fifth of a mile ahead, but Ford began to improve and cut down Winton's lead on the straightaways. Winton began to have trouble with a hotbox and slowed down. Ford shot ahead, rapidly outdistanced Winton, and won. The crowd went into an uproar. This win suggests that Ford had indeed carefully seen to the fabrication of his racer.

Another Business Failure

That the racer produced the effects for which Ford and Murphy had hoped was shown by the fact that the new Henry Ford Company was incorporated on November 30, 1901, with prominent Detroiters Clarence A. Black as president, Albert E. F. White as vice president, William H. Murphy as treasurer, and Lem W. Bowen as secretary. Ford was listed as engineer and was superintendent of production. Ford, though, evidently did not want to settle down to facing and overcoming the problems of car production. He apparently wanted to design cars and to build another racer. Murphy brought in Henry M. Leland, a successful engineer and machine shop owner. Car development proceeded slowly. A clash developed between Ford and participants in the company, particularly Leland, who criticized Ford's empirical make-and-try approaches. Ford was fired and left the Henry Ford Company on March 10, 1902, after being given $900, the uncompleted drawings for a new racer he wanted to build, and an agreement that the company would discontinue use of his name. However, Ford left his designs for the car he was to have placed in production. The company became the Cadillac Company, and under Leland, the Cadillac car was developed and produced using mainly Ford's car design and Leland's single-cylinder engine.

More Racing

Despite failures with two companies in less than 2 years, Henry still wanted to build another racing car. Before leaving the Henry Ford Company, he had made an arrangement with Tom Cooper, a famous, skilled, and successful racing cyclist who had become modestly wealthy, under which he and Cooper agreed to build two racers, the Arrow and the 999, the latter name being taken from a famous express train of the period. A shop was set up at 81 Park Place. Cooper made useful design suggestions, and the machines were built by Ford and C. Harold Wills, a hardworking, able mechanical designer and trained toolmaker. Also helping were Ed Huff, August Degener, and, near the end of construction, John Wandersee. The 999 was completed first. Its engine had four cylinders, each 7 by 7 inches, and was noisy when operating. In trials, the 999 was indeed found to be fast by Ford and Cooper. However, neither Cooper nor Ford would race the car.

Cooper got Barney Oldfield to race the 999 at Grosse Pointe on October 25, 1902. Four drivers and their respective cars started the 5-mile race:

Oldfield, Winton, Buckman, and Shanks. Oldfield promptly took the lead; Winston dropped out. Oldfield then lapped Buckman and defeated Shanks by a lap at the wire to win. The crowd's response was overwhelming. Oldfield, in later races, proceeded to establish several American speed records with the 999. For reasons that are not quite clear, Ford soon sold both these racing cars to Cooper, and he and Cooper parted.

THE AFFORDABLE AUTOMOBILE

One More Try

In spite of his two previous unsuccessful automobile manufacturing attempts and his demonstrated capacity to build winning racers, Ford still wanted to design and manufacture a lightweight, relatively simple, low-priced automobile for general usage.

In the spring or summer of 1902, Ford had begun to design a low- or medium-priced car with the feature of a vertical engine aimed at reducing vibration, noise, and wear. This engine would be mounted in front instead of the rear, as was found in, for example, the contemporary Oldsmobile. As work on the 999 progressed, Wills and Spider Huff found some time to work with Ford during the summer of 1902 to develop some drawings and probably some blueprints for this new car. Ford's skills included excellent technical expertise and innovation.

Concurrently, Ford succeeded in progressively arousing an enthusiasm for automobiles and in gaining a new investor, the successful Detroit business man, coal merchant Alexander T. Malcomson.

Before the 999 performed in October 1902, on August 20, 1902, Ford and Malcomson had signed an agreement under which Malcomson contributed $500 immediately and agreed to supply further sums as needed. All assets, including patents, models, drawings, supplies, tools, and dies were held by them in partnership. Once a model car had been built and capital had been obtained for manufacture, these assets would be assigned to a corporation that, they agreed, would be called the Ford Motor Company. Since Malcomson was overextended at banks where he had been doing business, all bills were to be paid from an account at a new bank. That account was in the name of James Couzens, the clerk, office manager, and business adviser of Malcomson.

This agreement marked the beginning of what was, in retrospect, Ford's most important and greatest adventure. The pathway thus opened was truly fraught with danger and threats of extinction, but Ford was now determined and motivated to succeed.

He continued to collaborate with the able C. Harold Wills, who functioned as his main assistant. Wills proved to be a gifted mechanical designer. To Ford's general idea and suggestions regarding a simple, low-priced machine, Wills added suggestions, helped make specific plans and blueprints, and worked with others to make and fit parts. Innovations were introduced, but exact inventorship determinations cannot be made since no records were kept. It was Wills who created the famous trademark with the blue oval lozenge containing the word "Ford" and the flowing "F."

Though James Couzens was at first very cautious and without faith in the venture, he gradually became converted as matters progressed, and he eventually took over and provided very effective management of all the venture's business affairs, to which he applied intelligence, ability, method, and organization. Ford had charge of design, innovation, and manufacturing. Like Ford, both Couzens and Wills enjoyed much energy and worked seemingly without ceasing.

In October 1902, Ford and Wills began hiring workers for the venture's Park Place Shop, and these included mechanics John Wandersee, Harry Love, Walter Gould, and Fred W. Seeman (a Canadian); pattern maker Dick Kettlesell (on whom Ford liked to play practical jokes); draftsman August Degener; and blacksmith Charles Mitchell. Their shop was rudimentary, with only a few machine tools that were unspecialized. The car would not actually be manufactured there but would be assembled from components produced by outsiders.

The Model A

By Thanksgiving of 1902, the engine design and prototype for the model car had been completed by Ford and Wills. The engine had two vertical cylinders, evidently a pioneering and basic idea, since cylinders have usually been vertical or almost vertical (relative to the direction of car movement) ever since in car engines. The chassis was formed out of angle iron. The body with a wooden shell was similar to the one Ford had developed for what was now the Cadillac Company. Wheels and other parts came from various sources. Between Thanksgiving and Christmas, the men assembled the prototype vehicle and readied it for exhibition and the taking of orders. The vehicle was called the "Model A".

In November 1902, to attract investors, Ford and Malcomson formed the Ford & Malcomson Company, a corporation to which they transferred all of the partnership assets for 6,900 shares. In addition, they paid in $3,500 in cash for 350 more shares, leaving 7,750 shares to be sold. But who would buy these remaining shares and provide the necessary additional money for a plant, production, and marketing?

Potential new investors were wary and not easy to find. The infant automotive industry was a pioneering and risky business, known to have many—too many—undercapitalized enterprises, colorful promotions, speculators, failures, and crooked dealings.

With two prior failures, Ford, as a key man, did not look like a good investment risk. Moreover, Malcomson was not only overextended, but also continued during the winter of 1902–1903 to buy up more coal businesses. By April, he had ten coal yards and was struggling to keep more than a hundred wagons and more than a hundred horses busy. As a result, he gave the automobile project intermittent attention, leaving Couzens to pursue that business.

Couzens went from one prospect to another but could not raise capital. Ford's associates, including his former boss, Alexander Dow at EIC, were not interested. For months, the partners kept going on convictions and nerves only.

In December 1902, Malcomson, through purchase of a small coal business, became the leaseholder of a coal yard with an adjacent old wagon shop on Mack Avenue. The property was owned by Albert Strelow, a large painting and carpentering contractor. Malcomson persuaded Strelow to convert the wagon shop into an automobile assembly plant, which came to be called the Mack Avenue plant, using Ford's plant designs. Malcomson and Ford agreed to pay $75 a month in rent for 3 years to compensate Strelow for his estimated conversion cost of $3,000–$4,000. Ford was able to bring his equipment and men over from the Park Place shop into the Mack Avenue plant on April 1, 1903. For power, they connected their machines to a large stationary Olds gasoline engine they installed in the plant.

Though the first prototype of the car worked quite well, experimentation had indicated the desirability of changes. Ford and Wills worked during the winter of 1902–1903 on another, better one. Ford wanted a car that was as good as possible.

While Strelow was getting the plant ready, Ford and his cohorts began making contracts with parts makers, even though the company had no appreciable cash capital. Since the engine was the most important part of the new car, an engine contract was essential. Ford had been acquainted with the brothers John F. and Horace E. Dodge, for years. They had achieved one of the top machine shops in the Midwest. They had faith in Ford's engine designs and Ford proposed lucrative manufacturing terms.

Ford proposed that they supply 650 chassis (each an engine and transmission with axles and frame) for the first season. Even though they were considering contracts presented by the Great Northern and Olds companies, the Dodge brothers signed a formal agreement with Ford and Malcomson on February 28, 1903, to supply the chassis for $250 each, for a total of $162,500. Each individual chassis was to be delivered to the Mack Avenue plant ready for assembly with wheels, tires, bodies, and related parts that the Ford Motor Company would acquire.

To make the deal, Ford and Malcomson agreed to pay for the first one hundred chasses with cash on delivery. Thus, the first $5,000 for twenty delivered chasses was due on March 15, provided the Dodge brothers could then show that they had invested that sum in equipment and materials pursuant to the

Ford contract. A second $5,000 for another twenty chasses was due a month later, if the Dodge investment had doubled, and a third $5,000 for another twenty chasses was due when the Dodge investment reached $15,000. The $15,000 thus paid was to cover the first sixty chasses delivered, with the next forty to be paid in cash as completed, and thereafter semimonthly payments were to be made. If Ford and Malcomson failed to make these payments, then all unsold product machinery reverted to the Dodge brothers.

For auto bodies, Ford and Malcomson contracted with the C. R. Wilson Carriage Company for wooden bodies at $52 each plus cushions at $16 each. For tires, they contracted with the Hartford Rubber Company at $40 per set of four, and, for wheels, with the Prudden Company of Lansing, Michigan.

A little later in 1903, Ford and Wills found they did not have an adequate carburetor. Ford got carburetor specialist George Holley of Bradford, Pennsylvania, to design one for their new Model A car, which they now sometimes called the "runabout."

To meet the various contracts, the Model A would have to be assembled and sold rapidly. Malcomson and Ford planned to sell the runabout car at $750, or $850 with tonneau (a rear seat with sides), yielding a profit of $196 or $246 each. Sale of the season's output of 650 cars would result in a gross profit of nearly $100,000 if all buyers took only the runabout.

Gradually, one by one, capital investors were gained through Malcomson's efforts. Malcomson got his elderly uncle, John S. Gray, a successful banker, to invest. Malcomson's cousin, Vernon C. Fry, and Malcomson's two attorneys, Horace H. Rackham and John W. Anderson, each decided to risk some money.

When the first $5,000 became due to the Dodge brothers, a crisis occurred. After a contentious March meeting of Ford, Malcomson, Couzens, Gray, and the Dodge brothers, Gray agreed to advance the $10,000 needed for the first and second engine shipments, with an added $500, all provided that he receive 105 shares in the company, that he be made its president, and that he be repaid by Malcomson if he became dissatisfied with his investment within a year.

On June 16, 1903, incorporation papers were filed in Lansing, Michigan, for the Ford Motor Company, a name proposed by Malcomson to the stockholders and accepted. This new company received all assets of Ford & Malcomson, Ltd. These assets were transferred to the company for 510 shares to be divided equally between the two partners. One thousand shares of stock nominally valued at $100,000 were divided among twelve investors (including the Dodges, Ford, and Malcomson plus Couzens and Gray). Only $28,000 was paid in cash, but Malcomson and the Dodge brothers had already made significant cash expenditures.

The investors agreed, importantly for the future, on a policy of closed ownership: no shareholder could sell his shares without first offering them to the other shareholders, and no new stock would be issued without approval of existing stockholders.

It was a classic of shoestring financing. Though overhead costs were low and operations simple, the business had barely sufficient money to start. Yet by June, the Mack Street assembly plant was starting to operate. Ten or twelve workmen at $1.50 per day each had been hired, and, under Degener, were ready to mount tires on wheels, install the wheels and the bodies on each chassis, install the cushions, paint each vehicle, and test it. According to Couzens's cost analysis, to the cost of parts per car of $384, the cost of assembly of $20 per car was added. Sales costs were estimated to be $150 per car, making the cost $554 without the tonneau or $604 with it.

For a few weeks, the company had a harrowing race against bankruptcy. On June 26, with a bank balance of $14,500, disbursements were made: to Malcomson, $10,000 for payments to the Dodge brothers, to the Dodge brothers, $5,000 for chasses, to Hartford Rubber Works, $640 for 64 tires, and to others, some other, smaller sums. The company account thus would have been overdrawn except for the fact that one investor (J. W. Anderson) on that day paid in his $5,000. Five days later, payments were made for salaries and services, followed by payments between July 7 and 11 of $5,000 to the Dodge brothers plus checks for rent and other expenses.

The bank balance dropped to $223.65. However, on July 11, Malcomson and Ford were successful in getting another investor (Albert Strelow) to meet his stock subscription with a $5,000 check.

Not until July 15 was a single car sold and price paid, but on this date the first check from a buyer, Dr. E. Pfennig, of Chicago, for $850, was deposited. Within the next week, payments from five other buyers were received. Ford and the company had won the race against bankruptcy.

On July 14, the Dodge brothers were paid another $5,000, and on August 1, the bank balance was $3,831.67. Additional customer payments resulted in a bank balance of $23,060.67 on August 20. The Model A was being made and sold. Production proceeded, perhaps surprisingly, in view of Ford's previous manufacturing failures and lack of manufacturing experience, but perhaps Ford's prior efforts had been a needed learning curve.

From the first three and a half months of operations, the company made a profit of $36,957, and on November 21 it paid a 10% dividend. In the first nine months, from mid-June 1903 to the end of March 1904, the company sold 658 Model A's for a total of $354,190, with a net profit of $98,851. This was followed by April through June 1904, when sales reached almost $650,000. The total number of cars made and sold in the first 15 months was 1,700.

The profits distributed by June 2004 almost equaled the whole original capitalization. While Couzens and also Ford had, with strenuous efforts, guided and driven sales, the new company had obviously been fortunate in entering the car market with a car that customers liked at a time when a "car-mad" public demanded affordable vehicles.

From the start, the company was able to finance itself by revenues. Profits were large compared with the capital invested. A second story was added to the Mack Avenue plant.

The early Model A cars had a great many faults, and many dealers and their service personnel knew little about servicing the cars they sold. But the customers tolerated these problems. Nobody expected a car to be dependable in 1903 and 1904. Demand outran production. From the quadricycle to the successful Model A, about 7 years had elapsed. From the available information, and as reviewed above, it appears that Ford's main contributions to the success of the Model A were a mixture of some innovation, some design, manufacturing management, business acumen, and a determination to produce a low-cost car that worked.

Company Growth

Market momentum and sales needed to be maintained. The Model A was not to be just a flash. During the next selling season, 1904–1905, production of the Model A was replaced by production of three new, improved models designed by Ford and Wills: the new improved Model C, selling at $800 ($900 with tonneau); a new touring car, the Model B; and the Model F, selling at $1,000. Both new models, Models C and F, incorporated Ford's original two-cylinder engine. The Model B was a heavier, faster touring car, selling at $2,000, with a four-cylinder Ford engine. Malcomson had insisted on the Model B.

Ford disliked the Model B because it was moving away from his dream of a low-cost family car. Nevertheless, to promote the Model B, Ford drove the vehicle and set a new world speed record with the car. The 1904–1905 sales surpassed all expectations, and on April 4, 1904, the stockholders at a special meeting approved the construction of a new and larger plant, which became known as the Piquette Avenue plant and which began operating in late 1904.

In plant assembly, each chassis was still set upon wooden horses at a designated spot to which a body was brought and assembled with other parts, so that a car was built and completed by a group of workmen. Initially, there were about four such car assembly sites, and at the Piquette Avenue plant, perhaps 12 or 15 such assembly sites were concurrently functioning.

In this period, Ford and Wills designed two new models, a moderately priced Model N as an improvement on the Models C and F, and a costlier Model K that Malcomson wanted that had six cylinders. Ford wanted to make only one type of car, selling at below $650, then the factory price of an Oldsmobile. The Model N is believed to have approached Ford's inclinations. It weighed about 800 pounds, had a front-mounted four-cylinder engine of 15–18 horsepower, and eventually used vanadium steel extensively. It was tougher and more reliable yet lighter than cars of much greater price.

When, increasingly, Ford and Malcomson could not agree, Ford and Couzens established a second corporation, the Ford Manufacturing Company, in November 1905, to make finished engines, parts, and other components at a site called the Bellevue shops, all for sale to the Ford Motor Company. From Bellevue, finished engines and parts were hauled to the Piquette plant by horse-drawn trucks. Malcomson received no stock in the manufacturing company, but Ford wound up owning 2,900 shares, and seven other men each had 350 shares. One objective of this second company was to make important parts at controlled quality and cost for a low-cost car, then the Model N. Another was to improve the manufacturing process by better volume and quality of components. But a main objective was encourage Malcomson to dissociate from the company. The basic idea was that payments made by the Ford Motor Company to the Ford Manufacturing Company for component manufacture would reduce the money available for dividends to stockholders (including Malcomson) of the Ford Motor Company. Malcomson had meanwhile formed a new car company, the Aerocar Company, with Malcomson as president and majority stockholder, and he wanted the income from his Ford Motor Company stock dividends to finance his new company.

The Model N reached full production in the fall of 1906, but the combined sales of the Models C, F, N, and K in 1905–1906 were lower than total car sales in 1904–1905. Profits slumped and dividends fell; only $10,000 was distributed on October 24. With no dividend money from his stock to further his Aerocar Company, Malcomson chose to exit. He negotiated with Ford to buy his stock, and Ford wound up paying him only $175,000 for his 255 shares in the company.

Within a year, three more original stockholders, friends of Malcomson, sold out to Ford and Couzens, while a fourth, Strelow, being then interested in a Canadian venture, sold his stock to Couzens. A fifth, Gray, died suddenly. The result was a shift in company ownership to eight stockholders, with Ford (as president) and Couzens (as treasurer) in control. Ford was in charge of design and production, Couzens of business matters.

For the 1906–1907 season, the low-priced Model N, cosmetically upgraded versions identified as the Model R and the Model S, and the Model K, all designed by Ford and Wills, were produced. The Model K sold for $2,800, the Models R and S apparently for about $650–$800, and the Model N for $500–$600. Led by sales of the very successful Model N, 8,243 cars were sold from October 1906 to September 30, 1907. Ford products developed a reputation for holding up well in use. Dealers gradually were able to offer improved vehicle servicing.

In the spring of 1906, to meet the growing demand, a bold decision was made to build a new and much larger plant in Highland Park, Michigan. The

Ford Manufacturing Company was absorbed into the Ford Motor Company early in 1907. To lessen Couzens's load, an advertising manager, Le Roy Pelletier, and a sales manager, Norval A. Hawkins, were taken on in 1907, and both became key figures in operations and marketing. Also, in about 1907, under Ford's direction, the company began extensive use of steel alloys, such as vanadium steel, in its automobiles. Variations using the Model N chassis, including a taxicab, a landaulet (a car having an open driver's side and a collapsible roof), and a touring car, were introduced. The company continued to display excellent applied engineering skills and to build durable cars. Important for sales, Couzens developed a national marketing and dealership organization. Expansion without outside capital continued.

Some thought that the company was expanding too fast after the Panic of 1907, but the Panic had no adverse effect on Ford sales, which increased in late 1907, climbed in 1908, and exceeded all previous records, including significant sales in Canada and England. However, at least twice in the fall of 1907, paydays at the company had to be postponed for lack of funds.

In contrast, five Detroit area automakers became bankrupt in 1907: Malcomson's Aerocar Company, the Detroit Auto Vehicle Company, the Marvel, the St. Clair, and the Huber. Packard and Pierce-Arrow were seriously affected but survived. Buick, Reo, Cadillac, Oldsmobile, Oakland, and Maxwell-Briscoe, along with Ford, survived as the leaders. As recovery progressed early in 1908, Ford had more than 500 employees in the Highland Park plant. A great automobile boom followed the Panic of 1907.

The Model T

Meanwhile, since early 1907, Ford and about a dozen men had been working at the company to develop a new model. Ford was pursuing his wish of many years to design and build a cheap, efficient "family horse" that could be made with various body styles and that many people could buy and use. He felt that more money could be made by selling more models of a cheaper car than by selling fewer models of a more expensive car.

Ford was now nearly 45 and had his full mental powers. Also, he had achieved control of his company, he was affluent, and he had the money to do almost anything within mechanical capabilities. For this new car, called the "Model T", he used wooden models, special forgings, and impact tests. He plunged himself into developing this new car. He reviewed the smallest details, was involved in everything, and strove to get everything right. He innovated where necessary, but where possible, he improved on existing components. He maintained his vision of what he thought this new car should be. He had demonstrated an ability to proceed from concept through implemented details to maintain or reach an objective and to materialize particular components. Many hours each day and into the night he worked with his team.

In general, in designing and developing the Model T, Ford provided management, leadership, insights, and many general ideas and innovations that he communicated to those on his team. The team, in fact, was made up of gifted men with technological training and experience who worked out Ford's ideas and brought them into specific reality and practice, and added their own. The exact origin and inventorship of many innovations in the initial and subsequent Model T autos, as in other cars produced under Ford, will undoubtedly never be known because of inadequate records. Written records, such as entries in dated laboratory notebooks, periodic reports, or research and development summaries, were never kept. However, but for Ford, like the Model A, the Model T probably would never even have come into existence or been manufactured.

The Model T, with subsequently and gradually incorporated improvements, became probably the most famous of all automobiles. In March 1908, circulars describing the Model T were released to Ford dealers and received enthusiastic responses. The car was unconventional; it had a four-cylinder engine of 20 horsepower, usually with a 10-gallon gasoline tank, strong wheels, and springs. It had a completely enclosed engine and transmission, a body located high above the road, extensively used vanadium steel, had enclosable passenger areas, and had various other features. The initial price ranged from $825 for the runabout to $1,000 for the landaulet. The circulars produced a flood of orders. Production seems to have begun on September 27, 1908, but deliveries could not begin until October 1, 1908. The advertising emphasized low cost, durability, simplicity, standardization, and interchangeability of parts. Speed, endurance, and hill-climbing power were also featured. The planned and achieved production of 25,000 cars for the 1908–1909 model year was inadequate to meet the demand.

Several faults with the early Model Ts were gradually corrected. Among the national and local publicity given the car, the New York to Seattle race in the summer of 1909, arranged by mining magnate Robert Guggenheim, was especially appealing to many people. West of Missouri, most of the distance was roadless. Two Model Ts were among the six cars entered. When the Ford drivers reached the fearsome Snoqualmie Pass, they found Henry already there to cheer them. In awarding the trophy cup to the winning Ford, Guggenheim said:

> Mr. Ford's theory that a light-weight car, highly powered for its weight, can go places where heavier cars cannot go, and can beat heavier cars, costing five and six times as much, on the steep hills or bad roads, has been proved.

The winning Ford was driven down the Pacific coast and then back east, providing further publicity for individual dealers along the way.

As purchased, all original Ford cars came bare, without speedometer, windshield wipers, or doors. The gas gauge was a long, slender stick that the user had to find on the car and insert into the tank. Supplying accessories for bolting, screwing, strapping, or inserting into or onto the car became an industry.

The Alleged Selden Patent Infringement

The famous Selden patent offered a potentially very significant impediment to the Ford Motor Company and to American automobile development generally. On September 15, 1909, the Federal Circuit Court of the Southern District of New York held that the Ford Motor Company infringed George Selden's U.S. Patent 549,160, which was issued November 5, 1895, and was based on a patent application filed May 8, 1879, that Selden had maintained in a pending status for about 14.5 years. The court's decision was widely publicized. The complainants in that suit had maintained that this patent covered the Ford automobile, as well as all other automobiles then being made with gasoline-fueled internal combustion engines.

The Selden patent in fact discloses and claims broad combinations of internal combustion engine with a carriage. George Selden, a patent attorney with engineering training, thought his combination back in 1879 when he filed his patent application was ahead of its time from invention and commercialization standpoints. Selden had intentionally kept his patent application from issuing into a patent for many years until he perceived, in 1895, that the time of his combination's commercialization had arrived. The patent issued then would be effective for 17 years from its issue date under the law then in effect, and he thought that he could receive a lot of royalty money because his patent, in its claims, intentionally covered virtually all automobiles with internal combustion engines.

George Selden, having little money himself, needed strong financial support to exploit and enforce his patent. In November 1999, he licensed the Columbia and Electric Vehicle Company of New York, later called the Electric Vehicle Company (EVC), for $10,000 plus a percentage of whatever royalties that company could collect.

In July 1900, the EVC and Selden, pursuant to their agreed strategy, sued two concerns for infringement of Selden's patent: a leading car manufacturer, the well-known Winton Motor Carriage Company of Cleveland, and also a leading parts manufacturer and supplier, the Buffalo Gasoline Motor Company of Buffalo, New York. The idea of the EVC and Selden was that if these concerns could be forced to take a license, then the rest of the industry would follow.

Some prominent makers of gasoline-powered cars, especially Henry Joy of Packard and Frederick Smith of Olds Motor, negotiated with the EVC and formed the American Association of Licensed Automobile Manufacturers (ALAM) in 1902. In a written agreement between ALAM, EVC, and

Selden, the EVC was to collect royalties and fees; it would pay two-fifths of them to the ALAM, pay one-fifth to Selden, and retain two-fifths for itself.

As the ALAM was being formed, the Winton Company began to negotiate a settlement. Later, in March 1903, having apparently received favorable terms, Winton Motor and Buffalo Gasoline for the record in the lawsuit conceded validity of the Selden patent and accepted licenses. The result, as anticipated, stimulated car manufacturers to join the ALAM and pay the royalties and fees set. Twenty-six concerns did, but not the Ford Motor Company and perhaps as many as a hundred other companies.

In June or July 1903, with the big early success of the Model A, Henry Ford seems to have approached the president of the ALAM about a license and membership but was told that his organization was unfit because it was nothing but "an assemblage plant." In fact, all ALAM members and other car manufacturers obtained some components from suppliers and carried out assembly. The ALAM had simply decided to deny an immediate license and membership to Ford. The Ford Motor Company was even then moving into having the largest share of the American car market.

In 1903, Ford and Couzens, after receiving advice from their Detroit patent attorney, took the position that if the Selden patent was broad enough to cover their cars, then it was invalid as being anticipated by prior art, and also, if it was not that broad but valid, then their cars did not infringe the patent's claims. Ford and Couzens also evidently gambled that even if a court finally found the Selden patent to be valid and infringed by Ford (a worst-case scenario), then the Company could probably pay the royalties owing because of the company's great profitability.

Advertisements by the ALAM, which were intended to inhibit the Ford Motor Company's commercialization efforts, stated that only members were authorized to make and sell gasoline-powered automobiles. The Ford Motor Company responded with its own advertisements promising to protect "dealers, importers, agents, and users" against prosecution for alleged patent infringement. The company led the independent car makers in opposing the ALAM and the EVC. The company tried to convey the impression to the public that Ford was a small, struggling businessman while the ALAM was a large and powerful group of monopolists.

The ALAM, the EVC, and Selden concluded they had no choice but to sue the Ford Motor Company and did so in what they undoubtedly thought was the then most favorable jurisdiction for finding patent infringement: the New York lower federal circuit court (now identified as a federal district court). The lawsuit against Ford was filed October 22, 1903, and for the next 6 years, the company had bursts of adverse publicity, time-consuming and costly legal maneuvers, and the threat of ultimately receiving an adverse federal court decision (which did happen). Despite these problems, the lawsuit actually did little or nothing to impede the growth and development of the company.

Two days after the Selden patent suit was filed, Couzens told the Detroit *Journal* that although it "may take years to thresh the matter out in the courts," the company was enlarging its plant and "increasing its 1903 output fivefold." Subsequent pretrial discovery procedures and the trial resulted in the buildup of an extensive record. Despite the fact that the lower court had received evidence showing that in fact the broad Selden combination represented prior art and was previously known before the application was filed, the lower Circuit Court held that the Selden patent was valid and was infringed by Ford.

The ALAM, the EVC, and Selden were delighted with their lower court win and with the publicity generated. After some soul-searching, Ford and Couzens decided to appeal the adverse lower court decision.

On January 9, 1911, the Second District Appellate Court, after review of the lower court record and oral arguments, reversed the lower court's holding. This was a great victory for Ford and Couzens.

The appellate court in effect agreed with the initial position of Ford and Couzens and concluded that since every element in the claims of the Selden patent's broadly claimed combinations, as well as the basic main combination itself, were all shown in prior art, the broad combination of each of the Selden patent claims was old and unpatentable.

The appellate court, in its opinion, proceeded to explain that the only engine actually disclosed in the Selden patent was a compression engine of the type known as a Brayton engine, which was a constant pressure, two-cycle engine, not a four-cycle Otto engine like those used by Ford and other car manufacturers.

The appellate court observed that the original Brayton engine type was in the prior art when the Selden patent application was filed, but it found that Selden disclosed in his patent a new type of Brayton engine that the court held was a narrow improvement over the original or prior art Brayton engine. The Court concluded that Selden's variously claimed combinations were novel and narrowly patentable if considered just with Selden's improved Brayton engine. However, since the Ford automobile, which incorporated an improved Otto engine, was distinctly different from, and not equivalent to, the (narrow) patentably novel combination claimed by Selden, Selden's patent claims did not apply to (that is, "read on") the Ford automobile. Seldon's broadly claimed combination was old and unpatentable, and Selden could not enlarge the scope of any of his claimed (narrow, novel) combinations to include internal combustion engines of the Otto type or cars using such engines.

Hence, the Ford Motor Company and other car manufacturers won a decisive victory. The court's decision was accepted by all parties as well reasoned and fair. No appeal was taken to the U.S. Supreme Court. No car manufacturer needed a license under the Seldon patent.

After the appellate court decision regarding the Selden patent, work at the company continued on erecting the architect-designed, 62-acre, state-of-the-art Highland Park Ford plant, and manufacturing shifted to this new plant at the beginning of 1910 from the Piquette Avenue plant. The company purchased the John R. Keim Mills, a company that made pressed and drawn steel, to incorporate that company's forming technology into Ford's production of automobile parts, and moved the Keim Mills from Buffalo to the Highland Park plant in 1912. The company could now itself fabricate most of the components used in its cars.

By 1910, as the automobile was becoming the primary mode of transportation, Henry Ford thought to reduce the price of the Model T with each new model year. Profits were seemingly unbelievable. By the spring of 1914, there were 7,000 Ford dealers handling only Ford cars. They were apparently the largest single-product sales group in the United States. The company asserted that no other industry in which money had been legitimately invested and where work was actually done ("not mere high financing") could show such a result.

Within 10 years of its founding, Henry Ford's company had become the most successful of automakers and had the biggest automobile manufacturing facilities in the world. The factory could turn out and sell completed automobiles faster than competitors, including all European car manufacturers, could manufacture some components. Ford's cars were reliable, by the standards of the time, and able to withstand hard service. His cars and their customers made Henry one of the world's richest men.

Continuous Mass Production

Time and motion studies begun at the Piquette Avenue plant led to the assembly of a plurality of vehicles, sequentially and progressively, at the Highland Park plant. The old procedure of stationary sites or stationary group centers was abandoned. From the assembly of subcomponents on an assembly line, a chassis assembly line gradually emerged that greatly reduced the time needed to assemble a car.

Under the stimulus of demand for the Model T, ways to increase production were sought. Thousands of machine tools were purchased and operated, and, as new machine tools became available, the old tools were replaced without consideration of cost.

The placing of men and their machines sequentially in a well-planned arrangement with localized stockpiling of materials along a moving assembly line was a practice that was already well developed in many industries, but such lines involved a conveyor on which products being progressively assembled moved in a stop-and-go manner (i.e., by jerks, not continuously).

The innovation of continuous assembly line movement and production first appeared in the Highland Park Ford plant in 1912–1913. Its evolution and application proceeded in relatively small, incremental developmental steps. Carefully planned time and motion studies were developed and carried out, and were followed by plant-scale trial-and-error experiments involving assembly procedures. The prior teachings of men such as Charles Babbage in 1832 and Frederick W. Taylor in 1911 were reviewed. The basic Eli Whitney elements of mass production—interchangeable standardized parts, close tolerances, and efficient handling—had to be combined with a moving, accurately timed line whose speed was appropriate to the needs, skills, and tools of assemblers along the line. The goal was to have only work in process of assembly move on the belt. First, conveyors, slides, and rollaways appeared; then came material handling and delivery to continuous assembly lines.

Many company employees participated in data accumulation and feasibility development, leading to the achievement of actual continuous mass production. Ford biographers indicate that Henry Ford was not himself the sole innovator of the continuous form of mass production. It is not known who the actual innovators were. Exact inventorship records were not kept.

Continuing Growth

Employee Benefits

On January 6, 1914, Ford announced that the company had then adopted a plan of sharing equally its profits with its employees. Briefly, under the plan, the company would take 50% of its net profits, but the other 50%, estimated to be $10,000,000 for 1914, would be distributed to the men over 22 years of age working for the Ford Motor Company. This money would be given in the regular pay envelopes handed out every week in proportion to the daily wage received, but would be paid at the rate of at least $5 a day. Ford's announcement was dramatic and well publicized, but it horrified other companies. The *New York Times* warned of labor unrest at other companies. The announcement improved company worker morale. For years thereafter, the workers were proud to be Ford employees. The company then had what was probably the most advanced labor policy in the world.

A company Sociological Department was established about this time and operated for a few years. It was aimed at worker welfare and reducing employee turnover. It verged on paternalism and included a program involving company investigators. Each investigator had an assigned Detroit district and was given a list of employee subjects for inquiry under which individual workers and members of their families underwent searching interviews and detailed inspections, including inspections of their homes. The inspectors filed reports of their findings and conclusions with the company. Using the reports, the department fed, housed, trained, and educated company employees and their families. The program was considered successful because employee turnover was much reduced.

World War I

Until U.S. entry into World War I, Henry was an outspoken pacifist and against preparedness and conscription. Ford's sponsoring of a "peace ship" in 1915 was a failure and an embarrassment.

Couzens, though a man of strong character and unshakable honesty, with strong organizing capacity and great commercial acumen, had a negative disposition and was generally not well liked, but he is given credit, along with Ford, for achieving the company's success. So long as Couzens operated separately from Henry, things were stable. However, when in 1915 Couzens objected to a pacifist statement by Henry as being injurious to company interests, Henry forced Couzens to resign. John Dodge also resigned.

After Germany announced on February 1, 1917, its intention of resuming unrestricted submarine warfare on shipping suspected of delivering war material to the Allies, diplomatic ties were severed between the United States and Germany on February 3, 1917. Henry Ford suddenly changed from a pacifist to an industrial warrior. He promised to operate his factories for war production "without one cent of profit." Although his factories produced military goods, his biographers were later able to suggest that this promise was evidently not kept.

Highland Park turned out a variety of war materiel. Ford's American plants produced about 2,000 ambulances and about 39,000 other motor vehicles. The Model T was adapted for use as an ambulance, but Ford's proposal to use the car as a sort of disposable tank was never accepted. Two tank prototypes were developed but were never produced because of the coming of peace.

Henry thought it would be a good idea for the company to make its own steel to better control quality. He liked the idea of a new large Rouge manufacturing plant. In the spring of 1915, the company began to buy large tracts of land along the Rouge River southeast of Detroit. Plans for a huge industrial complex on the site were announced.

The Dodge brothers, desiring to use Ford dividends to finance their own expansion into car production, brought a lawsuit against Ford to stop diversion of profits into the Rouge plant. Although Ford was permitted by a court in 1917 to continue his Rouge expansion, in 1919 another court required Ford to distribute a dividend that amounted to well over $19 million plus interest. Though Ford as principal stockholder gained much of this dividend himself, he then resolved to buy out all minority stockholders, including the Dodge brothers, and did so using $75 million borrowed from a consortium of banks. Work on the Rouge plant construction continued.

In the 1920s, the vast Rouge River plant was built, occupying almost 2 square miles, with large buildings and more than 100,000 employees. Experimentation and growth continued to be main ideas of Henry.

Foreign competitors seemed dumbfounded that Ford could turn out and sell cars profitably at a fifth or even a sixth of the price at which they could afford to sell theirs. By 1920, the Ford Motor Company owned not only the Highland Park and Rouge River plants but also branch plants throughout the world. Henry Ford had achieved an industry of unprecedented size.

In 1920–1921, Ford got into a financial scrape from which he extricated himself by his own techniques. Between the adverse decision in the Dodge brothers' lawsuit regarding profit distributions (see above), the outlay of capital for development of the Rouge and for the purchase of coal and iron mines in northern Michigan for his new blast furnaces, and the money needed to buy out the minority stockholders, Henry then reportedly had about a $60 million debt, which was associated with a payment due date at lending banks. Henry's initial response was to cut car prices drastically, but, after an initial surge, Ford car sales fell drastically, as did the car sales of other manufacturers. General Motors was driven into real difficulties. Cash flow became dangerously low. Ford shut down the company on Christmas Eve, 1920, and did not reopen until February 1, 1921. In the interim, in manufacturing areas and offices, all nonessential equipment, including office equipment such as desks, typewriters, pencil sharpeners, and telephones, was stripped and sold. Staff was drastically cut, various departments were merged or eliminated, and capable executives were fired or resigned (including Knudsen and Hawkins, who both went to General Motors). Parts inventory was reduced, new inventory was bought only as required, and payments to suppliers were lengthened to 90 days. Excessive consignments of new cars were shipped to dealers, forcing dealers to borrow from their local banker to make their contractually required payments to Ford in full upon delivery. As a result, Henry paid off all his debts and even had a surplus without borrowing further and without losing control of his company to lenders.

Ford then, in early 1922, bought the Lincoln Motor Company out of bankruptcy for $8 million from the Lelands (Henry and his son Wilfred). Henry Ford evidently considered that it was Henry Leland who had been responsible for his being fired in the early days. Soon both Lelands were fired by Henry. Henry gave Lincoln to Edsel to operate. Lincoln designs under Edsel Ford and Eugene Gregory then gave the company good visibility at the upper end of the automobile market.

The End of the Model T and the New Model A

In the midst of his various activities, production and sales of the Model T were not forgotten. Production of the Model T slowed during 1917 and 1918, but they gained in 1919 with an improved Model T that included an all-electric starter.

Over the preceding 12 months ending July 31, 1916, the Ford Motor Company reportedly made more than 500,000 Model T cars, sold nearly all for about $440 (the touring car price; the runabout was less and the closed town car more), and achieved profits of almost $60 million. Some Model Ts were reportedly sold for only $290. In the year ending in December 1919, Ford made and sold more than three-quarters of a million Model Ts. By 1919, one in three cars sold in the U.S. was a Model T. In 1920, the postwar boom abruptly ended, and car sales generally dropped and then slowly improved.

The all-time peak year for Model T sales was 1923, but Henry was opposed to updating and making needed improvements in the car, such as hydraulic brakes. Sales of Model Ts slowly began to slump. On May 26, 1927, Ford announced that the 15 millionth Model T had been made and that production would stop at the end of May. Ford also announced that a new Model A (which he had named himself) would succeed the Model T, which caused many buyers to hold off purchasing a car until they had seen the new Ford car.

Times and tastes of car buyers were changing. A car customer wanted his new car to have the newest comforts and technology and the latest in fashions. Planned obsolescence became a reality. Ford had to adapt, but never again would there be a car like the Model T that was so low cost, so durable.

Six months later, after the end of Model T production, in November 1927, a hastily designed new Model A automobile was unveiled. Within a day and a half, more than 10 million people had made the effort to see the car. By Christmas of 1927, dealers had almost half a million orders, each supported by a cash deposit.

Wills was gone, but young Lawrence A. Sheldrick worked with Henry designing a new engine for the car. Edsel worked on styling.

The new Model A had many features, including balloon tires; a transmission with strong gears operated by a stick shift; a reliable electrical system; a self-starter; a strong, peppy, rapidly acceleratable, four-cylinder engine; hydraulic brakes; hydraulic shock absorbers with rubber cushioning; and quieting insulation. The entire mechanical system was enclosed in a rounded body shell that Edsel had developed. The passenger compartment was enclosed with access doors. Henry Ford apparently managed the engineers and the designers, but (as far as is known) except perhaps for some aspects of the engine design, no specific technological innovations in the car are attributable to Henry Ford.

Ford achieved the new Model A without extensive research, development, innovation, or testing facilities, such as were characteristically then utilized by General Motors. Henry's research and development work was comparable to the same old cut-and-try, pragmatic, experimental approach he had used at the Bagley Avenue shop and earlier. He insisted on absolute control yet provided divided attention. Ford's greatest difficulty with the new Model

Henry Ford's Other Endeavors

Ford became involved with other endeavors besides automobiles.

Tractors

Henry liked farm machinery. He desired to reduce "farm drudgery" and wanted to produce a low-cost, efficient tractor. As soon as his income from car sales permitted in 1905, he started serious tractor development.

Testing continued for years. In 1915, Ford established, initially apparently at his own expense, a tractor plant in Dearborn, and in 1916, he began commercializing his "Fordson" tractor. The name Fordson resulted from shortening Henry Ford & Son; he could not use his name alone because a Minnesota entrepreneur had already registered the name Ford as a trademark for a tractor.

After the United States entered World War I in April 1916, Ford, in response to a British request, agreed to supply engineering drawings and personnel (including Sorensen) for producing the tractor in England. These manufacturing plans were abruptly changed after the Germans bombed London's Fleet Street financial district, and the originally proposed tractor plants were rushed into airplane manufacture. The tractors that were needed would have to come from America. The British government ultimately ordered 8,000 Fordsons, but Sorensen was able to ship and deliver only about 3,600 Fordsons by March 1918. Fordsons for U.S. consumption were available in April, 1918. The Fordson did reform the tractor and the tractor business, and American production ultimately exceeded 750,000.

Ships

The Navy agreed to have Ford build a specialized, fast, lightweight, submarine chaser that the Navy had designed and called the Eagle. Henry assigned supervision of this project to William Knudsen, a manager acquired in the Keim Mills purchase. Henry got the U.S. government to fund $3.5 million for a huge Eagle ship assembly building. After the war, about 60 ships were delivered to the Navy.

Aircraft

In late 1922, Henry invested in the Stout Metal Airplane Company. The Stout-designed trimotor, approved by Ford, was the first American multiengine, all-metal aircraft. Embodiments were built by the company and are among the most durable aircraft ever built. Ford was the first company to use radio to guide a commercial airliner, the first to establish regularly scheduled passenger flights (between Cleveland and Detroit), and the first to

(*continued*)

establish a regular airmail service. The Great Depression forced Ford to concentrate, about 9 years later, on automobiles. Ford was never an innovator, only a promoter, in early aviation.

Hospital Development

In 1914, when it appeared that a Detroit-area hospital, for which Ford had previously agreed to be chairman of the fundraising committee, would collapse, Henry paid back every dollar in contributions previously made by others, assumed all liability for the hospital's debts, and made the project all his own. The hospital was renamed the Henry Ford Hospital. Some Detroit doctors considered the place a "human garage," but under Ford, the hospital introduced a systematic line examination and diagnosis procedure.

Village Industries and Education

In 1916 for a brief period, Henry formed and supported the Henry Ford Trade School for training boys from poor families in various trades. Later, he funded a private school system at Greenfield Village emphasizing vocational training.

For a brief period in about 1918, Henry supported the unsuccessful development of "village industries" in southeast Michigan.

Soybeans and Chemicals

In 1916, Henry set up a company with himself and his son Edsel as stockholders that was called Henry Ford and Son Laboratories. The two main projects attempted were fuel alcohol from agricultural plant wastes and soybean culture. He hired his boyhood friend, now Dr. Edsel Ruddimen (for whom Henry's son Edsel was named), away from university research. Later, these projects were abandoned.

Greenfield Village

In 1919, Henry started a project that became known as Greenfield Village. It now occupies over 250 acres and contains what is essentially Henry Ford's tribute to the past that was forever changed by his development of a successful automobile.

Thomas Edison was persuaded by Henry to lay the foundation stone of the Henry Ford Museum in Greenfield Village on October 21, 1929, the fiftieth anniversary of the date of Edison's invention of the incandescent light bulb.

A involved production. Although Ford sold 1.5 million new Model A's in 1929, Chevrolet overtook Ford again the following year.

The V-8 Engine

Perhaps the last significant engineering challenge faced by Henry Ford was the V-8 engine.

At the end of 1929, Ford's former employee, William Knudsen, now in charge of General Motor's Chevrolet Division, introduced the six-cylinder Chevrolet. A few weeks later, Henry Ford decided that, because Chevrolet was going to a six, "we are going from a four to an eight."

Eight-cylinder engines were then well known, including those with two four-cylinder groups, each inclined side by side in a V-shaped configuration, but all had blocks with multiple sections. Henry Ford envisioned a V-8 engine costing only a fraction of that of other V-8 engines that would use a relatively simple, single-cast V-8 cylinder block that could go down the Rouge production line at forty units per hour or more. He was raising problems that no one had previously contemplated.

By December 1931, Ford engineers, working on Henry's orders, had designed, built, and tested perhaps thirty different V-8 engines, and all had been rejected by him. Henry and Edsel then conferred and decided to make a V-8 engine for a new Model B as a sequel to the lagging sales of the four-cylinder Model A. Every Model B customer would have the option of either a four-cylinder engine or an eight-cylinder engine. Charles Sorensen estimated that the new investment needed for the V-8 engine would be as much as $50 million, but Henry authorized the expenditure.

News of the new Ford car with a V-8 engine was carefully leaked to the public, and 100,000 advance orders for the new Model B with the V-8 engine were received. Orders doubled after the product appeared on March 31, 1932. This first low-cost, truly hot rod car was a delight to many, especially to gangsters like John Dillinger and Bonnie and Clyde Barrow, who sent written testimonials to Henry. In the early 1950s, Fords were still being made with V-8 engines that were essentially the same as designed about 21 years earlier.

Ford's Other Endeavors

With his position in automobiles established, Henry Ford turned to other projects. He used his wealth to support a variety of projects, most of which can be loosely classified as nonautomotive. Ford certainly displayed management and direction in these projects and spent some of his money well, but except perhaps for the V-8 engine, all seem to have involved little or no personal innovation, and none reached product innovation levels like those associated with the Model A and the Model T. A highly abbreviated review of these projects, within space limitations, is provided in the sidebar for comparison and information purposes.

Ford as a Person

Nature and Character

Henry Ford functioned as a creative managerial person who provided direction and produced innovative development. Particularly in his early automobile years, he had the personality and ability to get willing associates to understand and pursue his communications. Ford would move among his men, both those on his immediate development team and also the car assembly laborers. He was perceived to be kind and casual, radiating good cheer, conveying an infectious sense of achievement, inspiring loyalty, and inducing men to carry out and achieve his ideas and concepts. In, for example, *Reminiscences*, one George Brown comments:

> He could get anything out of the men because he just talked and would tell them stories. He'd never say, "I want this done!" He'd say, "I wonder if we can do it. I wonder." Well, the men would just break their necks to see if they could do it.

Particularly in his early automobile production years, Henry was more concerned with development, production details, and improvements than with details of business, sales, and money, which he was happy to leave to Couzens and others, to whom much of the credit for the success of the Ford Motor Company must fairly be given.

Generally, as he grew older, Ford changed. He tended to be outspoken, and he elicited from others a full spectrum of responses, ranging from admiration and approval to hostility and disapproval. As he advanced in years, he became increasingly contradictory, controversial, changeable, seemingly unable to achieve a balanced perspective on events, and even able to say one thing while meaning or thinking another.

In 1918, Henry campaigned for the U.S. Senate and narrowly lost by about 2,200 votes to his Republican opponent. He made an unsuccessful presidential bid that began in 1922 and that was linked to his unsuccessful attempts to buy for $5 million Muscle Shoals, a 75-mile-long complex of dams and generating plants that had been developed with about $85 million in federal funds along the Tennessee River and that later became the core of the Tennessee Valley Authority.

A friendship developed between Henry Ford and Thomas Edison. In 1918, Henry Ford, Thomas Edison, Harvey Firestone, and the naturalist John Burroughs began an annual tradition, which continued until 1924, of going camping in the Adirondacks of upstate New York. The men had chauffeured cars and trucks with attendants, and each man emerged from his 10-by-10 foot tent in the morning in a three piece suit with collared shirt and tie. One year, President Warren G. Harding joined them. The occasions were well covered by the press. Henry Ford and Thomas Edison established adjoining vacation residences in Florida.

Henry was, at least for a time, anti-Semitic. Through the Dearborn *Independent*, a publication that Ford bought in 1919, highly anti-Semitic material was collected and published first fortnightly, then weekly, beginning with the issue of May 22, 1920. A book was published by Ford reprinting this material. A lawsuit by Aaron Shapiro resulted in a settlement agreement under which Henry published a retraction in the *Independent* on July 7, 1927. Thereafter, the *Independent* seems to have gone out of business. However, Henry seems to have always had many Jewish friends and business associates.

Edsel and Henry II Succeed Henry as President

On December 31, 1918, Henry Ford formally left the presidency of the company and placed his son Edsel, then 25 years old, in the spot, where Edsel remained for the rest of his life. Edsel displayed some mechanical and managerial ability and had a good capacity to design vehicles. He was a patient, even-tempered, kindly man.

Henry never gave the reins of control to Edsel, and Edsel was never able to grasp them. Henry thought he needed to actually run things. Edsel remained ever submissive to his father.

Unfortunately, Edsel became ill and died in 1943, four years before Henry's own death. Henry, though mentally enfeebled, then reassumed the presidency of the company. Edsel's son (Henry's grandson), Henry II, then 26, was called home from the Navy and subsequently became president of the company.

Henry II, before his own death in 1987, removed and destroyed a large quantity of material on his grandfather and father from the Ford Archives in Dearborn and from the Henry Ford Hospital in Detroit, which handicaps scholars and researchers today.

Ford's Demise and Achievements

When Henry Ford died on April 7, 1947, at age 83, he had sent over 20 million automobiles and 1.7 million tractors out of his factory doors with his name on them. Ford thought that a low-priced car of standardized components and standardized production would expand sales and increase profits, and his thought proved to be correct. Ford and his associates built cars that developed and fitted the market, promoted road improvements, developed industry and commerce, established many new jobs, and created billions of dollars in new capital.

FURTHER RESOURCES

Henry Ford. *My Life and Work*. Sioux Falls, SD: NuVision Publications, 2007.
Jacqueline Harris. *Henry Ford (An Impact Biography)*. New York: Franklin Watts, 1984.
Robert Lacey. *Ford: The Men and the Machine*. Boston: Little, Brown and Company, 1986.
Allan Nevins and Frank E. Hill. *Ford: The Times, the Man, the Company*. New York: Charles Scribner's Sons, 1954.
Allan Nevins and Frank E. Hill. *Ford: Expansion and Challenge, 1915–1933*. New York: Charles Scribner's Sons, 1957.
Allan Nevins and Frank E. Hill. *Ford: Decline and Rebirth, 1933–1962*. New York: Charles Scribner's Sons, 1962.
Michael Pollard. *Henry Ford (Giants of American Industry)*. Woodbridge, CT: Blackbirch Press, 2003.
Steven Watts. *The People's Tycoon: Henry Ford and the American Century*. New York: Vintage Books, 2005.

Library of Congress Prints and Photographs Division

The Airplane and the Wright Brothers

That the brothers Wright, Orville and Wilbur, succeeded in creating and building a powered, controlled, aircraft with practical flight characteristics, which had never before been accomplished, was an important accomplishment. Done with much applied intelligence and intuition by men with little formal education, their achievement must be regarded as great history. Aircraft development thereafter proceeded rapidly everywhere, but the brothers' technology was soon outmoded.

BACKGROUND

In antiquity, humans had created wonderful fables about flying, but all were not only fictional but also physically impossible: Daedalus and his wax-bonded feather wings, King Kai Kaus and his throne transported by eagles, and Prince Ahmad and his flying carpet, to name some well-known examples.

Not until the seventeenth century did humankind achieve flight, when the Montgolfier brothers in France succeeded with large balloons charged with hot air (see sidebar). In following years, proposals for various flying machines appeared, including powered balloons (airships), ornithopters (flying machines with flapping wings), and helicopters (aircraft having a powered, large, horizontal propeller mounted above the fuselage). By the nineteenth century, balloons were used as circus attractions, as air exploration vehicles, and as military observation posts. Count Zeppelin, after observing American Civil War observation balloons, returned home to Germany and later, in the 1890s, created the zeppelin, a cigar-shaped, rigid, dirigible airship internally holding a row of bags filled with a lifting gas, and externally supporting a suspended gondola that housed the powering engines, a control cabin with pilots, and sometimes passengers. Though some embodiments of the zeppelin survived and were used until about World War II, by then, the evolution of airplanes had reached a level where the zeppelin was only a curiosity. Contemporarily, occasionally, a blimp, or relatively small, nonrigid, powered, controllable airship, can still be observed as an advertising or attention-getting structure moving slowly over the vicinity of a crowd.

Balloons

Before the powered, pilot-controlled airplane, the balloon and gliders were invented. In 1783, the French Montgolfier brothers, Joseph (1740–1810) and Jacques (1745–1799), two of sixteen children of a paper manufacturer, filled a linen bag about 35 feet in diameter with hot air. In the first flight, in June 1783, their balloon, unmanned, lifted about 1,500 feet and floated about 1.2 miles in about 10 minutes. Later, in November 1783, in Paris, before an estimated crowd of 300,000 that included Benjamin Franklin, they

(*continued*)

achieved a manned flight of over 6 miles. Their balloons rose because of the expansion of hot air, which is less dense and thus lighter than the surrounding atmosphere.

In Paris, Jacques Charles (1746–1823; a French physicist and teacher at the Sorbonne), upon hearing of the Montgolfiers and their balloons, immediately realized that hydrogen, the light gas discovered in 1766 by Henry Cavendish (1731–1810; an English chemist and physicist), would be a much better buoyant force, though substantially more costly. By late August 1783, Charles had constructed the first hydrogen balloon, complete with the devices needed to operate it. He himself went up several times and reached a height of over a mile.

In August 1804, Joseph Gay-Lussac (1788–1850; a French chemist) went up with Jean Biot (1774–1862; a French physicist) in a hydrogen balloon. Their balloon was left over from Napoleon's Egyptian campaign. They loaded it with instruments and small animals. In their descent, which was tricky, Biot reportedly panicked. Later, in his own balloon, Gay-Lussac reached a height of about four miles. The measurements made when airborne showed no change in air composition or in the earth's magnetic force.

Apart from their work on balloons, Charles and Gay-Lussac each independently discovered that different gases expanded or contracted by equal amounts with a rising (increasing) or cooling (decreasing) temperature, the amount being 1/273 of the volume at 0° on an absolute scale. Gay Lussac made various other discoveries in chemistry.

Beginning mainly in about the mid-eighteenth century, the science of flight developed from experiments, though Giovanni Borelli is credited with having observed about 1680 that the arm muscles of humans are not strong enough in proportion to total weight to support flying by flapping as birds do.

The experimental and fundamental discoveries by the Englishmen George Cayley (1773–1857) and John Stringfellow (1779–1883) illustrate the important developments in flying the century before the Wrights and that prepared the way for manned, powered, controlled flight. Cayley learned, inter alia, that slightly arched (cambered) wings provide greater lift than flat ones, that a region of low pressure that develops on the upper surface of a moving, cambered airfoil aids lift, that upwardly tilted wings provide lateral stability, that rear rudder and elevator surfaces provide directional and longitudinal stability, and that streamlining reduces drag. Stringfellow, at first working with engineer William Hensen, succeeded in 1848 in getting a small model to fly and in 1868 produced a light, powerful steam engine that generated more than 1 horsepower yet weighed only 13 pounds. Stringfellow even built a bird-like flying machine with a 10-foot wingspread and with airscrews powered by a small steam engine that flew about 40 yards.

Clement Adair in France and Hiram Stevens Maxim in England each built full-sized, steam-powered flying machines, but when they were tested, each elevated a few inches, moved forward a few yards, and crashed.

In the 1890s, gliding superseded ballooning as a great aeronautical sport. German Otto Lilienthal, trying to imitate birds, produced arched wings and in 1891 made his first glide, but after more than 2,000 flights, while testing a new rudder design in 1896, he crashed and died. His work, though, did encourage the Wright brothers: they thought they had deduced why he had crashed.

Samuel Pierpont Langley, an American astronomer who became secretary of the Smithsonian Institution, worked out aeronautical principles. He made an unmanned heavier-than-air powered flying machine that flew in 1896 and then tried to construct a machine that could carry a human. Encouraged by President William McKinley and granted $50,000 of government money, he made three unsuccessful trials between 1897 and 1903. The money had been granted because the Spanish-American War had stimulated interest in potential military applications, but he could not obtain further money. Langley is thought to have failed because of the insufficient strength of the structural materials he used and/or the limitations of the engines he used. Nine days after Langley's last attempt, the first successful flight of a powered, controlled aircraft was made by the Wright brothers at Kitty Hawk, North Carolina. Subsequently, Langley was friendly with the Wrights. He died in 1906.

Langley had failed where the Wrights had succeeded with far fewer resources. It has been proposed that Langley proceeded in an incorrect manner: He had first built a powered flying machine in the belief that the pilot would learn to fly it after it was airborne. The Wrights, in contrast, had first learned to fly in gliders and then had installed an engine into a flying machine that they knew how to fly.

Before December 1903, various surprising full-size machines were created and fabricated, but these either did not have an airframe and associated engine sufficient to sustain independent flight, or did not have, perhaps most importantly, a pilot-controlled system allowing a course or changes in course while airborne. Also, these prior machines were not well documented by actual witnesses and/or were not repeatable in aerial performance. However, apparently some still argue that the Wrights were not the first to fly a powered, controlled airplane.

On December 17, 1903, though, the Wright brothers succeeded in flying a powered, heavier-than-air, pilot-controlled airplane that they had created, developed, and produced following careful experiments and testing. Their flying was repeated and documented.

Their first flight achievement was not a situation where the brothers had suddenly and miraculously succeeded in doing what many previously considered impossible. Previously, others had carried out tests and experiments and had gained knowledge sufficient for them to conclude that flight was an achievement that would inevitably occur.

How and why did the Wright brothers succeed in producing the world's first powered, pilot-controlled, sustainable aircraft? The answer seems to be clear from their biographies.

THE WRIGHT BROTHERS

Early Years

The brothers Wilbur and Orville were the sons of a bishop of a minor Christian group, the United Brethren Church. In 1884, the family moved from Millville, Indiana, to Dayton, Ohio, then the center of their church. Their mother died of tuberculosis in 1889. Their father, Milton, raised his family.

One day in 1878, when the boys were young, their father brought home from one of his frequent trips a flying toy. It was a small model made of bamboo and cork with paper airscrews driven by rubber bands. The model had been invented by Alphonse Penaud, a French invalid, who called it a *helicoptere*. The bishop operated the toy and sent it flying around the house. The boys at once were fascinated. When the toy became broken, the boys made others and experimented.

Wilbur and Orville had two older brothers and a younger sister, Katherine, all of whom had been sent to college. The older brothers, Loren and Reuchlin, married and left home. Katherine took care of her younger brothers, who it seems, at times, almost considered her their mother.

Wilbur and Orville decided not to go to college. Both were proficient with their hands and had mechanical aptitude. Before 1890, they built a covered, wrap-around porch on two sides of their family home in Dayton. They joined kite clubs and began to make kites they sold to other boys. Orville was a champion bicyclist. At age 13, Orville built a printing press and his 17-year-old brother Wilbur made improvements to it. In 1889, Orville opened a print shop, and his brother joined him in 1890. The brothers became inseparable in their teens. It seems that they quit high school before finishing. In 1892, the so-called safety bicycle appeared and became popular, and the brothers opened their own bicycle shop at Dayton, where they not only repaired bikes, but also, in 1896, designed, fabricated, and sold bikes. The shop became successful and profitable.

Wilbur (1867–1912) was the dominant one of the pair and was tall, thin, narrow-lipped, and hawk-nosed with gimlet-like eyes. Orville (1871–1948) was smaller, rather common in appearance, shy and even retiring, and liked his mustache. Wilbur was serious and thoughtful; Orville was more fun-loving and happier in outlook. It seems that each made inventive contributions regarding their work. They were known to argue intensively, but they enjoyed doing so. Both were stubborn regarding what they considered to be matters of principle. As sons of a minister, they lived the most proper of lives. Neither smoked, drank, or married. Both usually wore conventional business suits complete with neckties, even when tinkering with machines and tools. Orville conceded years later that they should have gone to college, but by then it was too late. Orville commented that it would have saved them "much fumbling."

They pursued gliding as a hobby, and they became well educated regarding aviation. Wilbur seems to have taken the initiative, using "I" and "my" a flying machine and working on plans in 1899–1900 before Orville became fully involved. In late May 1899, Wilbur had written to the Smithsonian Institution in Washington, D.C., for "papers" regarding flight that the Smithsonian had published and "if possible a list of other works in print in English," and in reply was sent a number of pamphlets and books. The brothers decided they should read more about aviation. They read publications by and about Otto Lilienthal (1848–1896), including accounts of his gliders and manned gliding efforts, and data about flying he had gathered. Lilienthal had, as mentioned above, made more than 2,000 glider flights. He controlled his gliders mainly by body swings below the glider, but he had died in 1896 from injuries sustained in a crash with a glider having a new rudder design.

They read Octave Chanute's "Progress in Flying Machines," which had been written in 1891–1893 and had first appeared in the *Railroad and Engineering Journal*. Chanute (1832–1910) was a French civil engineer who then lived in Chicago and who had taken up gliding as a hobby at age 60. Chanute, perhaps considering himself too old for the body movements required to control a glider, provided his gliders with control surfaces that he or coworkers could move. Chanute's first glider had five wing pairs, one on each side and one above another.

The brothers also read Samuel Pierpont Langley's 1891 *Experiments in Aerodynamics*, and anything else they could acquire on flying. Langley (1834–1906) had even received government funding for his manned flying machine, but it was unsuccessful. They were knowledgeable about Cayley's and Penaud's machines. They were aware of the death in 1899 of the British aerialist Percy Pilcher, also through hang gliding. The brothers were well read regarding prior aeronautical science.

Progressive Experiments and Developments

From the start, the brothers proceeded deliberately and carefully, step by step. Their mechanical aptitude and manual training in the bike shop were assets. At first, for control, they liked the idea that the pilot would use his own balance and muscle control to keep the winged surfaces of a glider facing into the wind, but they soon decided that the deaths of Lilienthal and other glider pilots were caused by a lack of lateral control. They observed that previous flying machines had rudders for left and right steering and elevators for up and down movement, but had no control for preventing a machine from tipping over laterally on one side or the other. They came to think that an aircraft should have control means for each of the three axes.

Wilbur and Orville began aeronautical experimentation in 1899. In December 1899, Wilbur wrote to the U.S. Weather Bureau for information about where steady winds could be found. In May 1900, Wilbur wrote to Octave Chanute, starting a correspondence with him that continued until Chanute's death in 1910. The brothers were seeking suggestions and comments from Chanute. A reply letter from Octave Chanute indicated that steady breezes and soft landing surfaces existed along the Atlantic coast.

The first proposed solution to the perceived problem lack of lateral control was Orville's. He suggested mounting the outer wing sections on shafts that pivoted in opposed relationship so that each of these sections would be rotated about a lateral axis. When one wingtip was turned up, the other would be turned down, and the first wingtip would give more lift than the other. In effect, he had invented ailerons.

After some thought, they found this solution to be not practical because it was associated with structural weakness and mechanical complexity. A few weeks later, Wilbur proposed another solution: make the wings flexible and twist them by control cables. He illustrated this idea by twisting the opposite ends of a cut-open cardboard box between his hands. The wing on one side would meet the oncoming air at an angle different from the wing on the other side. They decided to try out the wing-warping system.

To test wing warping, in the summer of 1899, Wilbur, along with Orville, made a kite with a span of 5 feet that resembled a biplane. Attached cords enabled the opposed wing tips to be manipulated in opposite directions, from the ground while the kite was flying. The twisting made the kite turn and bank left or right, so it was clear that the wing warping idea worked.

(Later, it was learned that, 2 years previously, Professor Edson Gallaudet of Yale had experimented with such a wing-warping system using a similar kite. However, he dropped his experiments and even quit his job when told that his "flying gimcracks" were "making of an ass of himself and a laughingstock of the faculty.")

Next, the Wrights designed a glider that could carry a human; it could be flown tethered, it had a wing area of a little over 150 square feet, and it used the wing-warping system. The various wing and other design modifications achieved by the brothers became part of the disclosure in their subsequently filed original U.S. patent.

The brothers thought that maybe they could learn something by watching birds and decided to lie on their backs outside. This exercise was not particularly useful. As Orville later said, "Learning the secret of flight from a bird was a good deal like learning the secret of magic from a magician. After you once know the trick and what to look for, you see things that you did not notice when you did not know exactly what to look for." However, the brothers did note how birds soar on the wind and face into the wind, which creates a lifting force on the wing underside. They noted how birds twist their wing tips and changed the angle at which the leading edge of the wing met the moving air.

Apparently they then built a model glider that weighed 52 pounds and had a wingspan of 17.5 feet. The opposite ends of the wings were attached to cords that were manipulated from the ground. Their glider had been designed based on an interpolation of data published by Lilienthal and on information previously published by Chanute. The brothers liked to equip their models with a canard. A canard is an attached structure located in front of the fuselage and any propeller and which is provided with controllable surfaces for air. Later, the canard was omitted and became obsolete.

The brothers decided that strong, steady winds would be desirable for their experiments. They studied weather information received from the U.S. Weather Bureau and learned that such winds could be found at Kitty Hawk, North Carolina, a sparsely settled area. The ground there was also softer than it was in Dayton (desirable for soft landings). Train transportation from Dayton, though not direct, was available.

In September 1900, when the bicycle business became slack for the winter, they went to Kitty Hawk, taking the glider (disassembled) they had built. They camped, evidently, in a hand-built shack on the sandy, windy dunes at a site called Kill Devil Hills. Here they made camp for each of the next 3 years, during their visits for testing. Their housekeeping-type chores were divided between them. It seems, for example, that Orville cooked lunch and Wilbur washed the dishes.

They flew this glider structure both as a kite and as a glider. They found that the lift was not what they had desired, although the wing-warping system seemed to work well. The glider was tested unmanned as suspended from a little homemade tower, and on one day, Wilbur made about a dozen glides while lying flat on the lower wing to reduce drag. They used this pilot position for the next 5 years.

The following year, 1901, they returned in July with a new, much larger, improved glider having a wingspan of 22 feet, an increased chord (a straight line extending directly across an airfoil from the leading to the trailing edge) of 7 feet and a total lifting wing area of about 300 square feet. They were unaware that a low aspect ratio (the ratio of wingspan length to wing chord width) can reduce lift and cause control problems. Pitch control could be achieved by varying the camber (or arch) of the canard achieved by pushing down on a pitch bar. Wing warp provided roll control using foot pedals.

Chanute had joined them for a part of this period. Again they flew their structure both as a kite and as a glider. Proceeding cautiously, they kept the glider close to the ground as they learned their new art of flying. They found that the 1901 glider performed worse than the 1900 glider. Lift of the

1901 glider was less than Lilienthal tables would indicate, and the glider did not glide well. It achieved only about a third of the expected (calculated) lift. Sometimes it failed to respond as expected to wing warping and the craft turned opposite to the intended direction, a problem later termed adverse yaw. To try to improve performance, they associated the 1901 glider with a complex system of wire and rope that reduced the camber to 1/19 from 1/12. Performance improved some, but control was not as good as with their 1900 glider. With the 1/12 camber, a large nose down pitching moment resulted, so that the pilot had to move back to avoid nosediving into the sand. The brothers were disappointed. They returned to Dayton in August. Wilbur, in particular, was dejected and said to Orville that man might fly but not in their lifetimes. The brothers considered abandoning their ambitions regarding flying.

Chanute cheered them up. When they got home, they found an invitation from Octave Chanute in Chicago to deliver an address and a paper there in September 1901 at the Western Society of Engineers about their gliding work. The brothers, though flattered by the invitation, wondered what they had done worth attention, and they decided that their measurement of actual performance of their 1901 glider compared with Lilienthal's data, which had apparently been developed in a laboratory, might be of interest. Their sister Katherine "nagged" (in Wilbur's word) them to accept. They accepted the invitation, and Wilbur developed a speech and paper on this comparison.

On September 18, 1901, Wilbur gave a speech in Chicago to the Western Society of Engineers describing his and Orville's experiments and tests at Kitty Hawk until fall 1901; this was the first public report of their work.

The poor lift of their gliders caused the Wrights to question both Lilienthal's data and the so-called Smeaton coefficient, which had been used for over a century in the standard equation for lift. Wilbur suspected that the value of this coefficient was actually about 0.0033 on the basis of measurements that he and Orville had made, not the value of 0.0054 previously used, which in effect caused calculated lift to be greater than actual value. To test, the brothers attached a freely rotating, horizontally mounted bicycle wheel before the handlebars of a bicycle that was associated with a Lilienthal airfoil and a counter plate. While operating the bicycle on local streets, they found that the third wheel rotated in response to the airfoil instead of remaining motionless, as expected from calculations. The brothers' suspicions that published data on lift were inaccurate was verified. They decided to carry out some more research.

Wilbur with Orville started tests in their shop using a little 6-foot wind tunnel that incorporated a soapbox. From October to December of 1901, after Wilbur's return from Chicago, they systematically tested various miniature wings using supports they called "balances" that were mounted in the tunnel to hold the wings so that they were able to balance lift against drag. Thus, they were able to calculate the performance of each wing. There was a viewing window in the top of the tunnel, through which they could watch the wings and observe which wings worked well. The brothers used straightforward trigonometry to study the forces, which was the best methodology available then. (Today, vector analysis with calculus is used to perform such a study.)

Their preliminary data had suggested that Lilienthal's data were wrong. They discovered that all known published tables of lift and drag, even Langley's, were too inaccurate to be useful. So, in Dayton in 1901, upon Wilbur's return, the brothers, using their more precise wind tunnel and associated equipment, developed their own lift and drag information.

They thought out the forces acting on a wing in flight and decided on what they should measure. They fabricated simple equipment that was actually a sophisticated mechanical analog of the forces acting on a wing. During the last quarter of 1901, the brothers tested wing surfaces. They evaluated the action of wind pressure on more than two hundred miniature wing surfaces. They found that the leading edge should not be sharp, as had previously been thought; that highly cambered wings were inefficient; and that unless a glider aircraft could glide and soar without an engine, it would not be stable in flight with an engine. They calculated their own tables of lift and drag. Lift and drag forces act at 90° relative to each other. From their measurements, the brothers applied their data to aircraft design and achieved an understanding of the relationships between lift, drag, and wing shapes. From the discouragement of September 1901, the brothers advanced to confidence in aeronautics by February 1902. They realized that their lift and drag measurements had accomplished in a few months what had previously escaped scientists for over a century.

The lift coefficient and the drag/lift ratio were derived from the raw data. They found that the wing of their 1901 glider had an aspect ratio (defined as the square of the wing span divided by the wing area) of about 3.14, so that such might only lift about 75% as much as a wing with an aspect ratio of about 6. The wind tunnel experiments conducted from October to December 1901 provided the Wrights with answers regarding the low performance questions from their 1901 flight trials. However, problems with control and stability would have to wait until the next round of flight tests to be addressed. Design and construction of a new glider, known as the 1902 glider, proceeded in Dayton.

In late August 1902, Wilbur and Orville returned to Kitty Hawk with their important new 1902 glider. This glider had an aspect ratio of 6.4, a wing area of slightly more than 300 square feet, a wing camber of 1/24, a canard of increased area and aspect ratio, and a fixed, two-surface vertical tail located only 3.5 feet back from the wing trailing edge. The tail was intended to compensate for the adverse yaw produced by wing warp and perhaps possible adverse effects from their slightly anhedral wing design. For control, wing warping was achieved by the pilot shifting his hips in a cradle in the direction of turn, while the canard was now operated by a twist bar, with twisting upward providing increased pitch.

At Kitty Hawk, the brothers tied off the controls of their glider and flew it as a kite with the tether ropes attached to the leading outboard struts. The tether ropes stretched up at an angle that corresponded to the result of lift and drag, and the rope in effect was the hypotenuse of the lift-drag triangle. The angle was measured with a clinometer (a surveyor's instrument given to them by Chanute), and the tension on a rope was measured with an ordinary spring scale. Using trigonometry, plus the angle measured and the tension measured, they could calculate lift and drag. From the numbers for lift and drag, the brothers could and did determine from other formulas whether their wings were developing the desired lift and drag. From their equipment, called a lift balance, the coefficient of lift (a dimensionless factor relating airfoil lift, air pressure around the airfoil and planform area of the airfoil) was determined from which they could determine, by mathematics, the coefficient of drag (a dimensionless quantity used to quantify drag and to model the complex dependencies of shape, inclination, etc.).

They were thus able to build a table of coefficients suitable for many different airfoils at various angles of attack. The efficiency of wing shapes could be determined by plotting the lift and the drag of each foil against the angles of attack. The Wrights sought an efficient wing shape that would produce the most lift and the least drag. A high lift-to-drag ratio would produce a shallow gliding angle and extend a flight. The Wrights were beginning to realize that the aircraft design problem was more complex than they had first surmised.

To design their aircraft, they needed one more piece of information. Using their table, they compared the shape of the wing on the glider with the shape of the airfoil and used the standard lift formula and the drag formula to calculate the coefficient of air pressure (previously called the Smeaton coefficient). They determined the coefficient of air pressure to be 0.0033. (Later developmental work has placed this value at 0.00339 for low-speed aircraft. With their crude and primitive equipment, the Wright brothers were off by less than one ten-thousandth.) Using their own data, they made the airfoil flatter and reduced the camber (the wing curvature depth divided by the wing cord).

They had been corresponding with Chanute. He visited their camp in each of two different years, and he expressed wonder at their progress. Chanute cheered them up when they were discouraged. He also paid the railroad fare of a young man with medical training to be on hand with them in case they needed some broken bones set.

The Wrights designed and built their improved 1902 glider using their own data. It was the first plane to actually achieve the predicted lift. For this glider, they had decided to use separate control surfaces in front and a fixed, vertical tail fin in the rear, in contrast to the cross-configured control surfaces that combined rudder and elevator that Chanute and Langley had used.

The Wrights made more than 2,000 glides with their 1902 glider. It proved the success and correctness of their three-axis control system, but they discovered a new and hazardous problem: once in a while, the glider would tilt to one side and descend with a corkscrew motion. It turned round and round on one wingtip. The pilot could not control the event. Later this action was called a tailspin.

Much thought by the brothers at first produced no solution. Orville finally analyzed and solved the problem: if and when the glider tilted sideways, then if the pilot did not immediately correct the tilt, the glider would slide somewhat sideways in the air. The sideways component of the airstream struck the tail fin and caused the glider to turn in the direction it was sliding. Orville thought the remedy was to make the tail fin (or rear rudder) movable. The remedy proved to be successful. Their tests showed that the trailing edge of the rudder should be turned away from the wing end that had more drag (or lift) due to wing warping. During a turn, rudder pressure overcame differential drag and pointed the aircraft nose in the direction of the turn. The rear rudder prevented adverse yaw.

In late October 1902, the brothers left Kitty Hawk returning to Dayton. Their 1902 glider was very successful. In their final week, on just two days they made over 250 glides. At least one expert has maintained that the success of controlling the 1902 glider was actually the invention of the airplane. On June 24, 1903, Wilbur gave a second paper to the Society of Western Engineers.

Powered, Controlled Flight

Back in Dayton, the brothers decided to proceed as rapidly as possible with construction of a powered aircraft, to be called the *Wright Flyer I*. They needed an engine, and they wanted one that would develop 8 or 9 brake horsepower, weigh no more than 180 pounds, and be substantially free of vibration. When various engine manufacturers could not supply such an engine, they designed and built their own.

It was difficult to make the engine light in weight yet powerful. Their bicycle shop employee, Charlie Taylor, was helpful. After some struggles, they succeeded. They brought the engine weight down to about 7 pounds per horsepower. The engine had 12 horsepower and weighed 170 pounds.

Propellers presented another problem. They found they had to design their own laminated wood propellers. From more wind tunnel tests, they concluded that a propeller was basically a wing rotating in a vertical plane. They used twin "pusher," counter-rotating (to neutralize torque) propellers, each having three lamination layers. They mounted the engine and propellers in an improved version of their 1902 glider having a movable rudder.

These 1903 propellers were later tested and, remarkably, found to be more than 75% efficient under the conditions of the first flights.

The *Flier I* was a biplane that had white muslin wings about 40 feet across and weighed 625 pounds. The two pusher-type propellers were driven by their engine using automobile drive chains.

Back again at Kill Devil Hills in the fall of 1903, broken propeller shafts had to be replaced by two trips back to Dayton. Also, they were stalled at first by accidents and bad weather. By December 14, they were ready. The airplane had sled-runner-like skids in place of wheels and was launched from a little two-wheeled car that ran along a track defined by two-by-fours. There was then no strong wind, so it was decided to launch the plane down the side slope of Kill Devil Hill to gain speed for takeoff. Wilbur won the toss for who should first pilot the machine. On takeoff, the craft climbed too steeply, stalled, descended, and landed hard. Several parts were broken.

By December 17, repairs had been made and they would try again. The white sand was blowing in a 27 mile-per-hour wind, and at 10 a.m. it was useless to wait longer in hopes that the wind would lessen. Though local people had been invited to come and watch, only five came on that cold, windy morning: three men from a nearby lifesaving station, another man, and a boy. The witnesses apparently helped lay the track from which the plane would take off. On the track rode the little two-wheeled truck that was made from bicycle wheel hubs that fitted about and rode on a pair of laterally spaced rails made of two-by-fours. Frequent warm-ups for all in their stove-heated shack were needed.

Orville, whose turn it was, would pilot. He set up his cameras pointing at the location where he thought he would take off and asked a volunteer witness to press the button at the right time. He then boarded the plane and assumed the planned, prone position. The engine had been warming up, and the engine, through the chain drive, began rotating each of the two propellers, located behind the wings. The plane moved forward into the wind about 40 feet. Wilbur, apparently holding one wing tip to balance the craft, moved alongside until the craft accelerated and he began to stumble. The plane took off, lifted about 10 feet above the ground, advanced shakily into the wind at a ground speed of apparently about 10 miles per hour (an airspeed of 30 miles per hour), flew about 120 feet, and settled down upon the sand after about 12 seconds. Their plane had really flown!

At about 11 a.m., it was Wilbur's turn, and the second flight was made. Then, third and fourth flights were made. In the fourth flight, Wilbur had flown about 852 feet. Then, the plane was damaged while on the ground by a gust of wind. The damage was so severe that the *Flyer I* never flew again, though it was shipped back to Dayton and restored years later by Orville. However, not only had they flown the plane, but they had flown it with a complete, effective control system.

Aftermath

Orville wired their father with the news of their success. Lorin Wright, the elder brother, gave carefully worded, accurate statements to the newspapers. However, the Norfolk telegraph operator, disregarding a request from the Wrights, tipped off a local reporter, Moore, for the Norfolk *Virginia Pilot*.

Moore, unable to obtain details, created a story with imagined facts: The plane had flown 3 miles, and it had a propeller that pushed it from behind and another that raised it from below. The story was published in the *Virginia Pilot* and sent out by the Associated Press. Because of Langley's recent failure, this story about unknown adventurer's alleged exploits was viewed with skepticism. Only three papers published accounts, and they were short and inaccurate. During the succeeding few years, occasional short press accounts about the Wrights appeared commonly accompanied by editorials stating that the whole thing was presumably a hoax. The October 1905 issue of the *Scientific American* magazine said the alleged event must not have occurred because if it had, American journalists would certainly have reported it.

Developing a Practical Airplane

Back in Dayton, there was much to do. The brothers made a decision to begin withdrawing from the bicycle business and to pursue making and marketing practical aircraft. The brothers were not satisfied with the *Flier I*—it was underpowered and difficult to control—but by the following spring, they had built another airplane, the *Flyer II*, that they flew from a field called Huffman Prairie near Dayton (now part of Dayton Aviation Heritage National Historical Park, adjacent to Wright-Patterson Air Force Base). Twice they invited reporters, and each time the plane's engine refused to operate, so the reporters left further convinced that the thing was a hoax. Perhaps the brothers had intentionally feigned trouble to discourage reporters and publicity. Of course, when nobody was around, the engine worked and the machine flew well.

Flyer II in 1904 did demonstrate progress, but it frequently was out of control. In 1905, they built *Flyer III*, which had a separate rear rudder control not linked to the wing-warping cradle, so that each axis each had it own independent control, but *Flyer III* still had only marginal performance. First flown on June 23, the craft crashed on July 14 when flown by Orville, who was shaken but fortunately not injured. As rebuilt, the forward elevator and the rear rudder of *Flyer III* were both enlarged, and each was located longitudinally several feet further away from the wings. With these modifications, control and stability were improved. Longer flights became possible. These flights convinced the Wrights that they had achieved a practical machine they could make and sell. *Flyer III* in 1905 not only could fly in a straight line for 30 minutes (fuel permitting) but also could fly figure eights over Huffman Prairie.

On the borders of the field were two busy roads and a streetcar line. The local people seem mainly to have discounted what they were seeing, to have become accustomed to the flights, and to have been unaware that the flights were not only occurring but also lengthening.

Evidently, the Wrights made no flights whatever during 1906–1907. The brothers did not try to keep their successes secret but did not seek publicity. They used money earned in their bicycle business to pay their own way, and they did most of their mechanical work themselves. They later indicated that the cash cost for their first airplane was under $1,000, but cost accounting would have produced a larger number.

Their basic U.S. patent, number 821,393, was issued May 22, 1906, and related mainly to the wing-warping system of lateral control. In 1906–1907, the brothers built several airplanes. They wrote their Ohio congressman, Representative Robert M. Nevin, asking whether the government was interested. The inquiry landed in the Army's Board of Ordnance and Fortification, which wrote the Wrights that their apparatus "has not yet been brought to the stage of practical operation" and they could not "make allotment for the experimental development of devices for mechanical flight." A reply from the Wrights induced this august board to respond that it could do nothing "until a machine is produced that by actual operation is shown to be able to produce horizontal flight and to carry an operator."

After 1905, the brothers stated they would make no demonstration or public flights unless they had a firm contract to sell an airplane. Efforts by the brothers to interest governments, concerns, and individuals in the United States, England, France, Germany, and Russia in buying or having a demonstration of one of their airplanes were unsuccessful, but in late 1907, the U.S. Army Signal Corps expressed interest in purchasing a plane, and in early 1908, a French syndicate expressed a similar interest. Both asked that the aircraft be capable of carrying a passenger. President Theodore Roosevelt asked Howard Taft, secretary of war, to investigate the Wrights' success.

The Wrights hastily modified the *Flier III* to have a second seat and a more powerful engine. They secretly tested these modifications back at Kitty Hawk (there for the first time in several years). They also practiced their flight demonstrations. Curiously, each brother preferred a different arrangement for the lateral and yaw controls on the pilot's right stick. Orville's method of rudder control differed from Wilbur's method. Then the brothers separated.

Orville, in the United States, flew at Fort Meyer, Virginia, with several different Army officers, who were greatly impressed. Orville offered to sell the Army a plane for $25,000, but then, as he was flying Lieutenant Thomas Selfridge on September 17, 1908, a propeller blade split, and a portion separated and connected with a guy wire. The plane crashed, killing Selfridge and substantially injuring Orville, who in time recovered. As matters developed, the U.S. government had no money for such a purchase, and Congress would presumably refuse to provide it. Selfridge was a member of a group of four that Alexander Graham Bell had assembled and was himself financing in Nova Scotia. The other members were Glenn H. Curtiss, a motorcycle manufacturer in Hammondsport, New York, and two Canadians, Frederick W. Baldwin and J. A. D. McCurdy. Bell had named the group the Aerial Experiment Association and had paid salaries to the three civilians to develop airplanes. Baldwin was the principal designer. The group's third machine, which actually flew, embodied Baldwin's invention of a hinged aileron in place of the Wright's wing-warping arrangement. After Selfridge's death, the group disbanded, but Curtiss built airplanes at his Hammondsport factory using the hinged aileron.

In 1908, Wilbur took a plane to France for exhibition flights. He began demonstration flights near Le Mans, France, on August 8, 1908. His short first flight of only one minute, 45 seconds, involving banking turns and flying in a circle, amazed a group of very skeptical French onlookers. Subsequent flights involved more challenging events that demonstrated the aircraft and Wilbur's piloting skills. Previous doubters gave apologies and offered much praise.

The recovering Orville and his sister Katherine joined him. In Europe, the three were much acclaimed. In June 1909 they returned to Dayton, where Orville and Wilbur enjoyed fanfare as the finally discovered local eminent sons.

Commercialization

Two of the army officers who had flown with Orville had gone to the White House in 1908. Theodore Roosevelt produced on the spot $25,000 from an emergency appropriation provided by Congress during the Spanish-American War. Orville finished purchase contract qualification tests in July 1909, delivered to the Army its first airplane, and was paid a total of $30,000, of which $5,000 was a bonus for exceeding the speed specification. He then left for Germany to provide exhibition flights.

The Wright Company was incorporated in November 1909 with money from investors. The Wrights sold their patents to the company for $100,000 and received one-third of a million dollar stock issue plus a 10% royalty on every plane sold. The company set up a factory in Dayton and a testing and a flight school at Huffman Prairie, with a headquarters office in New York City. The company began to produce aircraft.

In 1910, the brothers changed the design of their craft by moving the horizontal elevator from the front (the so-called canard design) to the back because a rear elevator made the airplane easier to control, particularly at higher speeds. They also added wheels.

The brothers also began suing a large group of infringers of their patent. In the midst of these events, Wilbur died of typhoid fever in 1912 at age 45.

Orville continued to pursue the lawsuits. Orville and Katherine felt Curtiss was partly responsible for Wilbur's death, which they indicated was brought on by his exhausting travel and stress during the Curtiss patent litigation.

Glenn Curtiss was considered by the Wrights to be the most significant infringer. Curtiss thought his assembly of positioning ailerons between the upper and lower wings would circumvent the Wrights' patent, but after protracted litigation the federal court held the patent valid and infringed. Curtiss had to pay royalties, but he continued to seek ways to attack or circumvent the Wrights' patent.

Curtiss obtained, evidently on loan, the wrecked but repaired and reassembled Langley aircraft and transported it to Hammondsport, where he rebuilt it. He made many changes, including moving the wings and changing the bracing. He added seaplane pontoons (Curtiss invented the seaplane). As thus rebuilt, the plane made several flights of a few seconds each on Lake Keuka. Curtiss hoped to weaken the Wrights' patent by indicating that the Langley machine could have flown so that the Wrights were not the first to fly and be airplane pioneers. These efforts by Curtiss were without consequence.

The Airplane and Lindbergh

Charles Lindbergh (1902–1974) was the first to make a solo, nonstop flight across the Atlantic Ocean from New York to Paris; he accomplished this on May 20–21, 1927, in 33.5 hours in a single-engine plane he had named *The Spirit of St. Louis*. He had obtained the backing of some St. Louis businessmen, purchased the plane, and won the prize of $25,000 offered to whoever made that first nonstop transatlantic flight. In the United States, the public response was tremendous.

His flight was more than a stunt. Since the 1903 achievement of the Wright brothers in flying their plane, flying had remained little more than stunts and thrills. The dogfights during World War I and some airmail service had not resulted in the public taking airplanes seriously as a means of transportation. Lindbergh's flight, though, brought the airplane into public awareness and expanded the potential for commercial flight.

Lindbergh, the son of a Minnesota congressman, had entered the University of Wisconsin in 1920 with the goal of becoming a mechanical engineer. Two years later, he suspended his university education and attended a flying school. In 1925, he purchased his own plane and became an airmail pilot.

After his solo flight, he worked in the early 1930s with Alexis Carrell (1873–1944; a French American surgeon) in designing an artificial heart intended for tissue perfusion. He twice thereafter entered the news. In 1932, his first son was kidnapped and murdered in a sensational crime, and in the late 1930s he was a leading isolationist opposing entry of the United States into World War II.

Entering History

Orville bought out the original backers and then sold the company to other backers and gained a fortune. He spent the remainder of his life in Dayton researching and developing aircraft improvements, such as an automatic stabilizer and a split-wing flap air brake. The successor company merged with that of Curtiss to form the Curtiss-Wright Company in 1929, ending the threat of additional lawsuits. It seems that the lawsuits scarred the reputation of the Wrights, but supporters argued that the Wrights were protecting their interests after years of work.

Orville also had a long-enduring feud with the Smithsonian Institution. Charles D. Walcott, a paleontologist, succeeded his former chief Langley as secretary. Walcott wanted to prove that Langley, and not the Wrights, invented the first actual flying machine. When the Langley machine, as rebuilt and changed by Curtiss, was returned to the Smithsonian, Walcott exhibited it at the Smithsonian with a legend identifying it as "The First Man-Carrying Aeroplane in the History of the World Capable of Sustained Free Flight." Whether or not it was so capable has never been resolved. The legend angered Orville since it implied that that the aeroplane could have flown before its reconstruction and before the Wrights' aircraft.

The dispute with Walcott over the legend regarding Langley's efforts and their meaning remained unresolved. Orville sent out the original rebuilt Wright aircraft on loan to the Science Museum in South Kensington, England, in 1923. After Walcott died in 1927, his successor, Charles Greely Abbott, a meteorologist, changed the legend on Langley's aeroplane in the exhibit to be innocuous. Orville kept the Wright aircraft in England until the Smithsonian acknowledged its error. Abbott finally issued an apology to Orville in 1942, whereupon Orville agreed to bring the aircraft back, but this could not be done immediately because World War II was in progress. The aircraft spent the war years in a bombproof British cave. Just after Orville's death in 1948, it arrived back in the United States and is now on view in the National Air and Space Museum.

The brothers succeeded in their efforts to achieve controlled, powered flight because, very briefly, they achieved three-axis control and through their own scientific research and development gained good aeronautical technology, which they applied to their Flyers. They had team skills, developmental plans, persistence, and organization. However, from their success, they promptly focused on achieving revenue. They did not try to undertake further research and development that might keep them ahead of the vigorous, capable, extensive competition that instantly resulted when their success had shown the way. Moreover, the competition was more moneyed, larger, better connected, and technically more capable than the Wrights.

By as early as about 1910, the Wright aircraft technology was outmoded. However, they earned an important, charming, and even unique place in history by succeeding in becoming the first to achieve controlled, powered flight.

In the United States, both governmental and private limited activity regarding aircraft existed prior to the outbreak of World War I. The U.S. Army then reportedly had only fifty-five planes, of which fifty-one were obsolete and four obsolescent; of sixty-five officers, just thirty-five could fly and only five could operate in combat conditions. Only about two hundred aircraft had been built in the United States since 1903. In contrast, the German army had 1,000 planes and a civilian reservoir of 450 more, and the French army had 1,500 planes and could requisition 500 more from civilians. By the end of the war, though, the U.S. Army had forty-five squadrons and 767 pilots who had flown 35,000 hours in missions over enemy lines.

By 1920, the airplane was far advanced from the reconnaissance craft of 1914, with design improvements by such men as De Havilland of England, Anthony Fokker of Holland, and a group of French designers. The Wright brothers were history.

FURTHER RESOURCES

Tom D. Crouch. *The Bishop's Boys. A Life of Wilbur and Orville Wright.* New York: W. W. Norton, 1989.

Fred Howard. *Wilbur and Orville: A Biography of the Wright Brothers.* New York: Alfred A. Knopf, 1987.

Peter L. Jakab. *Visions of a Flying Machine: The Wright Brothers and the Process of Invention.* Washington, DC: Smithsonian Institution Press, 1990.

M. W. McFarland, ed. *The Papers of Wilbur and Orville Wright, including the Chanute-Wright Letters and Other Papers of Octave Chanute.* New York: McGraw Hill, 1953.

G. D. Padfield and B. Lawrence. The Birth of Flight Control: An Engineering Analysis of the Wright Brothers' 1902 Glider. *The Aeronautical Journal*, 2003: 697–718.

The Wright Brothers & The Invention of the Aerial Age. Washington, DC: Smithsonian Institution, 1908.

AP Photo

Radio: Fessenden, de Forest, and Armstrong

Although Marconi had successfully generated and interrupted radio waves to transmit Morse code, it remained for Fessenden, de Forest, and Armstrong to succeed in modulating these uninterrupted waves with sound waves. Modulating radio waves with sound opened the way for wide usage of radio as a medium for the transmission and reception of information.

BACKGROUND

As previously related (see the chapter on Marconi), Marconi and apparently a few of his contemporaries succeeded in generating (by a spark-gap transmitter) electromagnetic waves, interrupting them in a telegraphic pattern that conveyed information, and then transmitting them so as to convey the information wirelessly to a receiver. At the receiver, the information was then separated to provide audible signals that telegraphically corresponded to the originally conveyed information. With experiments and a few developments, Marconi was able to demonstrate significant applications and uses for wireless radio telegraphy.

Historically, the commercial field of wireless telegraphy (also known as radio telegraphy) is considered to have begun effectively with Marconi's wireless transmission of information using Hertzian waves produced by, and employing, his commercial radio telegraphy equipment, the early manufacture of which seems to have been carried out in England. Marconi's public demonstrations, particularly around 1899–1901, plus the publicity associated with the East Goodwin lightship accident in 1899, promoted wide public interest in and use of wireless telegraphy, though there were others who were contemporaneously inventing, making, and promoting apparatuses involving electromagnetic waves.

Marconi and his contemporaries, though, were unable either to modulate a continuously transmitted electromagnetic signal with sound or to detect and reproduce (or demodulate) sound from a received, modulated electromagnetic signal. The term "modulate" here refers to changes wrought in the amplitude or envelope of a generated and continuously transmitted electromagnetic (or radio) wave by mixing that wave with a variable electric signal corresponding to a sound (acoustic) wave. The electromagnetic wave itself is characterized by having a frequency substantially above the level of the range of audible sound frequencies (or variable electric signals corresponding thereto). Additional inventions were necessary before wireless transmission of uninterrupted radio waves modulated with voice and other sounds, especially music, could be possible.

The telephone had been in extensive and expanding commercial use almost since its introduction and public availability beginning about 1876, and by the end of the nineteenth century millions of folks were communicating with each other over wires carrying electrically transmitted information corresponding to actual sound, not interrupted electrical signals corresponding to Morse code. Particularly to inventors, it seemed clear that there had to be a way to create and transmit (broadcast) radio waves modulated with sound or audible information, and also a way to reproduce the original information at a receiver from the modulated radio waves.

A simple solution was not possible, but ways to achieve the solution were worked out. The present account essentially begins where Marconi's work with wireless telegraphy ended. Various people worked to solve the problems and achieve the solution. Careful review and study disclose that there was no single originator or inventor of the modern modulated radio waves, their wireless transmission, or their reproduction at a receiver. Rather, it was the outstanding and cumulative contributions of a few inventive individuals that enabled suitable radio waves to be generated, modulated with audio information, transmitted, received, and reproduced in a practical manner.

These inventors found ways to do the following:

- Achieve practical modulation of radio waves with sound.
- Broadcast the modulated radio waves.
- Capture (receive) and amplify the faint (as received) modulated radio waves.
- Reproduce the original sound information.

As a group, they were not only bright, resourceful, innovative, and independent but also persistent, aggressive, and highly competitive. Looking back over the history of modern radio technology, three iconic inventions stand out, each achieved by a different inventor: Fessenden, de Forest, and Armstrong.

REGINALD AUBREY FESSENDEN (1866–1932) AND RADIO WAVE MODULATION

Summary

The first inventor to provide a practical solution to the problem of associating (modulating) a radio wave with a sound wave was Reginald A. Fessenden. In 1906, he succeeded in generating and transmitting a continuous radio wave of almost constant, relatively high frequency, whose amplitude he varied with the electrical equivalent of the lower frequencies of sound waves. The resulting radio wave had an envelope or periphery whose amplitude incorporated variations that corresponded to the frequency of sound waves whose electrical equivalent he had admixed with the continuous radio wave. Although, characteristically, sound waves usually vary greatly from one time interval to another, the constant frequency radio wave readily incorporated these changes as successive amplitude changes. Fessenden thus achieved amplitude modulation (AM) of radio waves.

For this successful experimental work, Fessenden used a single or fixed frequency radio wave that had a substantially constant frequency of about 50,000 cycles per second (or Hertz, in the current nomenclature). Separately but concurrently, he converted starting ordinary sound waves, which characteristically have frequencies below about 20,000 Hertz, to a corresponding audio wave electrical signal that was a variable and alternating electric current signal. He then mixed the relatively high fixed frequency radio wave signal with the relatively low-frequency variable current signal, incorporating the two waves together and modulating the amplitude (but not the frequency) of the radio wave signal. The resulting composite was a radio wave signal that retained its frequency. He then transmitted this composite radio wave signal, which thus comprised an electromagnetic wave signal of constant frequency whose amplitude varied in accord with the variations in the incorporated audio wave signal. Fessenden was the first to thus modulate and transmit radio waves with sound waves, and he patented this system.

To detect and reproduce the modulated radio wave signal at a receiver, Fessenden used the so-called heterodyne principle, which he had previously worked out. He was the first to apply the heterodyne principle to receiving modulated radio transmissions. The idea of mixing two different, independent, simultaneous sounds, one having a higher frequency than the other, to produce a resulting difference signal had been appreciated since antiquity and was called heterodyning. The application of the heterodyne principle to radio frequency waves had to await the development of suitable fixed frequency generating equipment, such as low-power radio frequency (RF) sources, such as small oscillator devices, including vacuum tube oscillators (in Fessenden's time). With electromagnetic waves, in the heterodyne situation, the frequency difference between two waves of different frequency when mixed together results in a product frequency that can be above or below each of the two mixed frequencies. Fessenden accomplished his heterodyne receiver invention with electromagnetic waves of frequency below the mixed frequencies, for which he received U.S. Patent 706,740 on August 12, 1902.

Fessenden's Early Years

Reginald Fessenden was the eldest of Elisha and Clementina Fessenden's four children and was born in East Bolton, Quebec, Canada, in 1866. Reginald, as a youth, was a bright and diligent student, particularly in mathematics. Elisha, as an Anglican priest, during Reginald's youth, was moved with his family to various parishes in the province of Ontario, Canada.

A little later, Reginald was a student at Trinity College School in Port Hope, Ontario, for 2 years, and at the age of 14 he received a mathematics mastership to Bishop's College (subsequently Bishop's University) in Lennoxville, Quebec. Although he had accomplished "substantially all the work necessary," he left the Bishop's College at age 18 without having received a degree and worked for the next 2 years as the principal and sole master at the Whitney Institute in Bermuda. While in Bermuda in 1885, he met Helen Trott, and in September 1890, they married in New York City. Later they had a son, Reginald Kennelly Fessenden.

Though in his formal education he had received only limited scientific and technical training, he became interested in the electrical field and wished to exploit his mathematical abilities. Hoping to gain employment with Thomas Edison, in 1886 he moved to New York City. He called at Edison's electric lamp factory at Harrison, New Jersey, and sent in his card and application for Edison to review. Edison wrote a note back, "Am very busy. What do you know about electricity?"

Though Fessenden knew from his own efforts much of what was generally known about the subject, in his reply note he said in a conservative understatement, "Do not know anything about electricity, but can learn pretty quick."

Edison responded, "Have enough men who do not know anything about electricity."

Fessenden persisted, and later did get a job with Edison. He subsequently showed the notes to Edison, who grinned and said, "Things must have been going pretty badly that day."

Fessenden was a large, burly, bearded, robust, talented young man. For Edison, he worked as a tester, inspector, and repairer of defective electric mains. The latter he did without permits in New York City to avoid delays and graft. In making a repair, his procedure, with crew members, was to dig up a street, replace a section of main, and then place the street back into its original condition before the police next routinely arrived. On one occasion, J. Pierpont Morgan rushed out of his house, which had been one of the first electrified, wanting to know why his house power had been shut off. After Fessenden explained, Morgan invited him into his house and asked him how to prevent the fires that kept occurring from wiring defects. Fessenden suggested enclosing the wire in galvanized iron pipe, and this became a routine procedure.

Fessenden proved valuable, soon advanced through various promotions and became chief inspector. In late 1886, he began directly working for Edison at his new West Orange, New Jersey, laboratory. There, Fessenden successfully worked on a range of problems involving electricity, physics, and chemistry, but in 1890, because of financial problems, Edison was forced to lay off most of his laboratory employees, including Fessenden.

Because of his practical experience with Edison, Fessenden obtained positions successively with various manufacturing concerns, and in 1892, he obtained an appointment as professor for the newly formed Electrical Engineering Department of Purdue University in West Lafayette, Indiana. While there, he aided the Westinghouse Corporation with the installation of lighting at the 1893 World Columbian Exposition in Chicago and did much of the engineering development work on Westinghouse's stopper lamps, which were designed to circumvent the key Edison patent. Somewhat later, George Westinghouse himself recruited Fessenden to chair the newly formed Electrical Engineering Department of the Western University of Pennsylvania (which later became the University of Pittsburgh). Fessenden never graduated formally from a university but, because of the posts he held at Purdue and the University of Pittsburgh, was sometimes called "Professor." He always remained a Canadian citizen.

Fessenden was not an easygoing or agreeable person and seemed to have strong opinions about things. He was independent, headstrong, impatient, and easily offended, and he was recalled as one lavish with his associates' money. He talked a lot but paid little attention as to whether or not his companions understood him. His career is marked with controversies, quarrels, and litigation, but he was a prolific inventor.

Invention of AM Radio Waves

When Marconi's success with radiotelegraphy appeared, Fessenden started in the late 1890s some experimentation and soon decided he could achieve a much more efficient system than the existing basic combination of spark-gap transmitter (which only produced a group of electromagnetic waves, not a single frequency) and coherer receiver (which used an imperfect contact between conductors that functioned to produce an improvement in conductance by passage through them of a high-frequency current). Oliver Lodge and Guglielmo Marconi initially used the spark-gap transmitter and the coherer receiver.

In 1900, Fessenden left the University of Pittsburgh and obtained a contract with the U.S. Weather Bureau to develop a practical network of coastal radio stations to transmit weather information that would avoid the need for using telegraph or telephone lines. Fessenden would retain ownership of his inventions, and the Weather Bureau could use any devices Fessenden invented. Fessenden soon achieved major advances, including receiver structures, for which he invented and used first a so-called barretter detector and then an electrolytic detector. The latter utilized a fine wire dipped in nitric acid, and this detector soon replaced the coherer. In 1902, Fessenden quarreled with the chief of the Weather Bureau and resigned his post. Fessenden alleged that the Bureau Chief Willis Moore had attempted to obtain a half interest in his new patents, and he (Fessenden) had refused.

In this period, Fessenden invented the heterodyne circuitry in which two electromagnetic (radio) signals of differing frequency were combined to produce a third signal in the acoustic frequency range. This invention, though, was not practical for about a decade until stable, uniform (fixed) frequency radio signals of small or minimal bandwidth could be continuously generated, which was achieved with vacuum tube oscillator circuitry.

It occurred to Fessenden later in this period that radio frequency waves could be modulated in amplitude. His idea was to generate and transmit a relatively high-frequency electromagnetic wave of constant frequency (a carrier wave). That wave was admixed and combined (or modulated) with electrical waves of much lower frequency derived from and corresponding to sound waves. The amplitude variations of the resulting modulated carrier wave would correspond with the variations in the original sound waves. Then, in a receiver, he installed a heterodyne detector that responded to the resulting amplitude-modulated electromagnetic wave whereby the original sound waves were reproduced.

He tried at first to make this idea work using a spark-gap generator for the source of his high-frequency waves, but he found that these waves contained many different frequencies, some of which were in the range of sound. In a spark-gap, or Hertz-type, transmitter, an inductance and a capacitance are arranged in series, and a gap in the transmitter circuit was originally considered to be an essential element of the radiating system. When a modulated signal based on such a generator was received, the modulated signal was mixed with the noise of static from other signals generated by the spark-gap generator. Fessenden, though, thought he could hear from his receiver faint signals corresponding to the original sound.

Fessenden's radio work produced original results. He sought methods for achieving successful modulated radio waves that could be demodulated. On December 23, 1900, Fessenden succeeded in achieving what was apparently the first successful transmission and reception of an amplitude-modulated electromagnetic wave. The original sound waves were speech, and the experimental electromagnetic carrier wave was generated by a high-frequency spark-gap transmitter. The transmission distance was about 1 mile and occurred on Cobb Island, Maryland, which is located on the Potomac River about 50 miles from Washington, D.C. The sound quality of the received signal was distorted and commercially impractical, but as a test, this event demonstrated that, with improvements, radio waves could be modulated, transmitted, received, and demodulated, producing audio signals.

After leaving the Weather Bureau in 1902, he persuaded two wealthy Pittsburghers, Thomas H. Given, a banker, and Hay Walker, a soap manufacturer, to finance his development of so-called radiotelephony. They formed the National Electric Signaling Company (NESCO), with operations centered at Marshfield's Brant Rock, Massachusetts. Fessenden built three experimental stations on Chesapeake Bay. Although this company eventually came apart in 1912, it was during the term of this concern that Fessenden can be considered to have achieved his greatest inventions. He developed both a high-power, so-called rotary-gap (or rotary-spark) transmitter for long distance transmission and also a low-power, continuous wave alternator transmitter. These transmitters were based on a known electric alternator, which he speeded up to create currents of the order of tens of thousands of cycles per second.

In his alternating current (AC) generator, the output was coupled directly to a rotary spark gap so that sparks occurred at exact points in the input wave, the spark being between a fixed terminal on the stator and a terminal on the rotor; the rotor was, in effect, a spoked wheel rotating in synchronism with the AC generator. Not only did this arrangement achieve a higher spark rate, but it also achieved a sort of mechanically quenched spark-gap transmitter. The rotating gaps produced clear, almost musical, signals that were distinctive and came close to continuous wave signals. Fessenden's synchronous rotary spark-gap transmitter was completed on December 28, 1905 at his Brant Rock station and was coupled to a steam-driven alternator. This type of equipment was the only then known apparatus for generating radio wave power, and Fessenden regarded this equipment as research tools to be used until a continuous wave transmitter that produced a substantially pure sine wave, or single-frequency signal, was available.

Fessenden continued his development of a continuous wave generator, and by the fall of 1906, he had succeeded producing a continuous wave transmission using a high-frequency (HF) alternator that departed from the establishment view then that a spark-gap and damped wave transmission were required. With Fessenden and his continuous wave transmitter, frequency was determined by the speed of the HF alternator. Subsequently, vacuum tube transmitters replaced the HF alternator, and later they were replaced by solid-state transmitters, but the basic principle of operation of the Fessenden transmitter remains in use today.

In 1905, Fessenden set up the Fessenden Wireless Telegraph Company of Canada, but this venture never developed. The company never received support from his American venturers. The Canadian government authorized Marconi to exclusively install towers and operate radio equipment in Canada.

NESCO decided to establish a transatlantic radiotelegraph service. Apparently starting in about January 1905, employing his rotary spark transmitters, identical towers and stations were set up at Brant Rock and Machrihanish, Scotland, and Morse code messages were exchanged each way between those points. (Marconi at this time had achieved only one-way transmissions.) Because of atmospheric limitations, the distance could not be bridged during daylight hours or in summer months, so work was suspended until later in the year. However, on December 6, 1906, reportedly because of improper actions of contractors at the Scottish site relating to supporting cables, the Machrihanish radio tower collapsed, so the transatlantic radio communications ended. The Scottish tower was not rebuilt.

Subsequent Developments

Stateside development efforts continued. Fessenden now found himself in conflict with Lee de Forest, another aggressive and prolific inventor. Fessenden was conducting tests for the U.S. Navy involving signals sent between the U.S.S. *Topeka* and Sandy Hook. The de Forest Wireless Company installed a wireless shack adjacent to the Sandy Hook naval station. Fessenden maintained that de Forest told the shack operator to foul up Fessenden's tests, which the operator did by operating the shack's powerful radio transmitter while holding the signal key closed, so that static was transmitted. Fessenden's crew overcame the difficulty by bribing the operator with food and liquor to open the key.

As early as 1890, Tesla had built high-frequency AC generators. One produced an output of 10 kHz; a later one produced frequencies of up to 20 kHz. However, in the early days of radio, useful antennas were not able to radiate efficiency (and therefore usefully) at such low frequencies.

Fessenden, meanwhile, contracted with General Electric (GE) to design high-frequency transmitters. In 1903, Charles P. Steinmetz (the hunchbacked German genius of GE who, among other things, discovered the laws of magnetic hysteresis) produced a 10,000 cycle per second (Hertz) device, which proved to be of limited utility and unsuitable for use as a radio transmitter. Responsive to Fessenden's further request, the design work at GE was continued and assigned to Ernst F. W. Alexanderson, a young, brilliant Swedish engineer. After 2 years, in September 1906, GE's delivered effort was found by Fessenden to be a "useless machine."

Fessenden decided to rebuild the machine, and, apparently working day and night, by November 1906, he had a new machine that would operate at frequencies in the range of 50 to 90 kHz. His alternator had a fixed, thin disc armature and a revolving field magnet with 360 projections (or teeth). When operating at 139 revolutions per second, an alternating current of 50,000 Hz, having a maximum output of about 300 watts, was produced. The alternator frequency was determined by the speed of a well-regulated steam engine. Fessenden placed a carbon microphone in the transmission line to modulate the output radio frequency output carrier wave. (Subsequent years of costly research and development work by others occurred before practical versions became available.)

Fessenden, on December 21, 1906, made an extensive demonstration of his equipment at Brant Rock, showing its capacity for point-to-point wireless telephony and even demonstrating interconnection with telephone wires.

A few days later, Fessenden, having alerted nearby ships equipped with Fessenden wireless equipment to stand by, on December 24, 1906, Christmas Eve, Fessenden made the first radio broadcast of AM radio—audio signals that amplitude modulated the radio frequency constant signal that was transmitted. Instead of receiving Morse code signals with clicking and buzzing, the owners of Fessenden receivers heard voices and music coming from their apparatus. Another, similar broadcast was made on December 31, 1906, New Year's Eve. These two broadcasts are now seen as historic events and were believed to have been heard by radio operators on board U.S. Navy and United Fruit Company ships at various locations over the

South and North Atlantic and even in the West Indies. However, no acknowledging reports involving shipboard radio operators or others are known. The broadcasts were soon forgotten. A few of the audio test transmissions from Brant Rock were evidently received by a NESCO employee at the Scottish site (evidently before the tower there collapsed). (Previously, the U.S. Navy, beginning in 1904, had broadcast daily time signals and weather reports using spark transmitters and Morse code transmissions.)

Walker and Given had intended to sell NESCO to a large concern but were unable to find a buyer. The company Fessenden had formed in Montreal in 1905, called Fessenden Wireless Company of Canada, seems to have caused concern to Walker and Given. A long dispute arose between Fessenden and his backers over various matters. Fessenden was dismissed from NESCO in January 1911. He sued for breach of contract; he won after a trial and was awarded damages, apparently for $406,175. NESCO appealed and prevailed but Fessenden persisted. To conserve assets, NESCO went into receivership in 1912, and Samuel Kintner was appointed general manager of the company. In 1917, NESCO exited from receivership and then was renamed the International Radio Telegraph Company, which was sold to Westinghouse in 1920. The following year, the assets, including many Fessenden patents, were sold to Radio Corporation of America (RCA). On March 1, 1928, Fessenden settled his outstanding lawsuits with RCA and received a large cash payment. Walker and Given seem to have left NESCO without any return on their investment. Some historians think that the question of ownership of the alternator-transmitter patents led to the formation of RCA, because for national security purposes the federal government did not want the British-owned Marconi company to obtain control of them.

Developments apart from NESCO

Before leaving NESCO, Fessenden apparently had developed a high-frequency alternator to a level where it had an output of 50,000 kW. After Fessenden left NESCO, E. F. W. Alexanderson continued to work at GE on alternator-transmitter development. By 1916, the machine had a 200 kW output and was suitable for transmitting across the Atlantic. The Fessenden-Alexanderson alternator had become more reliable for transatlantic communication than spark-based transmitters.

By 1920, audio radio broadcasting had become well known and widespread, but vacuum tube transmitters had replaced the alternator. Fessenden's 1906 idea of using continuous wave amplitude-modulated signals had become standard procedure.

Fessenden's work in early radio had included the achievements of the first amplitude modulation and audio transmission by radio (1900), the first two-way transatlantic radio transmission using Morse code (1906), and the first radio broadcast of voice and music (1906).

After NESCO, Fessenden withdrew from radio and went into other activities. He pursued submarine signaling and detection. He developed a sonar system for submarines to signal to each other and a method for locating icebergs to enable avoiding another disaster such as that involving the *Titanic*. He also became involved in the Niagara Falls power plant of the newly formed Hydro-Electric Power Commission of Ontario.

Fessenden achieved a number of patents. Many he owned; others were assigned to his various employers. One set related to the use of microphotography and microfilm for keeping records. One patent related to a vehicle parking garage equipped with ramps and hoists, but this patent expired before autos became so common that his garage was needed. He also patented versions of reflection seismology, a method useful for locating petroleum. Other patents related to a fathometer sonar device for determining water depth relative to a submerged object.

He developed silicon steel, which reduced the hysteresis loss in the carbon steels and irons used in transformer cores and in iron pole pieces in motors. Fessenden evidently reasoned that the relatively large carbon atoms, when replaced by the relatively smaller silicon atoms, would reduce hysteresis loss.

Fessenden was actually second only to Edison among his contemporaries in the number and variety of patents in which he was an inventor, apparently holding, in all, about 500 by the time of his death.

Fessenden liked to relax and cogitate. Outside, he liked to float on his back on a lake or river with a hat pulled down over his eyes and a cigar projecting from his mouth. Inside, in his home, he liked to lie on his back on a carpet, a cat on his chest. In this prone state, he apparently could think and dream his way to new ideas.

Fessenden was made inactive by high blood pressure and a weak heart. After his settlement with RCA, he bought a small estate called Wistowe in Bermuda, where he passed his last years comfortably. He received various honors and recognition. His only son died in a boating accident off Bermuda in 1944. After Fessenden's death in 1932, his wife Helen wrote the book *Fessenden: Builder of Tomorrows*. She died in 1941.

LEE DE FOREST (1873–1961) AND THE TRIODE

Summary

De Forest invented, in 1906, a three-element vacuum tube (or triode) that enabled the amplification and even detection of relatively weak electrical signals. Depending upon the circuitry employed, the tube made practical radio transmitters and receivers plus a large variety of electronic equipment by amplifying weak signals without distortion. In 1910, de Forest used his triodes plus Fessenden's AM system to broadcast voice and music.

Until the invention of the transistor by William Shockley and others in 1948, and its development, the triode was a basic component in products of the roughly $90 billion electronic industry.

Early Years

Lee de Forest was born in Council Bluffs, Iowa, in 1873 and was the son of Henry Swift de Forest and his wife, Anna Robbins. Henry was a Congregational minister. In 1879, motivated by a sense of Christian duty, Henry accepted a position as President of Talladega College in Talladega, Alabama, a school for African Americans. Although the family at the college was ostracized by their white neighbors, young Lee found his friends among black children. He was one of very few white students.

Lee went to Dwight L. Moody's Mount Hermon Boy's School in Massachusetts before entering Yale University in 1893, evidently with a small scholarship. Lee had early exhibited interest and talent in science, traits he had seemingly inherited from his father, who had himself been an enthusiastic student of astronomy under Newcomb during his own earlier years at Yale. Although his father had hoped Lee would become a minister, Lee, with the reluctant acceptance and support of his father, enrolled at the Sheffield Scientific School at Yale.

At Yale, Lee studied intensively and also worked summers to earn his tuition. He made some mechanical and gaming inventions from which he earned some income that contributed to his tuition. Apparently pursuing some invention, one time he tapped into the Yale electrical system and blacked out the entire campus. Although he was suspended, he was later allowed to complete his studies. Typical of many scientific students, he soon lost the religious beliefs fostered in his youth. He grew to be a man of above-average size with features deemed blunt and with deep-set eyes that sometimes seemed to be odd. He received a bachelor's degree in 1896, and his class voted him "the Nerviest in the class and also the Homeliest!" He remained a hard-working, competitive, independent individual who tended to be outspoken and even assertive.

He volunteered for military service in the Spanish-American War and became a bugler, but he saw no action. He pursued graduate studies at Yale when he worked under J. Willard Gibbs, who is still considered by many to have been the greatest intellect in American science. Lee seems to have been much influenced by Gibbs. Lee gained a Ph.D. in 1899 from Yale with a doctoral dissertation on radio waves, which was probably the first dissertation in the United States that considered this subject matter.

His first job after Yale was with the Western Electric Company in Chicago. Though he was supposed to be a telephone engineer, he seems to have spent all the time he could find on a wireless detector that he called a "responder." This first job did not last long and he then successively held several other jobs, including teaching at night while he devoted days to his detector. Using the auditorium of the Columbian World's Fair at Chicago and a yacht owned by a friend as sending and receiving stations, respectively, he carried out wireless tests. In 1902, a Wall Street promoter named Abraham White worked with de Forest to set up a company called the American de Forest Wireless Telegraph Company.

Invention of the Triode

His triode tube invention emerged from his study of Englishman John Ambrose Fleming's 1905 published lecture to the Royal Society describing Fleming's diode tube, which functioned to rectify a high-frequency alternating current. In this diode tube, the alternating current was applied to a negatively charged, hot filament that, in response, emitted electrons that passed through adjacent tube space to a plate electrode. By controlling applied voltages, this tube enabled separating (called rectifying) about half of the received alternating current signal to produce a variable output signal having only a nonalternating, so-called variable direct, component, preferably a positive electrical component. This component could comprise a linear successive cluster or succession of electrical pulses that, when in the frequency range of audible sound and applied to a diaphragm, would produce audible sound.

De Forest performed various experiments involving the diode, and in 1906, inserted a small grid of wires between the filament and the plate in the tube. He achieved, from one group of experiments, a rectified positive current at the plate that he connected to the grid. He applied to the filament a much larger voltage. He experimented with gases in the tube envelope. He found that when the grid had a negative charge, it stopped the flow of electrons from the filament to the plate, and when it had a positive charge, it allowed electrons to flow. The grid thus acted as a sensitive valve. A small current could regulate a large current, with the variations in the large current corresponding to those in the small current. He called his triode tube the "audion."

If one wanted still greater amplification, then the large current produced could be applied to the grid of a second triode tube to achieve a still larger current. In early 1907, de Forest filed a patent application on his three-element tube and was granted U.S. Patent 879,532 in early February 1908.

Broadcasting Efforts

In about 1906, de Forest discovered that White and his directors were selling all company assets to a dummy company. De Forest resigned and took with him $1,000 plus pending patent applications, and he formed a new corporation called the De Forest Radio Telephone Company in 1907. De Forest constructed experimental stations, made and sold wireless equipment to the U.S. Navy and others (particularly in 1907), and reported Army maneuvers by wireless for the Signal Corps.

De Forest endeavored to make radio broadcasting a commercial reality. In 1910, he put together the first musical radio broadcast from the Metropolitan Opera House with Enrico Caruso. Daily programs of music were broadcast. In 1916–1917 in New York and in 1919 in San Francisco, he made broadcast demonstrations. He attempted to popularize radio for news and music and as more than a means for two-way personal communication.

Although White and others seem to have eventually gone to jail, de Forest and other associates reorganized and kept their concern operating for a few years. However, in 1912, de Forest was arrested and indicted for alleged mail fraud. Actually, he had just been trying to raise money to finance and commercialize his triode and his broadcasting efforts. It was not enough to be an independent inventor, one also had to be a promoter.

Out on bail and seeking money for his defense, he offered to sell certain rights to the audion to the American Telephone & Telegraph Company (AT&T). The company never responded directly, but it sent an agent named Meyers, who indicated only that he represented an unnamed and undisclosed group. Meyers offered $50,000, which de Forest accepted in 1913, since he was badly in need of funds to cover legal bills.

The indictment of de Forest and other defendants had charged them with seeking to exploit "a strange device like an electric lamp, which he called an 'audion,' and which device had proven to be worthless." In 1913, the trial for mail fraud took place, and de Forest was acquitted. In 1916, de Forest achieved the first radio advertisements (for his own products) and also the first presidential election reported by radio.

The Audion Development

It was a long time before de Forest profited from his audion. Improvements from many tests were needed. Also, public demonstrations were needed.

When his business began to grow, the American Marconi Company sought through prolonged litigation to terminate it. However, the federal court finally held the claims of the allegedly infringed Marconi patent to be either not infringed or invalid on the grounds of the prior art of Oliver Joseph Lodge, Alexander Steponovich Popov, or Edouard Branly.

Fessenden had a patent on his detector and sued de Forest for infringement. Fessenden won after 3 years, but by then de Forest was using a different detector, his triode tube. De Forest moved to San Francisco in 1910 and worked for the Federal Telegraph Company, which in 1912 started development of the first global radio communications system.

Big electronics firms expended much money developing the triode. It proved to be essential to the development of radio, long distance telephone, and all electronic/electrical amplification, including public address systems, sound motion pictures, electronic phonographs, radar, and the like. It was the only electrical amplifier in general use until the invention by Shockley, Brattain, and Bardeen in 1948 of the transistor (see "Transistors: John Bardeen, William Shockley, and Walter Brattain" later in this volume).

Later Life

AT&T pursued de Forest seeking to acquire all the remaining rights to the audion. De Forest sold these to AT&T in two packages reportedly totaling $390,000.

In 1916, de Forest claimed to have found that overloading his audion would make it behave as an oscillator. He alleged that he had done laboratory work in 1912 on this effect. This oscillator could generate a fluctuating current of any selected frequency. Edwin Howard Armstrong had independently discovered that the utilization of the audion could be greatly increased by feeding some of the energy of variations in the filament-plate circuit back to the antenna-grid circuit to increase the amplification of the latter, a feedback arrangement known as the regenerative circuit.

De Forest filed his own patent application on the regenerative circuit that used the audion. His 1916 patent application on the feedback circuit was placed into interference by the U.S. Patent Office with patent applications of three other parties: Irving Langmuir of GE, a German named Alexander Meissner, and a previously issued U.S. patent to Armstrong, then a recent graduate of Columbia. The Langmuir and Meissner applications were eliminated because of their comparatively later dates. Armstrong's U.S. patent on the regenerative circuit, though, had issued in 1914, before the interference started and also apparently before de Forest had determined (or confirmed) that his triode tube could be made to oscillate. The resulting litigation went on for about 15 years, long after de Forest had sold all his rights to the involved patents and had left work in radio. (See further discussion below, under Armstrong.) In 1934, the U.S. Supreme Court decided the issue of priority in favor of de Forest, but various experts believe this judgment was erroneous.

In 1919, de Forest left the radio field and began research and development work in sound motion pictures, where he had previously done pioneer work. In the early 1920s, he developed a "glow lamp" that functioned to convert sound into a corresponding irregular electric current that in turn could produce similar irregularities in a lamp filament. The filament could be photographed together with a motion picture and recorded alongside the film images in a motion picture film. Subsequently, the sound track could be converted into sound. Although he was perhaps one of the key inventors of the photographic motion picture soundtrack, and although he exhibited film incorporating a soundtrack in 1923 on Broadway, the movie managers did not pursue his development. Their opinion was that the public did not want talking pictures. However, by the late 1920s, when sound motion pictures came to predominate, people other than de Forest profited.

For income, de Forest managed to develop various manufacturing concerns in the radio field. In 1931, de Forest sold one of his radio manufacturing concerns to RCA. He made improvement inventions in television, diathermy apparatus, and cathode ray scanning (later used in radar). After World War II, he contracted with the Bell Telephone Laboratory and was able to operate a well-equipped research and development laboratory in California.

De Forest successively married four wives, each for a longer term. The first three each ended in divorce: to Lucille Sheardown in February 1906, to Nora Blatch in February 1907, and to Mary Mayo in December 1912. He finally married Marie Mosquini in October 1930. She was a silent film actress and Will Rogers's former leading lady.

De Forest reportedly received over 300 patents. His last patent was granted in 1957, when he was 84 years old. He was mainly a trial-and-error inventor and also a promoter, but the record suggests he was not a good businessman. Although solid-state transistors mostly replaced the bulky audion tubes originally used in electronics, Lee de Forest provided many inventions and aggressive enthusiasm on the road leading to the electronic age. His most important invention clearly was the triode.

He wrote an autobiography, in which he indicated that he considered himself to be the rather of radio. However, history suggests that radio was not a single invention and clearly had many fathers. He is reported to have died in California in 1961.

EDWIN HOWARD ARMSTRONG (1890–1954) AND FREQUENCY MODULATION

Summary

Armstrong invented independently the regenerative circuit (1912), the superheterodyne circuit (1918), the super-regenerative circuit (1922), and the complete frequency modulation (FM) radio broadcasting system (1933). These circuits are basic to all modern radio, television, and radar. Various parties contested his priority of most of these inventions. His most famous invention is probably a system of practical frequency modulation.

Early Years

Edwin was born in New York City to John Armstrong and his wife, Emily Smith Armstrong, in 1890. John was a vice president of the U.S. branch of the Oxford University Press, and Emily had been a teacher in the public schools. Soon after Edwin's arrival, the family moved to Yonkers, New

York, where they lived in a house on a bluff overlooking the Hudson River near the Greystone railroad station. Edwin's maternal grandparents lived next door to the north. In 1978, the Yonkers Historical Society declared the house a historical landmark.

The Armstrongs had a type of middle-class culture and were devout Presbyterians. Many relatives among the two families were teachers. The families were gregarious, valued education, and enjoyed their many gatherings, but it appears that the family members, except for Edwin, were not particularly creative. Edwin's childhood was pleasant and happy, but he was always serious. At age 9 he contracted rheumatic fever, which kept him out of school for 2 years and left him with a lifelong tic in his shoulder and jaw, but afterward he soon caught up. By his teens, he was accumulating, tinkering with, and building wireless gadgets and carrying out experiments in his third-floor cupola, a turret-like, attic room of his parent's home.

He mastered Guglielmo Marconi's work when a boy. By age 14, he knew he wished to be an engineer. He tried to overcome the poor reception strength of radio waves. Among other things, he tried various antennas, including kite-flown antennas from windows of his attic room. He built a 125-foot antenna pole for testing in the south yard. These efforts did not solve the poor reception problem. The neighbors watched him moving up and down the antenna pole with awe and apprehension, but his mother, in response to one who telephoned to advise that Edwin was at the top and that she was nervous watching, said "Don't look then."

After graduating from grammar school and then high school in Yonkers, he attended Columbia University, commuting from Yonkers to the university by a red motorcycle he received as a high school graduation present. At Columbia, the famous inventor and professor, Michael I. Pupin, became Edwin's influential teacher, role model, promoter, and friend. Though a shy youth, Edwin seems to have developed into a rather unconventional, outspoken adult. Pupin and some of his colleagues needed to defend Edwin to irate faculty members who did not like this student.

Invention of Regenerative Circuit

Before his senior year at Columbia, Armstrong achieved his first significant invention. Edwin had long been seeking better reception means for sound-modulated, transmitted radio waves. During his student years at the university, he had made his own exhaustive investigation of the de Forest audion tube, and the regenerative circuit had come out of this investigation. As Pupin's assistant, he had the full use of Pupin's well-equipped laboratory.

Though invented in 1906 by de Forest and patented by him in 1907, the tube itself had been little used (mainly only as a detector) and was only slightly understood even by de Forest by the time Armstrong began his studies of the tube. Although de Forest had tried to use this tube for radio reception and signal amplification, he had himself observed only poor or small results: radio signals still were only faintly received.

Before summer vacation in 1912, Armstrong, from his investigation, understood and explained the operation of the audion. During summer vacation in 1912, Armstrong conceived the idea of a circuit that would feed back audion output signals to an input to the audion in a manner that would cause signal reinforcement ("positive feedback"). In autumn, back home in his attic room, he immediately tried out his idea. His younger sister Edith ("Cricket") later recalled how Edwin burst into her room late at night dancing around and shouting, "I've done it!" Loud received signals in his attic room were coming from the headphones left across the room on his work table. The regeneration from feedback had resulted in amplification by a factor of thousands.

Although others at about this time had observed feedback, Armstrong had characterized and applied it to achieve practical use. In addition to signal amplification by feedback, Armstrong discovered that with even higher levels of positive feedback, the circuit involving the audion became an oscillator that could transmit its own signal. He had invented both a regenerative receiver and a transmitter.

Edwin wanted to immediately patent his invention, but his father refused to give him the money, fearing that the extracurricular activity would conflict with his son's fourth year and graduation. Edwin proceeded to borrow the money from relatives and friends and even to sell his cherished red motorcycle. In 1913, he applied for a patent, which was issued the following year. The invention finally made radio practical, and receivers made using the invention soon appeared commercially.

Armstrong received a degree in electrical engineering in 1913 and was offered a position as an instructor and assistant to Pupin, which he accepted without hesitation. At Columbia, he strung a huge antenna between the campus buildings Philosophy (where Armstrong in the basement had his laboratory), Havemeyer, and Schermerhorn, and, using a regeneration circuit, was able to receive distant signals even from Honolulu, to the pleased amazement of Putin.

He explained his findings at a meeting of the Institute of Radio Engineers (IRE) in October 1916. These findings generally corroborated the previous theoretical mathematics worked out by Liebowitz. Analytical work had made Armstrong knowledgeable about heterodyne circuitry.

In one of the subsequent demonstrations, this one held at a conference by the Institute of Radio Engineers at Columbia, de Forest was in the audience, and, for the first time, he heard his audion amplify. De Forest immediately felt an animosity toward Armstrong that not only endured but also intensified with time. Litigation between de Forest and Armstrong over patent rights to the regeneration circuit was postponed by the entry of the United States into World War I.

World War I and Invention of the Superheterodyne Circuit

Armstrong in World War I was free from litigation; made loyal, long-lasting friendships; developed perhaps one of his greatest inventions, the superheterodyne circuit; and became known as "Major Armstrong."

When the United States entered World War I, Armstrong joined the Army Signal Corps, was commissioned as an officer (a captain), was sent to Paris, and was placed in charge of the Radio Group of the Research Section.

While traveling to France, Armstrong met Captain H. J. Round, an engineer with the British Marconi Company. Armstrong learned that the British were ahead of the Americans in development of vacuum tubes capable of handling high-frequency signals ranging from 500,000 to 3,500,000 cycles per second, a frequency range that, it was suspected, the Germans were using for communication. Round had devised a method of using such tubes to receive frequencies of up to about 1.5 MHz and was in charge of a group of direction finding and listening stations for the Admiralty. The British not only kept track of many German ships, but also, having broken the German codes, could read nearly all the messages. The Germans were confident that their shipboard intercommunication "buzzer" sets could not be heard more than a few miles on their 1.5 MHz wavelength, but Round's multistage amplifiers, using Round's V24 vacuum tube with low interelectrode capacitances, could pick up such signals and fix the position of German ships. A small degree change detected in a signal's bearing indicated ship movement and gave advance warning of German sorties. Armstrong sent the information back to Signal Corps laboratories, and the information promoted his interest in receiving and amplifying weak high-frequency spark signals.

Armstrong was assigned by the U.S. Army to detect shortwave enemy communications. The American Expeditionary Force was then using a French receiver, the L-3, which had no capacity to pick up the very short wave frontline communications rumored to be used by the Germans. Evidently, a regenerative detector was not satisfactory for various reasons. One reason was that the output was microphonic and caused the antenna to emit an output that seriously interrupted local direction-finding stations. The heterodyne technique was not considered suitable for receiving the signals then produced by spark transmitters because of their inherent large bandwidth that made tuning a heterodyne detector to a spark signal's central frequency impossible. To fulfill this assignment, Armstrong achieved his second significant invention: the so-called superheterodyne circuit.

Armstrong later related how his conception of this invention took place. Months after his meeting with Round, on Paris streets, he was watching a night bombing raid in Paris "wondering at the ineffectiveness of the antiaircraft fire." To increase the methodology of locating aircraft position, he wondered whether "the very short waves sent out from the aircraft from their motor ignition systems might be used." It was then, in March 1918, while walking back to his apartment in Paris after the raid, that he suddenly thought of the superheterodyne method.

From his previous work with the audion and regeneration, he had become knowledgeable and familiar with heterodyne circuitry. As his findings reported to the Institute of Radio Engineers in October 1916 had indicated, amplifications by heterodyne of 100 times or more were measurable, and the regenerative circuit could increase such amplification by 50 times.

Armstrong had at hand a tuned amplifier and detector for long waves. His concept in March 1918 was to bring the short wave signals down into the range of the long wave amplifier by heterodyning. The short wave signals were too weak to be amplified directly. As matters developed, this heterodyning did not alter the modulation content of the original spark signals. The large amplified signal available enabled final detection to be achieved by rectification.

To make this invention workable, experimentation and much development work by members of his group were necessary to reduce his concept to practice and to produce a prototype. It was not until about December 1918 that preliminary tests showed that the prototype achieved several thousand times the amplification of the French L-3. Additional tests and the addition of an audio frequency amplifier were carried out before the new receiver, a relatively complex eight-tube structure, was ready for a trial at the front, but by that time the armistice had been signed. The French government had allowed him to use the Eiffel Tower for experiments and testing. His receiver amplified weak signals to an extent not previously known. He was named by the French a Chevalier de la Legion d'Honneur. In December 1918, Armstrong filed a French patent application and later a corresponding U.S. patent application.

In 1918, he adapted the known basic technique of heterodyning that had been used by Fessenden in early wireless. In superheterodyne reception, information (sound, picture, etc.) is recovered from transmitted modulated carrier waves over a range of frequencies. In a receiver, the high-frequency current of an incoming wave is combined with a low-frequency current produced in the receiver to produce a beat (or heterodyne) frequency, which is the difference between the combined frequencies, now called the intermediate frequency (IF). The IF is above the audible range (thus, the original nomenclature, supersonic heterodyne reception) but can be amplified with higher gain and selectivity than can the original high-frequency current. The IF signal retains the modulation of the original high-frequency current. In a detector, the desired low-frequency information is recovered. The receiver is tuned to different broadcast frequencies by adjusting the frequency of the current used to combine with the carrier waves.

Armstrong appreciated that it is difficult to make an amplifier that will operate at high frequencies and difficult to make a tuning filter that can select a narrow band of frequencies and still be adjusted across a range of frequencies. The filter must tune in only one signal (station) frequency and reject all others. However, it is relatively easy to make a tunable oscillator. By adding the local oscillator IF signal to an incoming radio signal, a beat signal results, as indicated above. A fixed filter is built to narrowly select the beat frequency and pass it to a low frequency amplifier. As the local oscillator frequency is varied, correspondingly different radio frequencies are moved down to the beat frequency and thus selected. Thus, a variable local oscillator and a fixed, narrow filter can do the work of a variable narrow filter.

This arrangement is still employed in most radio receivers. Before the transistor arrived, the superheterodyne circuit was used in about 98% of all radio and television receivers and in microwave radar receivers developed in World War II. It is also used in space exploration.

Commercialization of Radio after World War I

After World War I, Armstrong returned to Columbia with the rank of major and the ribbon of France's Legion of Honor. He presented a paper in December 1919 to the IRE about his new superheterodyne circuit and ended by acknowledging the prior development efforts of others, including Meisner, Round, and Levy.

Radio technology had advanced to a point where radio broadcasting to receivers was practical. Though in 1922, 100,000 radio receivers had been produced, by 1923, 500,000 had been produced in the United States, and while in 1922 there were 30 operating radio stations in the United States, in 1923, there were 556 operating radio stations. As manufacturing developed, most radio units used the superheterodyne circuit and/or the regenerative circuit. In 1920, Armstrong sold rights to his two major circuits for $335,000 to Westinghouse Electric and Manufacturing Company.

Invention of the Super-Regenerative Circuit

In 1922, Armstrong invented an improved receiver that employed an innovative circuit called the super-regenerative circuit, which would drive an amplifier up to the edge of oscillation and then stop the gain to avoid a squeal (or howl). If it was operated at more than about 20 kHz, the ear could not detect the variation in gain. The circuit enabled a lot more gain to be achieved from one tube.

Although the super-regenerative circuit is said to be less profound than most of his other innovations, Armstrong sold the rights to the newly organized RCA for cash and a large block of stock (reportedly $200,000 plus 60,000 shares of RCA). From this invention, Armstrong made more money than from all of his other inventions. Having missed the opportunity to purchase Armstrong's rights to the regenerative circuit and to the superheterodyne circuit, and subsequently having to pay substantial monies to Westinghouse to use these circuits, RCA decided not to pass up this circuit.

After the war, Armstrong became a millionaire as early radio broadcasting became successful, but he continued at Columbia University as a professor and eventual successor to Pupin. Then, for being on the faculty at Columbia, Armstrong agreed to be paid a dollar a year; his patents gave him more income than the university could. He never taught classes and rented his basement laboratory space from Columbia. He never incorporated and did all his work with a few assistants.

In February 1923, Armstrong persuaded David Sarnoff, president of RCA, to produce and sell his new, simplified receiver model. Sarnoff canceled orders worth millions of dollars with its suppliers Westinghouse and GE, intending to have the latest receiver ready for the 1923–1924 market. Time was short. GE (but not Westinghouse) agreed to produce the new receiver for RCA. Production problems appeared. To solve them, Sarnoff, at the suggestion of his secretary, Marion MacInnis, hired Armstrong. For the success that Armstrong and his associate Harry Houck achieved, Armstrong received an additional 18,900 RCA shares. In February 1924, the new machines made their very successful and profitable market debut. Armstrong was now publicly established as the inventor of the superheterodyne receiver.

After a vacation in Paris, he courted MacInnis, and on December 1, 1923, they were married. During the courtship, Armstrong during the same day climbed to the top of each of two radio towers in New York City. One was RCA's 115-foot north tower, which extended from the roof of the twenty-one-story Aeolian Hall in midtown Manhattan. Armstrong sent photographs of the RCA tower exploit to Sarnoff and Marion. Sarnoff, angry, then banned him from the building. Edwin's wedding present to Marion was the first portable radio (which used superheterodyne circuitry). It was a monster, but they used it on the beach on their honeymoon in Palm Beach, Florida. Marion understood her creative, driven husband, and they were close companions for 30 years. Armstrong had very few close friends.

The Fight with de Forest over Regeneration

After learning in 1913 of Armstrong's breakthrough with amplification and regeneration using the audion, de Forest then carried out some research and an examination of his own prior research records. He found a 1912 entry in a laboratory notebook that a particular circuit produced a howl when tuned in a specified way. He then (based on this 1912 entry) filed, in 1916, a patent application claiming the regeneration circuit that included his audion and that was covered by Armstrong's 1914 patent. This 1912 notebook entry was used on de Forest's behalf as the sole basis to assert priority of invention for regeneration as shown and claimed in the 1916 de Forest patent application that had been placed in interference by the U.S. Patent Office with Armstrong's earlier 1914 patent.

The howl was caused by feedback that induced the de Forest 1912 circuit to oscillate, but from the available evidence, de Forest in 1912 and prior to 1916 did not pursue and did not understand this effect. In 1916, de Forest maintained that the howling observed with the particular circuit in his 1912 entry was the result of accidental positive feedback, but de Forest never mentioned, identified, or utilized feedback at or during that earlier time period prior to his 1916 patent application. He never took advantage of feedback, and in fact tried to suppress its effects in his amplifiers. As he had earlier done with his audion, de Forest sold the rights to his 1916 patent application to AT&T.

As indicated, in 1920, Armstrong sold his rights both to the regeneration circuit and also to the superheterodyne circuit he had invented in 1918 to the Westinghouse Electric and Manufacturing Company. Also, in about 1922, he sold his rights to the super-regenerative circuit to RCA, as indicated above.

Shortly after the end of World War I, protracted, complex litigation began regarding the basic issue of who first invented the regenerative circuit, Armstrong or de Forest. De Forest was backed by AT&T, Armstrong by Westinghouse and RCA. The litigation was conducted by the corporate owners of the patent rights and passed through a dozen courts between 1922 and 1934. Armstrong won a first round, lost a second, was checkmated in a third, and lost in a fourth and final round before the U.S. Supreme Court. In the course of the litigation, it was shown repeatedly that Armstrong was the actual, true, first inventor of the regenerative circuit. Also, among other facts, it was shown repeatedly that de Forest did not understand how his own audion worked and did not actually fabricate in the critical time period a practical regeneration circuit. In contrast, though, Armstrong had identified and was publishing this information. The courts decided in favor of Armstrong several times, but these decisions had been followed by reversals. In 1934, the Supreme Court, evidently without an understanding of the actual technical facts, found for de Forest on the basis of language detail.

AT&T's lawyers eventually got a favorable (to them) ruling from a judge in Philadelphia who was technically untrained and ignorant about the facts presented. However, from this ruling these attorneys were able to overcome appeals by maintaining that judgment had earlier been rendered. Before the Supreme Court, it appears that these attorneys were able to cloud and obscure the facts and the real issues so that the court refused to hear and judge the case on its merits, and judgment was rendered in de Forest's favor. The result was that Armstrong's U.S. patent on the regeneration circuit was withdrawn in favor of one based on de Forest's 1916 application.

The IRE had previously given Armstrong its Medal of Honor for the invention of the regenerative circuit and now not only refused to accept Armstrong's attempt to return the medal but gave him a standing ovation at an IRE convention. In 1941, the Franklin Institute gave Armstrong the Franklin Medal for the regenerative circuit, and in 1942, the American Institute of Electrical Engineers awarded Armstrong the Edison medal for the circuit. Armstrong was similarly honored in Europe.

The Fight with Levy over Superheterodyne

In the early 1920s, AT&T had bought for $20,000 Frenchman Lucien Levy's American patent application, hoping that it would be judged fundamental. AT&T in July 1920 had joined the cross-licensing radio group with RCA and GE. In 1921, Wireless Specialty and Westinghouse also joined.

In 1916 in France, Levy, then an officer in the Telegraphie Militaire, got the idea of modulating the RF carrier wave with a supersonic wave that was itself modulated with an audio signal. This idea was not practical or original, but Levy then thought of producing the supersonic wave in the receiver by heterodyning the received RF signal against a signal produced by a local oscillator. The resulting signal could be selected by a tuned circuit and then converted to a final audio signal. Levy obtained two French patents regarding this arrangement and the earlier one, then the subject of a corresponding U.S. patent application, was acquired together with the U.S. application by AT&T.

The original claims in this U.S. patent application differed substantially from those in Armstrong's U.S. Patent 1,342,885, which had been issued on June 8, 1920, but the claims in the Levy application were broadened (by patent counsel for Levy) by copying Armstrong's patent claims exactly. The Court of Claims for the District of Columbia held on appeal that Levy's original disclosure supported the new claims. Since Levy's filing date was about seven months before Armstrong's first date of conception, Levy was awarded a U.S. patent—number 1,734,038—that incorporated seven of Armstrong's nine claims. The remaining two claims were granted in respective U.S. patents to Alexanderson of GE and to Kendall of AT&T. Because AT&T was in the same patent pool as Westinghouse and RCA, the transfer of claims from Armstrong had no effect on the radio industry and was little noticed. Because of French patent procedures then in effect, no effect resulted in France, but in Germany a procedure approximately corresponding to that in the United States resulted (based on the same priority date) in a German patent being issued to Levy on October 1, 1931. Although Armstrong lost his patent rights to the superheterodyne circuit, Armstrong and his collaborators were responsible for making this methodology invaluable and universal in radio technology.

Invention of Frequency Modulation

One unsolved problem in radio was static in reception. Nature has many signals capable of modulating the amplitude of a radio signal, most famously thunderstorms. In AM, the strength of a radio signal is proportional to the strength of the modulated audio signal. Circuitry for the elimination of static in reception of AM radio signals is very difficult to envision. By the late 1920s, as a result of a thorough study with Pupin, Armstrong had concluded that the only way to avoid reception static in AM from natural phenomena, such as electrical storms, was to use a suitable FM system. If the frequency of the main signal is varied instead of its amplitude, then the potential for static interference is greatly reduced, because in nature there are few sources that can modulate the frequency of an FM signal.

In the 1920s, it was thought that FM might be a way of loading more signals into a particular frequency band than in conventional AM, but careful mathematical analysis had indicated that a narrow-band FM signal would always result in worse sound than that in an AM signal of the same power. Armstrong thought that an FM signal did not need to be confined to a narrow range of frequencies but instead could vary over a range that was, for example, five times larger than a conventional AM signal and have a much better signal-to-noise ratio. By reasoning and experiment, he moved beyond the mathematical equations. He established the foundations for information theory under which quantification of bandwidth can be substituted for noise immunity. Even though conventional opinion was that FM was useless for communication, Armstrong started creating an FM transmission and receiving system based on this quantification. Development timing of his idea for an FM radio system seemed to be appropriate. Because of the existing technological opinion regarding FM, no one was working in the same direction.

By 1933, Armstrong had achieved a wide-band FM system that in field tests demonstrated both clear reception even in the most violent electrical storms and also the highest sound fidelity known. However, the major producers of radio equipment, such as RCA, were unimpressed with his FM equipment or demonstrations. They had vast investments in AM and did not wish to be displaced. RCA, after several years of evaluating Armstrong's FM equipment and system, refused to use it on various alleged grounds, but Armstrong proceeded with licenses to smaller companies. He completed designs for a complete system of transmitter, receivers, and antennas. He set up pilot broadcasting in New York and New England in 1939. Apparently, he did these things using his own money. The public received FM broadcasting well and liked the high quality, particularly for music.

RCA decided to ignore FM radio broadcasting and to push the forthcoming television. Sarnoff requested Armstrong to remove his FM broadcasting equipment from their Manhattan transmitting tower at the top of the Empire State building. Armstrong built his own 425-foot tower in Alpine, New Jersey. He at length got an FCC license and began broadcasting in 1939.

During World War II, Armstrong allowed the military to use his FM patents royalty-free, a gift that he could scarcely afford with his various expenses. Mobile FM communications were of great value in Europe and the Pacific. Armstrong himself worked on continuous wave radar, but nothing was deployed before the war ended.

In 1945, a group of companies led by RCA got the FCC to move the FM band from 44–50 MHz to 88–108 MHz, where it has remained. The result was to render obsolete all existing FM transmitters and receivers. The FCC also sharply limited FM broadcasting power and disallowed many possible radio relays and relay sources, including central stations and mountaintops, thereby forcing FM broadcasters to send their signals over AT&T coaxial cables at significant rates. Thus AM survived, while FM was severely handicapped.

The Cellular Telephone

When they were first introduced, a portable telephone based on radio was commonly called a wireless mobile phone or a radiophone; they were in use from Fessenden's time into the 1950s. Such phones tended to be powerful, and they propagated signals over a wide area when in use. Though useful for military purposes, each phone tended to monopolize its operating frequency, effectively limiting use by others. In the 1920s, some German passenger trains were equipped with radio telephones. Also, in the same period, some passenger airplanes were similarly equipped.

Modern cell phone technology, with multiple noninterfering channels using a single frequency, multiple cell stations, automatic "handoff" of operating channels from one station to another as a caller's location changes, and other features, has been traced to 1979 U.S. Patent 4,152,647, to inventors C. A. Gladden and M. A. Parelman in Las Vegas, Nevada, although Martin Cooper of Motorola is commonly considered to be the inventor of the first mobile phone, which he demonstrated on April 3, 1973.

Though cell technology was conceptualized in the late 1940s by Bell Laboratories personnel, it was not until the 1960s that this group developed the electronics. However, the first fully automatic mobile phone system was introduced commercially by Ericsson in Sweden in 1956, and the first citywide commercial cell phone system was introduced in Japan in 1959 by NTT.

An automatic call handoff system was developed at Bell Laboratories in 1970, but it was not until 1982 that the Federal Communications Commission approved an analog system, which was superseded by a digital system in 1990. The first cell phones in the 1980s were large and were usually installed in vehicles ("car phones"). Second-generation cell phones appeared in the 1990s, along with digital cellular phones, and some analog systems were shut down to make room for digital systems. By about 2000, third-generation cell phones became available; they had various features, such as

(continued)

the ability to display video, but at a consumer cost. Cell phone technology and use continue to develop.

The cell phone industry has many inventors and manufacturers, with Nokia of Finland currently having the largest market share (about 40%). Usage varies from country to country. In the United States, cell phone penetration is about 81% currently; in Europe, some countries have substantially more cell phones per capita than people. In 2005, the total number of cell phone subscribers in the world was about 2.14 billion, and it reached about 3.3 billion by the end of 2007 (about half the earth's population).

Although cell phones are potentially useful in public-scale emergencies, because most networks operate normally near capacity, during an emergency a cell phone system can become overloaded and useless.

Armstrong proceeded to redesign all his systems and got them operating again by 1948. This was his last engineering achievement.

RCA had refused to pay Armstrong patent license fees, although RCA was making and selling FM equipment. With only about 2 years left before expiration of key U.S. patents, Armstrong brought a patent infringement suit against RCA in 1949. RCA made Armstrong their first witness and kept him on the witness stand for largely irrelevant matters for about a year. When RCA was required to disclose its work on FM in the 1930s, Sarnoff asserted that RCA had invented FM without Armstrong. Apparently, the dates and other supporting evidence, such as patents, that would support this assertion have never been disclosed. Armstrong proceeded to file lawsuits against 21 infringers.

By 1953, Armstrong's patents and licenses had expired. His legal bills and research costs had almost bankrupted him. It is speculated that he was depressed and that probably he wondered what on earth he had accomplished over his lifetime. His health and stamina had been deteriorating and were continuing to do so. On Thanksgiving Day, he had a monumental fight with his wife, and she left him and went to live with her sister in Connecticut.

On January 31, 1954, after writing a two-page letter to his wife that he left on his apartment desk, he dressed neatly; put on an overcoat, hat, scarf, and gloves; and walked out of a thirteenth-story window. He landed on a third-story overhang. His body was not discovered until the following day. Details regarding his suicide are not known, but friends and foes alike were saddened and horrified by the tragedy.

Marion continued his lawsuits. She seems to have first dealt with RCA and achieved a settlement of about a million dollars. She then used this money to finance the other lawsuits. She won or compromised all lawsuits and gained millions. The "last holdout," Motorola, did not give up until the Supreme Court ruled against it in 1967, 13 years after Armstrong's death.

Armstrong was a very talented individual. Three of Armstrong's four great inventions seem to have been challenged regarding the issue of priority. A priority challenge to the time of origin of a great invention is common. In Armstrong's case, two of his original inventions were later found through litigation to be subsequent in date to others who obtained the patent coverage. Priority regarding one of his original inventions—actually a group of component inventions relating to FM—seems never to have been fully tested through litigation. The fourth (the super-regenerative circuit) has no known public record regarding a challenge to priority. But priority is not everything. Apart from issues of priority, Armstrong and his circuit inventions appear to have been the leading contemporary force in promoting widespread commercial development and use of radio.

CONCLUSION

Development of broadcast radio, post Marconi, can be considered to have been enabled mainly by the inventions of AM, regeneration circuitry, superheterodyne circuitry, and FM.

FURTHER RESOURCES

Hugh G. J. Aitken. *The Continuous Wave: Technology and American Radio, 1900–1932*. Princeton, NJ: Princeton University Press, 1985.

Richard T. Ammon. The Rolls Royce of Reception: Super Heterodynes, 1999. http://www.superhets.info/page4.html.

E. H. Armstrong. A Method of Reducing Disturbances in Radio Signaling by a System of Frequency Modulation. *Proceedings of the IRE* 24:689–740, 1936.

Lee de Forest. *Father of Radio*. Chicago: Wilcox and Follett, 1950.

Lee De Forest, 87, Radio Pioneer, Dies; Lee De Forest, Inventor, Is Dead at 87. *New York Times*, July 2, 1961.

De Forest Sends Out the Opera from the Metropolitan. *New York Times*, January 1910.

Election Returns Flashed by Radio to 7,000 Amateurs. *Electrical Experimenter*, January 1917, p. 650.

Helen M. Fessenden. *Fessenden: Builder of Tomorrows*. New York: Coward McCann, Inc., 1940.

Reginald A. Fessenden. The Inventions of Reginald A. Fessenden. *Radio News*, 11-part series beginning with the January 1925 issue.

Sungook Hong. A History of the Regeneration Circuit: From Invention to Patent Litigation. University, Seoul, Korea, 2004. http://www.ieee.org/portal/cms_docs_iportals/iportals/aboutus/history_center/conferences/che2004/Hong.pdf.

Lawrence Lessing. *Man of High Fidelity: Edwin Howard Armstrong*. Philadelphia: J. B. Lippincott Company, 1956.

Tom Lewis. *Empire of the Air: The Men Who Made Radio*. New York: E. Burlingame Books, 1991.

James E. O'Neal. Fessenden: World's First Broadcaster?—a Radio History Buff Finds That Evidence for the Famous Brant Rock Broadcast Is Lacking. *Radio World Online*, October 25, 2006.

Radio Set-up Eliminates All Noise. *Ogden Standard Examiner* (United Press), June 18, 1936, p. 1.

Tsividis Yannis. Edwin Armstrong, Pioneer of the Airwaves, 2002. http://www.ee.columbia.edu/misc-pages/armstrong_main.html.

Rutgers University/AP Photo

Penicillin and Alexander Fleming

The discovery and early work on penicillin were achieved by Alexander Fleming, and the necessary but difficult research and development for manufacture of penicillin were carried out by Howard Florey, Ernst Chain, and Norman Heatley. Quantities of the first great antibiotic, penicillin, arrived in time for wide use in World War II (1939–1945). Penicillin has saved millions of lives, and its discovery and success led to the research and development of other antibiotics.

BACKGROUND

Ancient cultures are known to have utilized naturally occurring substances containing antibiotics for the treatment of diseases, but nothing was known then about the antibiotic agents in these substances. The Chinese achieved the first known use (over 2,500 years ago), but usage by the ancient Egyptians and Greeks, and apparently other ancient cultures, also occurred.

Technology for treating microbial infections was developed before modern usage of antibiotics. In 1796, Edward Jenner found that when people were inoculated with material taken from a pustule of cowpox, they were protected against smallpox. Techniques for producing vaccines had to wait almost a century until Louis Pasteur, from observation and experiment, proposed the germ theory of disease and proceeded to attenuate (weaken) an anthrax preparation by heat, which destroyed virulence but not immune response, leaving an inoculatable (or vaccinatable) material (1880). He then used and developed this procedure to also prepare inoculations for fowl cholera (1880) and rabies (1885).

It was then determined by others that some bacteria produce toxins for which antitoxins could be produced, which resulted in success against diphtheria (1891) and tetanus (1914). Even dead vaccines could be effective (against typhoid fever, 1897).

The next advance involved the discovery of agents for the killing of harmful bacteria. Frederick W. Twort of the University of London in 1915 and Felix H. D'Herelle at the Pasteur Institute in 1916 each independently discovered bacteria-killing viruses (called bacteriophages), but the discovery was not immediately considered or developed for usage as inoculatable substances for disease treatment in humans.

Ernest Duchesne in France in 1897 reported that mold of *Penicillium* species had antibiotic activity, but he had not further investigated the matter. His disclosure did not result in work by others. The discovery that harmful bacteria could be killed by a substance produced by a *Penicillium* mold (a fungus) without adverse effect upon humans was independently made by Alexander Fleming, who had pursued the matter.

ALEXANDER FLEMING (1881–1955)

Early Years

Alexander was the seventh of eight children of a sheep farmer, Hugh Fleming, and was born on August 6, 1881, at Lochfield Farm (800 acres) near Darvel, Ayrshire, Scotland. His father, after his first wife's death, had married Grace Morton, the daughter of a neighboring farmer, and Alexander was the third child of four surviving from this marriage. Hugh died at age 66, when Alexander was 7.

The Fleming children played and ranged through the countryside with its many moors, valleys, and streams. Alec (as he was known in his family) attended Loudoun Moore School and Darvel School and graduated from Kilmarnock Academy. The family was poor. The farm was inherited by the eldest brother upon the father's death, and at that time another brother, Tom, had completed study of medicine and was starting a practice in London. A sister and four brothers, including Alec, moved to London and lived together. Alec attended the Polytechnic School on Regent Street.

Tom encouraged Alec to enter business, and after completing school, he obtained employment in a shipping office as a shipping clerk. Alec worked there for 4 years, but he did not enjoy the work.

After the Boer War started in 1900 between the United Kingdom and its southern African colonies, Alec and two brothers joined a Scottish regiment, but though they were too late to see service in the war, they were able, while in their regiment to pursue shooting, swimming, and water polo. In his young days, Alec served in the Territorial Army, and he served as a private in the London Scottish Regiment from 1900 to 1914. He retained his membership in the rifle club. Afterward, the Flemings' uncle died and left each of the children 250 pounds. Now having a thriving practice, Tom encouraged Alec to study medicine.

Alec achieved high scores in the qualifying examinations and had his choice of medical schools. Although he lived about equally distant from three schools, he chose St. Mary's because he had played water polo against them, and he enrolled there in 1901. By 1905, Alec was specializing as a surgeon, but as a surgeon it would be necessary to leave St. Mary's. Alec, a great shot, was pursued by the captain of St. Mary's rifle club, who worked in the Inoculation Service. Alec decided to work in the captain's department at St. Mary's, pursue research, and join the rifle club. He became assistant bacteriologist to Sir Almroth Wright, a distinguished pioneer in vaccine therapy and immunology. For the rest of his career, Fleming stayed at St. Mary's. He received a bachelor of medicine degree and then a bachelor of science degree with gold medal in 1908. He specialized in bacteriological research and served as a lecturer at St. Mary's until 1914. St. Mary's was a teaching hospital. However, he also trained at the Royal College of Surgeons.

Fleming was a short man who usually wore a bow tie and who seems never to have followed all the conventions of society. Had he not discovered and proceeded to carry out tests with penicillin, he most probably would have been a quiet bacteriologist all his professional life.

In about 1909, German chemist-physician Paul Ehrlich with associate Sahachiro Hata discovered, after testing hundreds of arsenic-containing organic compounds, one identified as number 606, dihydroxydiamino-arsenobenzene hydrochloride, which killed the syphilis spirochete but was innocuous to humans. Syphilis was then a relatively common and dreaded disease with no good treatment. Erlich named the compound salvarsan (today, this compound is known as arsphenamine, but it is no longer used in medicine). Ehrlich visited London and met Alec, and Alec became interested in treating patients with the agent. Fleming became one of the few physicians to administer the compound using the new and risky technique of intravenous injection. Fleming soon had a busy practice and earned the name "Private 606."

Alec married a nurse, Sarah M. McElroy, from Killala, Ireland, in late 1915, and their only child, Robert, became a general medical physician. In 1949, Sarah died, and in 1953, Alec married Dr. Amalia Koutsouri-Vourekas, a Greek colleague at St. Mary's, who, although Alec died in 1954, lived till 1986.

When World War I began, most of St. Mary's bacteriology laboratory left for France to establish and operate a battlefield hospital laboratory and serve in battlefield hospitals at the Western Front in France. Fleming served as a captain in the British Royal Army Medical Corps. He was mentioned in dispatches. He made many innovations in treating the wounded. Though the infections they encountered were usually simple, some were so severe that wounded soldiers soon died from them. Suppurating wounds and fatalities due to gangrene and septicemia were common. Alec thought that there must be some chemical that would fight microbial infection.

He found that, unfortunately, known antiseptics destroyed a patient's immunological defenses even more effectively than they killed invading bacteria. In an article he submitted to the medical journal *The Lancet* during the war, he described an experiment that he was able to conduct using a glass apparatus he himself made by glass blowing. The experiment demonstrated why antiseptics were actually killing more soldiers than infection. The problem was that while antiseptics worked well on a wound surface, deep wounds tended to shield and shelter anaerobic bacteria from antiseptic. Antiseptics seemed to remove beneficial agents that protected patients and did nothing concerning bacteria that were not reached. The position of Alec was rejected—perhaps people found it hard to believe that pus helps fight disease—and most army surgeons continued to use antiseptics. He also provided an improved treatment for syphilis.

In 1918, he returned to St. Mary's Hospital. Recalling the deaths of many soldiers who died, particularly from septicemia from infected wounds, he actively searched for antibacterial agents. He authored numerous published papers involving bacteriology, immunology, and chemotherapy.

In 1921, he discovered a natural antiseptic he called "lysozyme," "the body's own antibiotic," an enzyme that is present in many body fluids, such as tears. It has a natural antibacterial effect and dissolves bacteria, but it is not effective against strong infectious agents. He continued his searching.

Discovery and Evaluation of Penicillin

Fleming liked to work and seems to always have been very busy in his laboratory. Though from his work he had a reputation for brilliance, in his own laboratory work he had a reputation for mess. It seems that he was reluctant to clean up and liked to pursue and have underway more than one project at the same time. His laboratory was not well organized or tidy.

In September 1928, after being appointed professor of bacteriology at the St. Mary's school, he went on a holiday (a vacation) thought to have been about two weeks and returned, apparently, to find a visitor to whom he wished to show what he had been researching. He noticed that many of the culture dishes (Petri dishes) that he had left in his laboratory, and in which he had been culturing staphylococci ("staph") bacteria, which can cause boils, carbuncles, abscesses, pneumonia, and septicemia, were contaminated with a fungus. He discarded the dishes in a disinfectant bath. However, to show the visitor what he had been working on, Fleming retrieved some of the not yet submerged but otherwise discarded dishes. He noticed then one of the retrieved dishes had been covered as usual but for some reason had not been put into the incubator. This dish had not grown cultured bacteria to an extent such as would have normally occurred under the existing summer conditions even outside the incubator. Instead, the dish contained, in addition to the bacterial culture, a mold culture. He saw what he believed to be a very unusual occurrence: some of the colonies of staphylococci bacteria that should have been growing near and around the mold had disappeared.

Fleming thought that the mold could be making something that was capable of destroying the bacteria. He did some experiments and found that this was indeed the situation. Even at a dilution of 1 to 800, the unknown compound produced by the mold prevented the growth of the staphylococci. He cultured the mold by growing it in a broth. He identified the mold as being from the *Penicillium* family, later characterized as being *Penicillium notatum*, and he determined that this mold had produced an unknown substance or chemical, which he tried to isolate. Since he was the first to attempt isolation, he had the privilege of naming it. He called the substance "penicillin." He confirmed that penicillin prevented bacterial growth for an unknown reason. Fleming reported his findings in 1929, but there was little interest.

Fleming continued to work with the mold. He attempted to grow various types of bacteria in the vicinity of the mold. Some grew well; others would not approach the mold beyond a certain distance. He further experimented on the effect of the unknown chemical on white blood cells. If the chemical was poisonous to human cells, then the material was not particularly useful in medicine. He found that the chemical did not affect white blood cells at concentrations that were highly inhibitive of bacteria. Further experiments showed that the chemical had a positive antibacterial effect on many microorganisms. Besides bacteria such as staphylococci, his tests showed that the compound killed generally all Gram-positive pathogens, such as pneumonia, scarlet fever, gonorrhea, meningitis, and diphtheria, but not typhoid or paratyphoid, diseases for which at that time he was seeking a curative agent, or cholera or bubonic plague. He found that penicillin given orally to animals did not work. He soon published another report in the *British Journal of Experimental Pathology* regarding penicillin and its potential uses.

By the end of this experimental work, Fleming, because of his training and experience, had about reached the limit of his technological capacity. He was not a chemist and he could not isolate or identify the chemical. He could not produce it in quantities sufficient for further experiments. Cultivating the mold on a laboratory scale was difficult, and he was unable to concentrate and stabilize the chemical for even short-term, shelf-life purposes. Though the chemical and his test results were unusual, he did not arouse much interest. Because its action appeared to be slow and because it appeared to be quickly eliminated from, or inactivated by, an animal body, Fleming was forced to conclude reluctantly that penicillin might not be practical for human treatment.

Fleming discovered early in his work that bacteria developed resistance to penicillin whenever either too little was used or it was used for too short a period. His own research on penicillin stopped, but apparently, upon request from other research organizations during most of the 1930s, Fleming supplied limited quantities of cultured penicillin material, but the results from most tests were inconclusive, probably because the material was used as a surface antiseptic. Sometimes, trials showed promise. He tried without success to interest a chemist in pursuing work with penicillin. Fleming kept, grew, and distributed the original mold for 12 years.

Making Penicillin

In 1938, at the Dunn School at Oxford University, Dr. Howard Florey (a group leader) with Dr. Ernst Chain and Dr. Norman Heatley began a systematic study of antibacterial agents produced by microorganisms. Florey was a pharmacologist from Australia who had become a professor of pathology at Oxford. Chain was a German-born, German-educated Jew who left Germany and moved to England in 1933 after Hitler came to power. Heatley was an Englishman.

Florey and Chain had been working on lysozyme (Fleming's earlier discovery). Chain supposedly came across Fleming's work on penicillin and brought it to the attention of the others. The researchers decided that Fleming's penicillin work offered the most promise among prior work in the field and should be investigated further.

They reproduced Fleming's work and confirmed Fleming's problems of isolating the active component, penicillin, from what Fleming had called the "juice" produced by the mold. Only about one part in 2 million was the active component. Chain worked out methodology for concentrating and isolating penicillin and theorized its structure, which was confirmed by X-ray crystallography by Dorothy Hodgkin. Heatley proposed that the active component, after separation by ether extraction, be transferred back into water by changing its acidity, which purified the material. This procedure at last enabled production of a sufficient quantity of the drug to permit testing on small animals. Chain also discovered penicillinase, an enzyme that catalyzed breakup of penicillin.

In 1940, the Oxford team published its first results. Shortly thereafter, Fleming telephoned Florey to say he would be visiting within a few days. When Chain heard this, he exclaimed, "Good God! I thought he was dead."

On Saturday, May 25, 1940, they inoculated each of eight mice with a lethal dose of a virulent strain of streptococci and then injected four of them with penicillin solution. Within hours, the mice receiving streptococci alone were dead. The four injected with penicillin were healthy.

Florey and his group immediately decided to make a stable form of penicillin and mass-produce the material. Various talented people comprised Florey's group, and much innovative work was needed to achieve a stable product and a practical preparation process. The Battle of Britain was then at its most severe, with nightly Luftwaffe air raids, and pharmaceutical companies were unwilling to use and divert their equipment and personnel capable of culturing microorganisms from war efforts when the product was still untested.

Florey could only try to produce penicillin on a pilot plant scale at the Dunn School. Florey appreciated that a human was about 3,000 times heavier than a mouse and would need a correspondingly larger dose of penicillin. He delegated the project to Heatley, who had been so far successful in cultivating penicillin, the objective being to acquire enough penicillin for human testing with just a few individuals.

The basic process the group decided to implement involved setting up cultures, harvesting them to separate the penicillin-containing culture fluid, and using an acidified extraction with ether and water. It was known that an acidified solution shaken with ether would result in penicillin passing into the ether. Then, the ether extract would be shaken with water at a neutral pH with buffer and/or alkali to bring the penicillin back into a clean water phase.

The high cost and the long delay found to be associated with procurement of specialized glass or glass-lined vessels that would be suitable for making a penicillin-containing product were deemed to be unacceptable. Heatley proposed using relatively low-cost porcelain and similar vessels. They considered using porcelain bedpans, but because of the wartime conditions, this essential article was in short supply. Heatley designed a modified version, of which several hundred were manufactured in the Potteries, a group of five towns in the English Midlands where fine tableware had been made for centuries.

Using these containers, the Oxford laboratory became the first penicillin factory; it used the entire laboratory staff. On Christmas Eve 1940, at the Sir William Dunn School of Pathology at Oxford, dozens of the new containers were washed, sterilized, and filled, and on Christmas Day, Heatley seeded each container with spores of the fungus *Penicillium notatum*. The containers were stacked for a 10-day incubation period, at the end of which it was hoped that enough penicillin would had been produced to begin human testing. Apparently a variety of other containers were also employed as fermentation vessels, including pie dishes, bedpans, and even biscuit tins.

Within a month, Heatley had about 80 liters of crude penicillin solution. There were about 1–2 units of penicillin per milliliter, or about 100,000 units total. A unit was measured on a custom culture plate Heatley had invented. Florey decided they had enough penicillin to try it on a human.

Lying in the Radcliffe Infirmary was a 43-year-old policeman who was dying of staphylococcal and streptococcal infection contracted when he had scratched his face on a rose bush a few months previously. His face was covered with suppurated abscesses. One eye had been removed. Not even sulfonamides had helped. After a single infusion of penicillin, he immediately started to improve. After less than a week, he was recovering. Penicillin recovered from his urine (a practice used to extend the readily excreted penicillin) was exhausted. The bacteria regrew and the man died a few weeks later.

The next patient was a 15-year-old boy who had undergone a hip operation to insert a pin. His wound had become septic, and sulfonamides again had not helped. Within two days after infusion with penicillin, his previously elevated temperature was normal. The boy fully recovered.

Other patients were successfully treated with the penicillin. Results were published in *The Lancet* in August 1941.

The value of penicillin had been proved, but Heatley and the group were unable to increase the penicillin yield. The Rockefeller Foundation had been supporting Florey and suggested that he seek help in the United States. Florey and Heatley flew to the United States and went to the North Region Research Laboratories of the Department of Agriculture in Peoria, Illinois.

An improved broth was developed. A worldwide search for a rapidly growing mold that would produce much penicillin ended when a moldy cantaloupe grown locally in Peoria was found to produce a rich yield of penicillin. The American pharmaceutical industry, with help especially from Heatley, solved the problems of large-scale production. In 1943, Florey carried out trials in North Africa on wounded soldiers and demonstrated excellent results. By June 1944 (the same month as D-Day), unlimited availability for treatment of wounded troops was available.

It was learned that penicillin works by targeting peptidoglycan chains (murein) in the cell walls of bacteria and inhibiting the transpeptidylation, or cross-linking, of peptidoglycan chains. Penicillin thus effectively halts bacterial cell wall expansion. When the cell wall fails to expand with the cytoplasm, the bacteria lyses (dies because of cell wall dissolution). Other factors contribute to this process.

Other scientists tried to repeat the work carried out by Fleming regarding penicillin. At first there was no success. Years later, the problem was solved by Ronald Hare, Fleming's research assistant. Penicillium mold grows best at about 20°C, and staphylococcal bacteria grow best at about 35°C. After the culture plate in Fleming's laboratory was contaminated, the temperature became unusually cool for about nine days, during which the *Penicillium* mold grew well but the staphylococcal grew very little. Then, the temperature rose and the bacteria started to grow, but by now the penicillin present was sufficient to destroy the bacteria. Had not Fleming forgotten to put the plate into the incubator before going on holiday, had not the weather not changed in the manner it did, had not the right *Penicillium* mold landed (presumably as spores) on the plate, penicillin would not have been discovered then.

Years later, Fleming, as a famous man visiting the United States, was shown a new laboratory in the United States that had fixed windows and air conditioning. Fleming is alleged to have remarked, "Ah, you realize that I would never have made my discovery in these conditions, they are too clean."

AFTERMATH

In 1944, Fleming was knighted. Ernst Chain, Howard Florey, and Alexander Fleming shared the 1945 Nobel Prize in medicine. Apparently, under the rules governing the prize, it can only be awarded to three individuals. Norman Heatley received the Order of the British Empire in 1978. U.S. companies and individuals received great profits from penicillin.

Several fables without substance emerged about Fleming. In one, young Winston Churchill was saved from death by Fleming's father, and Churchill's father as a result paid for Fleming's education. In another, Fleming saved Winston's life in Carthage, Tunisia, in 1943, with penicillin when Churchill was ill.

Fleming died suddenly in his London home from a heart attack in 1955 at age 74. He was cremated, and his ashes were interred in St. Paul's Cathedral. His widow presented the Nobel Prize medal awarded to him, Florey, and Chain to his London gentlemen's club, the Savage Club, which continues to display it.

Bacterial Antibiotic Resistance

Soon after discovering penicillin, Fleming discovered that bacteria developed antibiotic resistance if too little penicillin was used or if used for too short a time period. Almroth Wright, a former bacteriologist at St. Mary's hospital under whom Fleming had worked, even predicted antibiotic resistance before it was noticed in experiments. Fleming cautioned not to use penicillin unless there was a proper medical reason for doing so, and if used, never to use too little, or for too short a period.

The perception, even as late as the 1980s, seemed to be that bacterial infection had been overcome. Pharmaceutical companies then were not developing new agents. However, it was observed that bacterial resistance increased relative to a number of commonly used antibiotics. By about 1994, researchers were reporting bacteria in patient samples that were resistant to all currently available antibiotics. An antibiotic does not directly cause the resistance but rather leads to it by creating an environment where an already existing variant can flourish.

Years ago, some microbes began resisting penicillin. Sometimes related agents, such as methicillin or oxacillin, could be used instead. Sometimes other, unrelated antibiotics were available and could be used. However, there may be a limit on the number and variety of usable antimicrobial drugs. At present, the agents of last resort for many infections are apparently the antibiotics vancomycin, linezolid, or daptomycin.

Antibiotic resistance can have various causes. Penicillin, for example, operates by destroying a key portion of a cell wall. Microbes resistant to penicillin, though, either alter their cell walls so that penicillin does not bind or produce enzymes that break up penicillin. Another antibiotic, erythromycin, attacks the ribosomes of cells, ribosomes being the cellular mechanism that makes proteins. Resistant bacteria have somewhat altered ribosomes to which the antibiotic cannot bind.

Antibiotic resistance occurs because of gene changes in bacteria. Gene changes can occur in one of three ways: spontaneous bacterial DNA mutation; transformation, where one bacterium takes up DNA from another; and plasmid (a small ring of DNA) migration from one bacterium to another. Social and environmental factors can contribute to antibiotic resistance.

Measures to slow the rate of growth of antibiotic resistance include improving infection control, developing new antibiotics, and using pharmaceuticals more effectively. Hope has been expressed that bacteriophages having a lytic cycle may one day be practical for controlling bacterial infections.

His discovery of and experimentation with penicillin changed the world of medicine by introducing bacteria-controlling antibiotics and by stimulating their research, development, and manufacture. Penicillin has saved countless lives.

DEVELOPMENTS

Penicillin is considered to have changed the course of history. It was the first antibiotic widely used medically, saved countless lives, and stimulated worldwide research, development, and manufacture of various antibiotics. Although many antibiotics of different types are now known, relatively few are safe for human use. Because of its broad-spectrum properties, the antibiotic tetracycline has enjoyed extensive use and has replaced penicillin in many situations.

Now, however, many bacteria commonly called "superbugs," such as *Pseudomonas aeruginosa*, have been found that have developed resistance to many antibiotics yet are known to cause various life-threatening infections. Hospitals have established antimicrobial programs involving groups that include pharmacists, infectious disease specialists, and microbiologists, with the aim of monitoring antibiotic usage. The federal program Medicare apparently will soon refuse to pay hospitals for treatment of preventable infections that patients acquire while under the hospital's care.

As the *Wall Street Journal* recently reported, about 2 million people contract bacterial infections in U.S. hospitals per year, and about 90,000 of these people die. Antibiotics do function to kill or inhibit susceptible bacteria, but some bacteria survive and become resistant to the treating antibiotics. For example, increasingly, a form of drug-resistant staph identified as MRSA (methicillin resistant *Staphylococcus aureus*) is found in hospitals and places such as school-locker rooms; it has developed resistance to common antibiotics. Two leading hospital purchasing groups, VHA Inc., which involves more than 1,400 nonprofit hospitals, and Premier, Inc., which involves more than 2,000 hospitals, urge members to adopt programs that electronically track and monitor usage of certain pharmaceuticals. The objective is to induce doctors to use for an infected patient an antibiotic that attacks only the bacterium that is causing that patient's infection instead of a strong, broad-spectrum, commonly used antibiotic such as ciprofloxacin, which could treat any infection effectively. The theory is that avoiding the use of an overused antibiotic such as "cipro" preserves that antibiotic as a reserve for use in treating serious infections.

FURTHER RESOURCES

Isaac Asimov. Alexander Fleming. In *Asimov's Biographical Encyclopedia of Science and Technology*, 2nd ed., p. 1077–1079. New York: Doubleday and Company, Inc., 1982.

Alexander Fleming. On a Remarkable Bacteriolytic Element Found in Tissues and Secretions. *Proceedings of the Royal Society Series B* 93:306–317, 1922.
Andre Maurois. *The Life of Sir Alexander Fleming*. Penguin, 1959.
Nobel Lectures. *Physiology or Medicine 1942–1962*. Amsterdam: Elsevier Publishing Company, 1964.
Richard Tames. *Alexander Fleming*. London: Franklin Watts, 1990.

AP Photo

Television and Philo Farnsworth

Some will find it amazing, but the all-electronic television system used today was invented in 1922 by a 14-year-old, Philo Farnsworth, who came from a sharecropper's home near Rigby, Idaho. Television is probably the most widely used and pervasive telecommunications system for broadcasting and conveying information. In most countries, television has today become a primary medium and source for information, news, and entertainment.

HISTORY

Briefly, all-electronic television superseded its forerunner, electromechanical television, which reached its inherently maximum development and public usage in the late 1930s. Though available in the late 1930s, all-electronic television with black-and-white images was not a common domestic item until the late 1950s. Color television did not become universal until the early 1970s.

The first electromechanical television was proposed and patented by Paul Gottlieb Nipkow in 1884 when he was still a German student. Nipkow had then invented a scanning disk that is believed to be the first image rasterizer. The discovery of the photoconductivity of selenium, a nonmetallic element that acts as a rectifier and converts radiant to electrical energy, was made by Willoughby Smith in 1873. Together with Nipkow's scanning disk, the properties of selenium led to the ability to electronically transmit photographs and still pictures over wires. By the early 1900s, half-tone (gray-scale) photographs were being transmitted by a facsimile system over telegraph and telephone lines for newspapers. With developments in vacuum tube amplification technology beginning in about 1907, the system became practical for commercial purposes.

Scottish inventor John Logie Baird in 1925 first demonstrated televised silhouette images in motion and later demonstrated transmission of live, half-tone, moving images. Baird used a vertically scanned image and a scanning disk associated with a double spiral of lenses, but his system had only 30 resolution lines per frame, barely enough to produce a recognizable human face. Baird gradually improved his system, became involved in a German experimental electromechanical television system in 1929, and achieved an electromechanical system having 240 resolution lines per frame on British Broadcasting Corporation (BBC) television broadcasts in 1936. His system was discontinued when the Marconi-EMI company produced a 405-line all-electronic system.

Karl Ferdinand Braun (who had shared the 1909 Nobel Prize in physics with Guglielmo Marconi) had invented, in 1897, the pioneering cathode ray oscilloscope, which was suitable and adaptable for use as a receiver tube in the all-electronic television system. This basic electron beam scanning device is still widely used today.

In 1908 and 1911, Alan Archibald Campbell-Swinton in England proposed an all-electronic television system similar to that still used today using cathode ray tubes at each of the transmitting and receiving locations. This system was not developed by Campbell-Swinton.

In the United States, various research and demonstrations regarding electromechanical systems occurred in the 1920s. Though improved electromechanical television systems were still being introduced in the late 1920s, Philo Farnsworth and Vladimir Zworykin in the United States were then separately and independently seeking to produce all-electronic transmitting tubes.

The arrangement used today in the all-electronic television involves continuous electron emission from a transmitter with accumulation and storage of released secondary electrons during the scanning of a single frame. Such single-frame scanning is then immediately repeated to provide successive scans. The idea was evidently described by the Hungarian inventor Kalman Tihanyi in 1926, who achieved more developed versions in 1928.

Although there was a significant amount of electronic know-how available to well-educated specialists in the 1920s, it remains an almost incredible occurrence that the invention of all-electronic television was first achieved (embodied) by an American country boy in 1927.

PHILO TAYLOR FARNSWORTH (1906–1971)

Early Years

Philo was born August 19, 1906, to Mormons Lewis E. and Serena B. Farnsworth at Indian Creek, Utah, near the town of Beaver. His father was a sharecropper. In 1918, when Philo was 12, the family moved to a house on a ranch near Rigby, Idaho, where Philo was surprised to find that their house was wired for electricity with its own power plant. Their house was 4 miles from the nearest high school, which Philo attended daily on horseback.

His early interest in things electrical began with his first telephone conversation with an out-of-state relative. Then he discovered a bunch of technology magazines, probably left by a prior occupant, in the attic of the family's new home. Philo started to read in these magazines everything he could find about electrical and electronic things, especially about television ideas.

By age 13, Philo had taught himself much about electricity. He fixed the farm's generator when none of the adults could. He built motors out of spare parts. He fixed his mother's sewing machine, and he connected a motor to drive her washing machine. He won a first prize of $25 offered by the magazine *Science and Invention* for developing a theft-proof automobile ignition switch.

Invention of the "Image Dissector"

At Rigby High School, Philo excelled, especially in physics and chemistry. He had good skills in mathematics. In 1922, at age 16, he drew a diagram on a blackboard of his electronic "image dissector" for his high-school chemistry teacher, Justin Tolman, and explained that this was his idea for television. He had had this idea for about 2 years. Philo explained that his idea had come to him while he was tilling rows in a potato field with a horse-drawn harrow. He realized that an electron beam could scan images the same way, row by row, line by line, like reading text.

He persuaded Tolman to give him special instruction and allow him to audit a senior course. Tolman appreciated Philo and Philo appreciated Tolman. Philo always acknowledged his teacher's training and inspiration.

Philo had read in a magazine about electromechanical television that used a rotating disc as a scanner. The device was attributed to Baird. In 1922, Baird's system produced only blurs of light (though by 1926, Baird could produce a poor image of a face). Philo realized that there was a better way: use line-scanning electrons.

Philo thought in 1922 that only electricity could move rapidly enough to render pictures, and so he proposed to use electrons controlled in speed and direction and then diffused over a screen. Farnsworth had even then decided that he was going to make a working model of his television before anyone else did. As described below, the diagram that he drew on Tolman's blackboard was later the key piece of evidence in his winning a patent interference case set up by the U.S. Patent Office between himself and Vladimir Zworykin, an expert researcher employed by Radio Corporation of America (RCA) in its well-equipped laboratories.

Philo took some violin lessons in his high school years. The family moved to Provo, Utah, in 1923. After only 2 years of high school, though, Philo managed to demonstrate sufficient intelligence and education to be admitted to Brigham Young University, where he was a student from 1923 to 1925. There he continued his research into cathode ray and vacuum tubes.

In late 1923, his father died from pneumonia, which forced Philo to suspend his research, leave college, and find a job to support his family, including his mother. He had a brief stint in the Navy.

Achievement

Philo was then fortunate to meet George Everson, a community chest manager and professional money raiser, who employed Philo. Philo described to Everson and his coworker Leslie Gorell his idea for television. He acknowledged the work of others before him but thought he could go further. Everson, a chance taker regarding business matters, decided to back Philo and provided him $6,000 for research and development of this idea. Philo apparently had something like a partnership interest so he agreed to cofound Crocker Research Laboratories, which would be based in San Francisco. Before leaving Utah in 1926, he married at the age of 20, and he and his wife Elma Gardner ("Pam") traveled in a Pullman suite to California. Over time they had four sons.

In California at their new facilities, Philo started his research. When his preliminary results looked promising, Everson continued to back and fund Philo. To get corroboration and hopefully get other backers, Everson brought a professor from Cal Tech into Philo's laboratory, Dr. Mott Smith. Dr. Smith provided the desired corroboration and expressed wonder at "the daring of this boy's mind."

Backers came, and their funding allowed Philo to continue his research and development on an electronic scanning and receiving system. On September 7, 1927, Philo, age 21, was able to provide a demonstration at his laboratory at 202 Green Street, San Francisco, California. He transmitted a single white line from an early embodiment of his "image dissector" television camera to a television receiver screen. Seventy-five years later, his wife, Pam, was the only one then still alive who witnessed this demonstration. The backers agreed that this was a new invention worth investing in and provided further funding. A patent application identifying Farnsworth as inventor was filed in the U.S. Patent Office in 1927.

Philo's image detector employed a layer of the photoconductive, or photoelectric, element cesium. This layer emitted electrons in areas struck by light, producing a so-called electrical image. Philo created an electron beam that he caused to scan a light-imaged cesium layer rapidly and systematically in progressive lines at a rate of tens of thousands of times per second. A fluctuating electric current was formed that corresponded to variations in the electron beam scanning the imaged cesium layer.

He had little or no experience when he faced certain problems. For example, he had had no experience or training with high vacuum, but he came up with a method for sealing a flat lens on his image dissector camera tube and then for evacuating the tube to obtain a high vacuum.

By 1928, Farnsworth had developed his system further and had achieved a working video camera. He demonstrated for the press on September 1, 1928, a complete camera and receiver system in which a motion picture film was televised. His backers had demanded to know when they would see dollars from the invention, so the first image shown to them was a dollar sign. The image transmitted apparently used 60 horizontal scan lines.

In 1929, Philo eliminated a motor-generator, so the system then had no moving mechanical parts, and he transmitted a 3½-inch-square image of his wife. He apparently then had developed the image oscillator, a cathode ray tube receiver that could display images produced by the image dissector. His wife's eyes were closed because of the intensity of the light involved. Unfortunately, Farnsworth's image dissector camera required a lot of light to operate properly. He sought unsuccessfully to overcome this problem but was never able to. Farnsworth renamed his concern Farnsworth Television and Radio Corporation.

Zworykin

In 1930, RCA's president, David Sarnoff, having been informed about Philo's work, decided to send to Philo's laboratory someone knowledgeable to gather information for RCA. RCA and Sarnoff wanted to evaluate what Philo was doing, and see how far along he was.

Sarnoff had no intention of paying royalties to Farnsworth. He is alleged to have said, "RCA doesn't pay royalties, we collect them." At most, if necessary to permit RCA to manufacture and sell television sets, he might consider buying technology.

Sarnoff sent Vladimir Zworykin to Farnsworth's laboratory. Zworykin, an immigrant from Russia, was well educated and skilled in electrical subjects. He had been working on developing television for Westinghouse since 1923, but his work had previously been along lines similar to those being pursued by Baird and British scientists. Although these various other researchers had superior funding, their progress was slower than Philo's.

Appearing as an interested colleague, Zworykin visited Farnsworth's laboratory. Philo freely showed Zworykin what he was doing. Philo had always been generous with his knowledge and expertise. Zworykin stayed three days and found out that Philo had built an image dissector that achieved a satisfactory resolution of each individual image scanned with electrons. Witnesses recalled that Zworykin then said to Philo, "I wish that I might have invented it."

Soon afterward in 1930, at a Westinghouse laboratory in Pittsburgh, Zworykin substantially duplicated Farnsworth's camera. Reportedly, Zworykin found that Philo's camera was impractical because of the amount of light required for operation. One story told was that Zworykin did attempt, without success, to demonstrate this camera idea to executives at Westinghouse (where he was then employed), but their considered response was a request to him to find something "more useful" to work on. It seems that these Westinghouse executives thought that Zworykin's camera idea had no promise and was commercially inoperable.

RCA then became Zworykin's employer, and Zworykin, in Camden, New Jersey, with RCA engineers, then proceeded to try to develop Farnsworth's camera. Later they produced a Zworykin camera that was called the iconoscope and that operated with much less light. Also, RCA embarked on developing a commercially practical all-electronic television system. RCA had the money, the trained workers, and the equipment to undertake this development.

In 1931, Sarnoff offered Everson $100,000 for all of Farnsworth's television system, including not only all available patent coverage but also even

Farnsworth's own services. Farnsworth refused when Everson approached him with Sarnoff's offer. Farnsworth was independent and believed he would succeed in achieving commercially practical equipment. Everson then refused Sarnoff's offer. This refusal may have served to cut off further money for investment in Farnsworth's system from Everson and his investors. Probably they were not sure whether Farnsworth would achieve either anything more or gain any other customers. Therefore, later in 1931, Farnsworth accepted employment with the Philco company. Farnsworth moved both his laboratory and his family to Philadelphia.

Seeking to raise money, Farnsworth visited England in 1932 and conferred with Baird, who had developed an improved mechanical scan system. Farnsworth's account indicates that Baird, after Farnsworth demonstrated his system, "realized the futility of his [Baird's] efforts." The directors of Baird's company paid Farnsworth $50,000 to supply them with Farnsworth's equipment, and Baird and Farnsworth competed with a company named EMI, hoping to have the Baird-Farnsworth system adapted as the standard for television in the United Kingdom. However, EMI merged with Marconi's company in 1934, and Marconi then already had a patent-sharing agreement with RCA. The Marconi-EMI system was reportedly almost the same as the all-electronic RCA (Zworykin's) system. The BBC chose the Marconi-EMI system.

The Interference

By 1934, RCA had achieved its own all-electronic television system, which incorporated Zworykin's iconoscope television camera. Meanwhile, the U.S. Patent Office had set up an interference proceeding, identified as no. 64,027, between a patent application filed in 1923 by Zworykin and U.S. Patent 1,773,980 to Farnsworth that had issued in August 1930 from Farnsworth's original patent application filed in 1927.

Farnsworth believed that the iconoscope was covered by his patent. The issue involved in the interference proceeding concerned who had priority of invention regarding Claim 15 in Farnsworth's patent, that is, who first invented the subject matter involved in Claim 15, Farnsworth or Zworykin. The claim had been copied from Farnsworth's patent into Zworykin's earlier-filed patent application on behalf of Zworykin by his employer, RCA, and under Patent Office Rules of Practice, the interference resulted. Claim 15 concerned an "electrical image" achieved by the practice of the "image dissector" television camera that had been built by Farnsworth in his laboratory before his demonstration of September 7, 1927, and that had been fully disclosed in Farnsworth's original patent application.

In the interference, Tolman provided helpful testimony on Farnsworth's behalf, relating how young Philo, years previously (in 1922), had drawn a diagram on his classroom blackboard and had explained his idea for his "image dissector." Apparently hoping to show that Philo was then perhaps incompetent, RCA representatives questioned Tolman about other matters Philo had then considered. Regarding the theory of relativity, Mr. Tolman testified that Philo's explanation was the clearest and most concise he had ever heard.

Evidence brought out in the interference showed that the 1923 system disclosed in Zworykin's application did not work. Very little evidence was available to prove that Zworykin ever built, tested, or demonstrated a system such as that disclosed in his 1923 patent application. There were no laboratory documents, no corroborating eye witnesses, only seemingly indefinite accounts from Zworykin about his electronic camera. Zworykin seems to have testified that maybe sometime during "1924 or 25" his camera was demonstrated. During the interference, though, substantially no material evidence was provided. Two Zworykin associates provided confusing, even contradictory, verbal accounts.

The patent interference examiners concluded in 1934, "Zworykin has no right to make the count [that is, copy the Claim 15] because it is not apparent that the device would operate to produce a scanned electrical image unless it has discrete globules capable of producing discrete space charges and the Zworykin application as filed does not disclose such a device." The Zworykin device as disclosed in the 1923 patent application was inoperative. Farnsworth won the interference.

Regarding priority, Zworykin and RCA appealed, but they lost all appeals. Farnsworth succeeded on August 25, 1934, at the Franklin Institute in Philadelphia, in producing the world's first public demonstration of an all-electronic, complete, synchronized television system with coordinated television cameras that scanned images electronically, and television receivers that received and showed the scanned images. The system could operate with live, moving people and could produce half-tone images. Philo had invented the camera and the receiver. In 1934, the British company Gaumont bought a license from Farnsworth to permit them to make systems based on his designs.

The original Zworykin 1923 patent application, though defective as filed, was substantially amended and was allowed to issue, as amended, into a patent as a result of a decision by a court of appeals in 1938. Finally, in 1939, RCA in effect agreed that Farnsworth's camera was first, accepted a license from Farnsworth, and paid him reportedly $1 million, which ended the litigation between RCA and Farnsworth. Certain features of the iconoscope are regarded as improvements over Farnsworth's image dissector, but the iconoscope was not outside the scope of Farnsworth's patents.

Later, RCA and Zworykin developed their so-called Image Orthicon camera, an improved and commercially practical video camera. This camera, though, is in fact descended from the device used in Farnsworth's September 7, 1927, demonstration and is disclosed Farnsworth's U.S. Patent 2,087,683, which involves what is regarded as the first device to disclose the low-velocity method of electron scanning.

In March 1932, Philo lost his young son Kenny, but Philco refused to give Farnsworth time to travel to Utah from Philadelphia to bury the boy. The death strained Farnsworth's marriage and may be the beginning cause of Farnsworth's struggle with depression. Later, in 1934, Philco seems to have become dissatisfied with Philo's development work and terminated his employment, but apparently then Philco remained Farnsworth's only major television licensee in the United States.

In 1934, Farnsworth, traveling in Europe, negotiated an agreement with the German Goertz-Bosch-Fernseh interests. He returned to his own U.S. laboratory. By 1936, his company was transmitting regular but experimental television entertainment programs.

Working with University of Pennsylvania biologists, he developed a process for passing radio waves through milk to sterilize and pasteurize it. He also developed a penetrating beam for ships and airplanes.

In 1938, his company seems to have become consolidated with the Farnsworth Television and Radio Corporation of Fort Wayne, Indiana, where Philo was director of research and E. A. Nicolas was president.

In April 1939, the New York World's Fair exhibited all-electronic television sets, and shortly thereafter RCA began selling television receivers to the public.

World War II (1939–1945) interrupted television development and commercialization. In effect, WWII caused a freeze on all civilian television work. After the war, Farnsworth was not involved with television except for trying to improve the devices he had already designed.

After WWII, in the late 1940s, RCA became the dominant entity involved in production and sales of TV sets. RCA carried out a vigorous public-relations program that promoted both Zworykin and Sarnoff as the fathers of television.

Color Television

The desirability of color television seems to have been appreciated by researchers and developers from the beginning. Electromechanical efforts proved to be outmoded after the advent of all-electronic television systems. Philo Farnsworth's work was all black and white. Color television resulted from inventions by many others.

In 1940, a color television demonstration was made by RCA to Federal Communications Commission (FCC) members using a system in which color images were produced by optically combining images from two picture tubes into a single screen. The same year, another demonstration was made by CBS of its "field sequential" system, which used a spinning disc with red,

(continued)

blue, and green filters. However, the War Production Board stopped TV and radio manufacture for civilian use from April 1941 to August 1945. Then, three systems competed for FCC approval: CBS's field sequential system, RCA's dot sequential system, and Color Television Inc.'s system, which used complicated filters and electronics. All were initially incompatible with black-and-white reception, but in 1949 RCA's system became compatible.

The FCC found RCA's and CTI's systems unacceptable, but it approved CBS's system in October 1950. RCA initiated a lawsuit to delay commercialization to permit it to develop an acceptable system, but viewership of color was extremely limited because only prototype color receivers were available only in the New York area for public viewing, although 10.5 million black-and-white receivers had been sold. In October 1951, CBS withdrew its system, but the National Television System Committee (NTSC) from 1950 to 1953 worked to develop a compatible color system. In 1953, NTSC gained FCC approval for an all-electronic compatible system. On January 1, 1954, NBC (owned by RCA) broadcast coast-to-coast the Tournament of Roses Parade, with public demonstrations across the United States received on prototype color receivers made by many different manufacturers. By 1959, RCA was the only remaining large manufacturer of color receivers. Even by 1964, only 31% of television households in the United States had a color set. Not until 1972 did more than 50% of U.S. television households have a color set.

As late as 1964, reportedly only 3.1% of television households had a color receiver, but NBC promoted color usage when it made its fall 1965 schedule almost entirely in color programming. The number of color TV sets sold in the United States did not exceed the number of black-and-white TV sets sold until 1972.

Color TV developed independently in Cuba, Mexico, and Canada. European color TV developed a little later because of divisions regarding standards. The French used the Sequential Color with Memory (SECAM) system and involved the Russians in development. The Germans developed the phase alternating line (PAL) system in 1963, which is similar to the NTSC system and the PAL system, and which came to be used in most of Western Europe and in China. In Britain, color film for entertainment is commonly made for sale to U.S. networks. France was the only European country to use the broadcast standard of 819 lines; others use 625 lines. However, consistent with the European Comité Consultatif International des Radio Communications (CCIR) system, France reverted to the 625-line PAL standard. In North America, the 525-line NTSC standard, dating from 1941, was retained.

Nervous Breakdown

Farnsworth, suffering from depression, took up residence in a house in Maine and evidently made his condition worse by excessive alcohol consumption. He had a nervous breakdown, spent time in hospitals, and underwent shock therapy.

In 1947, his Maine house burned, but he survived. His public comments were few, but his son Kent years later recalled that Philo "felt he had created a monster, a way for people to waste a lot of their lives."

Other Inventions

In 1951, the Farnsworth Television and Radio Corporation, after reportedly making about $2.5 million, was bought by International Telephone and Telegraph Company (ITT). Apparently, Farnsworth rapidly spent his portion of this money. In 1951, he received an honorary doctorate in science from Indiana Institute of Technology. From 1951 to 1966, he was employed by ITT and he worked in a basement laboratory called "the cave" on Pontiac Street in Fort Wayne, Indiana. He originated and developed various concepts, including a defense early warning signal, radar calibration devices, submarine detectors, an infrared telescope, the so-called PPI Projector, an early air traffic control system, and controlled fusion in fusion reaction tubes.

In 1965, under a board of director's directive, ITT was to sell its Farnsworth subsidiary. Though Farnsworth was funded for 1966, the stress thus produced caused Farnsworth to have a relapse. In the spring of 1967, he moved with his family back to Utah. Brigham Young University gave him an honorary doctorate, office space, and an underground bunker for continuing his fusion research.

Facsimile

Farnsworth's original idea of scanning an image to reproduce it has had many applications. With a contemporary facsimile (or fax) machine, one can transmit over a telephone line (typically) or a coaxial cable carrier line an image of text, graphic art, or photo to another fax machine connected at a remote location. At the receiving machine, the image is reconstructed and usually printed on a sheet of paper or the like.

The transmitting fax machine has a scanning device that, line by line, traverses a predetermined field. The output from the scanning device is converted into a variable electrical signal corresponding to the scanned material. The signal is transmitted over the line to the receiving fax machine.

Invention of the fax machine is attributed to Alexander Bain, a Scottish mechanic, inventor, and amateur clock maker, in 1843. His fax transmitter

(continued)

scanned a flat metal surface with a stylus mounted on a pendulum. The stylus sensed images on the metal surface. He used parts from clocks and telegraph machines. The variable electric signal produced was sent over a wire, and at the receiver the original surface was reproduced. Since Bain, many inventors have worked to improve machines of the fax type.

Modern fax machines were not possible until about the mid-1970s because of technological advances, including digital facilities. Today, a fax machine is commonly a combination of an electronic image scanner, a modem, and a printer. The scanner converts the matter shown on a document page into a digital electronic signal output, the modem sends the signal over a phone line to a receiving device, and the printer reproduces the scanned document page.

Fax machines have become ubiquitous. Fax capabilities are expressed by group, class, data transmission rate, and conformance with the Telecommunications Standardization Sector of the International Telecommunication Union.

He invited his ITT staff associates to proceed with him to Salt Lake City and work in his projected company, called Philo T. Farnsworth Associates. However, sufficient funding could not be secured. Farnsworth liquidated and invested his assets in the project, but even with a contract with the National Aeronautics and Space Administration, this was insufficient to prevent closing down the project in January 1970.

The televised moon walk in 1969 utilized Farnsworth's image dissector. His wife, Elma Farnsworth, commented in a subsequent interview, "We were watching it, and when Neil Armstrong landed on the moon, Phil turned to me and said, 'Pam, this has made it all worthwhile.' Before then, he wasn't too sure."

Farnsworth contracted pneumonia and died of emphysema and in debt on March 11, 1971, at age 64. He was survived by his wife, who lived to be 98, and two sons, Russell and Kent.

Farnsworth was involved in research in many areas besides television, including the above-indicated fields, involving radar development, peaceful uses for atomic energy, electron microscopes, infant incubators, and aircraft guidance systems. He reportedly was the holder of about 165 U.S. patents plus many corresponding foreign patents. Though he was paid royalties, he never became wealthy.

Until the late twentieth century, all television cameras utilized the work of Farnsworth and Zworykin. At that point, improved technology appeared, such as charge-coupled devices. Farnsworth's fusor, a small nuclear fusion device, was later developed into a commercially practical neutron source and is manufactured for this use.

FURTHER RESOURCES

Biography of Philo T. Farnsworth. University of Utah Marriott Library Special Collections. http://db3-sql.staff.library.utah.edu/lucene/Manuscripts/null/Ms0648.xml/Bioghist.

Elma Gardner Farnsworth. *Distant Vision: Romance and Discovery on an Invisible Frontier.* Salt Lake City, UT: Pemberley Kent Publishers, 1989.

Russell Farnsworth. *Philo T. Farnsworth: The Life of Television's Forgotten Inventor.* Hockessin, DE: Mitchell Lane Publishers, 2002.

Paul Schatzkin. *The Boy Who Invented Television.* Silver Spring, MD: Teamcom Books, 2002.

Evan I. Schwartz. *The Last Lone Inventor: A Tale of Genius, Deceit & the Birth of Television.* New York: HarperCollins, 2002.

HEINKEL/AP Photo

Jet Engines: Hans von Ohain and Frank Whittle

Independently and concurrently, jet engines were invented and preliminarily developed by Hans von Ohain and Frank Whittle under circumstances suggesting that both should be recognized as inventors. Jet engines and associated aircraft constitute such a significant advance over the prior propeller-driven engines and associated aircraft that they fairly must be considered an iconic innovation. For most commercial and military applications, they soon replaced the prior propeller-driven counterparts.

BACKGROUND

A jet engine converts air and fuel into a hot gas and emits a jet of this gaseous material from the back end of the engine, thrusting the engine and an associated airplane forward. An aircraft may have one or more jet engines.

The jet engine had a profound impact upon aviation and civilization. Within 20 years of their advent in 1937, jet-powered aircraft had substantially replaced propeller-operating, piston engine-driven aircraft for military and commercial applications. The advantages of jet engines over propeller-driving, piston-operated engines include greater power and speed at high altitude, less weight relative to power produced, typically simpler construction with fewer moving parts, less and simpler maintenance, efficient operation, and cheaper fuel.

From their advent, jet engine-powered aircraft were usually capable of traveling much faster than piston engine powered aircraft. In the 1950s, a commercial transatlantic flight powered with four piston engine propellers took more than 15 hours, but in the 1960s that time was cut in half by such jet aircraft as Boeing's now classic 707, which used four jet engines. A little later, the French and British supersonic jet-powered Concorde made the commercial flight from New York to Paris or London in only 3.5 hours, but it enjoyed limited commercial success because of low payload and high fuel cost. Jet engines also made practical large jumbo jets, beginning with the Boeing 747 in 1969, which could hold 547 passengers and crew. Longer range was achieved by replacing turbojet engines with turbofan engines.

Historically, the invention of the rocket by the Chinese in the eleventh century resulted in a little use of the rocket in fireworks and in weapon propelling, but essentially no further technological advance in gas thrusting occurred for centuries. Meanwhile, the piston engine and propeller appeared and developed, but their efficiency seemed to approach performance limits by the 1930s. New power plants were needed and sought.

Early efforts to create jet engines were not particularly successful. Gas turbines had been proposed since John Barber in England had received a turbine patent in 1791. Aegidius Elling, a Norwegian engineer, had built an operating gas turbine in 1903, and patents disclosing jet propulsion issued in 1917, but engineering and metallurgical capability limited development.

In 1923, the U.S. National Bureau of Standards, through one Edgar Buckingham, issued a report indicating that jet engines would not be financially competitive with propeller-driven piston engines at the then-conventional airspeeds and low altitudes. The stage was set for those who would invent a better aircraft engine.

STRUCTURE OF A JET ENGINE

A brief summary of jet engine operation is desirable. All jet engines (sometimes called gas turbines) operate similarly. Very simply, at the engine front end, air is drawn in, usually with a fan, and passes into a compressor, which raises the pressure of the air. The compressor comprises several fans, each with many blades, the fans being attached, sometimes alternately, to a common axial shaft. The blades compress the air, and the air passes into a combustion zone where the compressed air is sprayed with fuel. The resulting mixture is burned (usually ignited with an electric spark), and gases form. The burning gases expand, develop more pressure, and pass outward as a jet-like stream at the rear end of the engine.

In some engines, some of the entering air flows medially through the engine and the remainder flows cylindrically and peripherally through side-adjacent interior regions of the engine. The medially flowing air is mixed with fuel spray and combusted, and so the resulting gas mixture is heated substantially and gases are pressurized. The cooler peripheral air and the heated medial air mix at the engine exit region. The mixture is ejected rearward through a nozzle. As a consequence, the engine and its associated airframe are thrust forward.

Before the gases leave the engine, they pass through turbine blades mounted about a common axial shaft. Alternate groups of turbine blades rotate and turn the shaft. This shaft also rotates the compressor shaft and also perhaps the fan shaft, which brings in a fresh supply of air through the engine front end.

Engine thrust can be increased by adding an afterburner section in front of the rear end of the engine, in which additional fuel is burned in the exhausting gases to give added gas pressure, power, and thrust.

At about 400 miles per hour (mph), 1 pound of thrust equals about 1 horsepower. A pound of thrust is greater than 1 horsepower at higher speeds and less than 1 horsepower at lower speeds.

Frank Whittle and Hans von Ohain each developed a turbojet, which can be defined as a jet engine comprising an air inlet, an air compressor, a combustion means, a gas turbine driving the air compressor, and a nozzle. A turbojet is regarded as simple, with a small frontal area and high exhaust speed. When flown below about Mach 2 (two times the speed of sound), a turbojet is relatively inefficient and noisy. Contemporary aircraft mainly use turbofan jet engines, but medium-range cruise missiles commonly use turbojet engines.

A turbofan jet engine has, very simply, a relatively large-diameter ducted fan followed by a smaller diameter turbojet engine that rotates the fan. Turbofans have a much lower exhaust speed than a turbojet and are more efficient at subsonic and even supersonic speeds up to about Mach 1.6. All commercial jet aircraft and many military jet aircraft currently use turbofan engines.

Early turbojet engines were very fuel-inefficient because their pressure ratio and turbine inlet temperature were limited by the available technology. The first turbofan was the German Daimler-Benz DB 670 (or 109-007), which was testbed operated April 1, 1943, but development was stopped during the war when problems could not be solved. The first production turbofan had a low bypass ratio of 0.3 and was the Rolls-Royce Conway. Improved materials and the introduction of twin compressors increased engine thermodynamic efficiency.

A turboprop engine has a propeller driven by a gas turbine. A simple turboprop has an air intake, compressor, combustor, turbine, and propelling nozzle. Most of the output from the turbine is employed to drive the propeller, with a little being used to drive the compressor. Some expansion of gases in the propelling nozzle provides thrust. The propeller is reduction-geared relative to the turbine so that high turbine revolutions per minute (rpm) is converted into propeller low rpm and high torque. The propeller operates at constant speed with variable pitch (similar to the piston engine of a large aircraft).

The first turboprop was the Jendrassik Cs-1, designed by Hungarian engineer György Jendrassik in 1938, and was intended to power a twin-engine heavy fighter, the RMI-1. The engine ran for the first time in 1940, but it had problems with combustion stability. The program was canceled after the Hungarian Air Force chose the Messerschmitt Me 210 for the heavy fighter role. The first British turboprop engine was the Rolls-Royce RB.50 Trent in 1945, which used a converted Derwent II fitted with reduction gear and a Rotol 7-foot 11-inch five-bladed propeller. The first American turboprop was the General-Electric T31.

INVENTION OF THE JET ENGINE

In 1929, in England, Frank Whittle, then a Royal Air Force (RAF) Flying Officer, submitted his ideas for a turbojet to his superiors, who rejected them, but on January 16, 1930, he filed a patent application on his engine, for which a British patent was issued in 1932. However, it was not until April 1937 that he had his first experimental and demonstration (or proof of concept) engine running. On May 15, 1941, a flyable version of his engine, fitted into a specially built aircraft, flew.

In 1935, in Germany, Hans von Ohain, unaware of Whittle's work, conceived of a similar engine structure. An embodiment of his experimental, concept-proving engine was demonstrated, and the aircraft industrialist Ernst Heinkel provided a developmental facility for von Ohain and his machinist, Max Hahn. They had a first engine ready by September 1937, and a second engine, fitted into a specially built aircraft, flew on August 27, 1939.

THE WHITTLE AND VON OHAIN STORIES

Frank Whittle's Early Years

Frank was born June 1, 1907, in the Earlsdon district of Coventry, Great Britain. His father was a foreman in a machine tool factory. When Frank was 4 years old, his father gave him a toy airplane fitted with a wind-up, clockwork-driven propeller and hung it from a gas mantle. From this beginning, Frank's interest in flying increased.

In 1916, his father bought the Leamington Valve and Piston Ring Company, and the family moved to Leamington Spa. The company's equipment comprised some lathes and other tools and a single-cylinder gas engine. Frank became familiar with the machines and carried out piece work for his father. During World War I, Frank saw aircraft being built at the nearby Standard works and was excited when a plane once force-landed near his home.

Frank was educated at the Milverton primary school and won a scholarship to Leamington College for boys, but he had to drop out when his father's business declined and there was not enough money available for tuition for him to continue. Frank proceeded to spend hours at the local library and to learn about steam and gas turbines.

After passing entrance examinations, he duly reported in January 1923 at age 16 to RAF Halton as an aircraft apprentice. He lasted only two days: he failed the medical examination because he was only five feet tall and his chest measurement was too small.

He pursued an intensive physical training program and a special diet. One instructor suggested a list of exercises for him to perform to promote height. Within six months and with a growth spurt he gained three inches in height. He reapplied, but was rejected again. Not defeated, he applied again under a different first name and passed the written examination again. He was ordered to RAF Cranwell, where he was accepted.

At Cranwell, he trained as a fitter/rigger for 3 years. Also, he maintained an interest in the Model Aircraft Society, where he built quality replicas. Also, he demonstrated substantial mathematical ability. He was strongly recommended by his commanding officer in 1926 for an officer cadetship and a chance to fly, an exceptional opportunity for a "commoner" in the class-oriented military organization. Moreover, of the very few apprentices accepted, only about 1% ever completed the course. Only the top five cadets from a course were given the chance to undergo flying training. Whittle finished his course as number six, but luckily for him the top candidate failed his physical, so Whittle was selected. In 1927, Whittle soloed after 8 hours of instruction in an Avro 504K biplane at Cranwell.

Whittle graduated in 1928 at age 21, ranking second in his class in academics and "exceptional to above average" as pilot, although his logbook contained red-inked warnings about overconfidence, low flying ("hedge hopping"), and performing to the gallery. Each cadet had to complete a thesis for graduation. Whittle chose future aircraft design developments involving high-altitude flight and speeds over 500 mph. He proposed using a "motorjet" in which a piston engine provided compressed air to a combustion chamber whose exhaust provided thrust, similar to an afterburner for a piston engine driving a propeller.

In August 1928, he was posted to operational fighter squadron 111, which used Siskins aircraft and was based at Hornchurch. He was next posted to the Central Flying School Wittering for a flying instructor's course. After successful completion, he was promoted to Flying Officer and posted to RAF Digby. Here, Whittle married his first wife, Dorothy, on May 24, 1930.

Whittle's Invention of the Jet Engine

Whittle later dropped the motorjet idea when further calculations indicated it would not have a weight advantage over a conventional engine of equivalent thrust.

He thought, why not use a turbine in place of a piston engine? A turbine could be used with some power from the exhaust and could be used to power a compressor, similar to a supercharger. In late 1929, Whittle, with the aid of his friend, sympathetic Flying Officer Pat Johnson, gained an interview with the commandant, and his concept was sent to the Air Ministry, but there Dr. A. A. Griffith, who was interested in gas turbines to drive propellers, rejected Whittle's proposals, and the RAF referred to the design as "impracticable."

Whittle nevertheless applied for a British patent on his jet engine in early 1930. Johnson, still convinced, then set up a meeting for Whittle with British Thomson-Huston Company (BTH) and the company's chief turbine engineer at Rugby, but the company did not wish to spend what was estimated to be 60,000 pounds on development.

In late 1930, Whittle was posted to the Marine Aircraft Experimental Establishment at Felixstowe to test floatplanes. There, his friend from Cranwell days, Rolf Dudley-Williams, who was with a flying-boat squadron, made efforts on Whittle's behalf. In the summer of 1932, after completing his tour of duty at Felixstowe, Whittle was sent to take an engineering course at RAF Henlow, where he did so well that he was invited to take a 2-year engineering course at Peterhouse College, Cambridge. In 1936, he took a First in the Mechanical Sciences Tripos. However, in this period, the renewal fee of 5 pounds for his jet engine patent was not paid by the Air Ministry, and the patent lapsed.

The start/stop development and funding problems for his engine had a negative impact on Whittle. He suffered from stress-related illnesses, such as heart palpitations, eczema, and overuse of alcohol. His weight dropped. To maintain his 16-hour days, he sniffed Benzedrine during the day and consumed tranquilizers and sleeping pills at night. He became easily irritated and developed an "explosive" temper.

In 1936, Whittle, working with Dudley-Williams and former RAF pilot Tinling, made a deal with O. T. Falk & Partners, a banking firm, and a new company, Power Jets, was incorporated. Whittle requested and received permission from the Air Ministry to be honorary chief engineer and technical consultant for 5 years provided that there was no conflict with his duties and that in this project he use not more than six hours a week.

At this time, Whittle's ideas and developments were not subject to the Official Secrets Act. One consequence was that Whittle retained his intellectual property rights.

In July 1936, in Germany, Heinkel and Junkers openly began turbojet experiments. It was a relief to Whittle when their HE178 engine was reportedly scrapped.

Whittle, seeking development of his design using the modest Power Jets' capital, returned to BTH at Rugby. The company agreed to build Whittle's first experimental jet engine, a WU (Whittle Unit). He also tried to interest companies to develop the special materials needed.

With limited funding, including 1,900 pounds from the Air Ministry, engine development proceeded, although the relationship between BTH and Whittle was difficult because of Whittle's ideas were new and BTH had previously established firm ideas in steam turbines. A concept-proving, experimental first embodiment of a jet engine was completed and ready for static (ground) testing in April 1937 at Rugby. The engine was liquid fueled and incorporated a fuel pump.

The first ground tests involved complications. During startup, with much shrill noise as the main engine shaft gained speed, the engine was observed to be continuously accelerating and out of control. All team members fled in haste except for Whittle, who cut off the fuel feed valve. (The incident is reminiscent of the scene at the startup of Edison's first commercial generator in New York in 1881.) After a little while, the engine died. Subsequent review showed that during the preliminary fuel line bleeding, fuel had leaked into the engine and accumulated in pools, so the engine could not stop until this fuel had burned off. The installation of a drain line stopped any future repeat of the adventure. The engine was successfully demonstrated.

In 1938, BTH moved the testbed, and the engine was reconstructed for a third time. The Air Ministry contributed a further 6,000 pounds. Engine testing continued. After World War II started in September 1939, the project developed faster. When Whipple demonstrated engine operation at high power for 20 minutes, the Air Ministry initiated a bigger, more powerful, full-scale engine development by Power Jets, but since Power Jets had limited capital and no manufacturing capability, it had to offer shared production and developmental engine contracts to BTH, Vauxhall, and Rover. Only Rover took the contract. An offered contract to build an experimental aircraft, which became the E28/39, was taken by Gloster Aircraft Company.

Even though financing was now more secure, Whipple now faced new threats. BTH said the jet engine would not compare with conventional piston engine power. Various government sources questioned Power Jets' capability. The government bypassed Power Jets and demoted it to a research group. Gloster was authorized to press ahead with a twin-engine jet interceptor that became known as the Meteor. Rover was unable to produce an engine by the time the E28/39 was ready, so Whipple attempted to convert the massive WU engine into a flyable structure that came to be called the W1.x ("x" standing for experimental). To make this engine, Whipple used various test parts, but he successfully got the engine constructed, and it first ran on December 14, 1940.

By April 1941, Whipple's still further developed W1.x engine was fitted into the completed Gloster E29/39, taxi-trialed at Gloster, and finally transported to RAF Cranwell. On May 15, 1941, the aircraft with Whipple's engine successfully flew with test pilot Gerry Sayer. Within a few days, the aircraft was exceeding the performance of Spitfires, reaching over 370 mph at high altitude. The jet age had begun.

Development

The success was clear. Nearly every engine-producing company in Britain started crash programs aimed at catching up.

Development of an improved, better engine, the W2, was undertaken. This engine, like the W1, used a "reverse flow" burner design wherein heated air from the combustor cans was conducted back through ducts to the front of the engine before reaching the turbine area. This arrangement allowed the engine to be made shorter. In May 1940, Power Jets created a design for a W2Y engine that had a straight-through air flow that made the engine longer but simpler, with a long driveshaft.

To obtain an operational jet aircraft, the Air Ministry authorized BTH to develop the twin-engine jet interceptor that became the Gloster Meteor. The arrangement between Rover and Power Jets did not work out, but Rolls-Royce became substituted for Rover. The engines powering the Meteor were much more reliable than those being developed concurrently by the Germans; they had much better power-to-weight ratio and specific fuel consumption.

In mid-1942, Whittle was sent to Boston, Massachusetts, to help the U.S. developmental jet engine program at General Electric. In autumn 1942, a modified engine based on the W2B design combined with a simple airframe from Bell Aircraft flew as the Bell XP-59A Aircomet.

Developments at Power Jets by Whittle continued resulting in improved jet engines. However, since almost every engine concern was producing its own jet engine designs, Power Jets was no longer able to generate sufficient income from innovations. Afraid that private industry would freely exploit the pioneering discoveries of Power Jets, Whittle proposed that Power Jets be nationalized. The Minister of Aircraft Production, Sir Stafford Cripps, accepted this proposal, so in April 1944, Power Jets was nationalized. It was eventually renamed National Gas Turbine Establishment, and in 1946 the Royal Aircraft Establishment divisions joined them.

After the War

In 1948, Whittle left Power Jets. Though originally a socialist, his experiences with nationalization of Power Jets caused him to change his mind. Later he campaigned for the Conservative Party. His friend Dudley-Williams, who was Managing Director of Power Jets, became Conservative Member of Parliament for Exeter. Whittle retired in 1948 from the RAF, blaming ill health, and leaving with the rank of Air Commodore.

He enjoyed the fact that Gloster Meteors of the No. 616 squadron were shooting down some German V1 flying bombs in the period from June 1949 to March 29, 1945. The Meteor was not yet fully operational, but it had ample speed. Its cannons, though, were found to jam frequently.

A little later, Whittle received 100,000 pounds from the Royal Commission on Awards to Inventors, partly to compensate him for turning over all his shares in Power Jets when it was nationalized. Much of this money he gave to his coworkers at Power Jets. He also was made a Knight of the Order of the British Empire that year.

Soon he joined British Overseas Airways Corporation (BOAC) as a technical advisor on aircraft gas turbine engines. Over the next few years, he traveled extensively, reviewing engine developments in various regions, including the United States, Canada, Asia, and the Middle East. In 1952, he left BOAC and wrote an autobiography, *Jet: The Story of a Pioneer.*

In 1953, he accepted a position with a Shell Oil subsidiary and invented a self-powered turbine drill that operated on the lubricating mud charged into a borehole during drilling. In 1957 he left Shell.

Hans von Ohain's Early Years

Hans von Ohain was born December 14, 1911, in Dessau, Germany. He earned a Ph.D. in physics and aerodynamics from the University of

Göttingen in 1935. The university was then one of the major centers for aeronautical research. In 1933, at age 22, during his studies, he conceived of "an engine that did not require a propeller" and formulated a theory of jet propulsion. After receiving his degree in 1935, he became junior assistant to Robert Wichard Pohl, then director of the Physical Institute of the university,

Von Ohain's Invention of the Jet Engine

In 1936, he received a first German patent on his turbojet engine, titled "Process and Apparatus for Producing Airstreams for Propelling Airplanes." Unlike the jet engine design in Whittle's first patent, von Ohain's engine used an assembly of centrifugal compressor, combustion zone, and turbine placed adjacent to one another longitudinally, with combustor cans located circumferentially around the outside of the assembly. His engine was larger than Whittle's in diameter but substantially shorter along the longitudinal (or thrust) axis.

While engaged at the university, von Ohain often took his sports car to be serviced as a local garage, Bartles and Becker. There he met an automotive engineer, Max Hahn, and eventually, in 1936, got him to build an embodiment of his proposed jet engine that was done at a cost of about 1,000 deutsche marks.

He took the completed engine to the university for testing. There, he experienced significant problems with combustion stability. Frequently, the fuel would not combust inside the combustor cans but would instead be blown through the adjacent rear turbine, where the combustible fuel/air mixture would ignite and shoot flames out the back end or nozzle of the engine. The electric motor powering the compressor was overheated.

Pohl, in February 1936, wrote a letter to Ernst Heinkel, the German industrialist, on behalf of von Ohain, describing the von Ohain engine design and its possibilities. Heinkel arranged a meeting between his engineers and von Ohain. They grilled von Ohain for hours, during which he stated plainly that his "garage engine" would never work, but that there was nothing wrong with the concept itself. The engineers were persuaded that von Ohain's idea was good. Hahn and von Ohain joined Heinkel, and in April 1936, von Ohain and Hahn were placed in a laboratory at the Marienehe airfield outside Rostock, in Warnemunde, Germany, and continued engine development.

There, a study was made of the engine's air flow, and over a two-month period, several improvements were made. The results were proven favorable, and it was decided to produce a new engine incorporating these changes and operating on hydrogen gas. The new engine was designated HeS-1 (for Heinkel-Strahltriebwerk 1, or Heinkel Jet Engine 1). Hahn would work on the combustion problem, an area in which he had some experience, while the new engine would be constructed by some of the best machinists in the company. Shop-floor supervisors of these machinists hated to lose their services.

The engine, as constructed, was very simple and made largely of sheet metal. Construction extended from late summer 1936 to March 1937. Two weeks later, the engine ran on hydrogen, but, though the tests were otherwise successful, the high temperatures caused considerable burning of the metal.

By September 1937, the combustors were replaced and the engine had been adopted to use gasoline as fuel. It was found that this fuel caused the combustors to clog up, so Hahn, using his soldering torch, designed a new version, which was found to work substantially better. Though this version was not suitable for use as a flight-quality structure, the tests established that the engine's concept was workable.

As work on HeS-1 progressed, work on the design of a flight-quality engine design, HeS-3, advanced. The main difference was that the bent and folded sheet metal pieces were replaced with machined compressor and turbine stages. Also, the component layout was changed to reduce the cross-sectional area of the engine, enabling the combustor cans to be located in a spacing between the compressor and the turbine. It was thought that on the original design the turbine area was too small to work efficiently. Increasing the size of the turbine meant that the combustor or flame cans no longer fit well into the spacing.

As development progressed, another new design, HeS 3b, was proposed and then utilized wherein the combustor cans were moved out of the gap and had a modified shape so as to permit the widest part of the cans to be positioned adjacent to the compressor's outer rim. In Hs-3b operation, compressed air was passed forward to the combustor can chambers, and from there the now hot air and exhaust gases flowed rearward into the turbine. Even though it was larger than the HeS-3, the HeS-3b was still considered to be compact.

The HeS-3b as first constructed was air-tested in the Heinkel He 118 dive bomber prototype, but it soon burned out. However, a second HeS-3b was constructed, tested, and flown in the new Heinkel He 178 on August 27, 1939, by Flight Captain Erich Warsitz and so became the first jet-powered aircraft to fly. The Luftwaffe beat the RAF into the air by about 9 months.

Work started promptly on larger, improved engine versions. Although von Ohain's HeS-8A was installed in the new Heinkel He 280 fighter in March 1941 and was subsequently flown in April 1941, progress toward manufacturing the HeS-8A engine was still not progressing well by early 1942. In the spring of 1943, work on the HeS-8A was abandoned. None of von Ohain's jet engines entered production.

Development

Austrian Anselm Franz of Junkers' engine division (Junkers Motoren, or Jumo) introduced the axial flow compressor, wherein air is compressed by rotor blades against nonrotating stator blades (the reverse of a turbine). To get the needed compression, a successive series of pairs of rotating fan blades (rotors) and stationary blade pairs (stators) was necessary, but the additional complexity still resulted in a smaller, more aerodynamic engine.

Production of the Franz axial flow engine was started in Germany in 1944 for installation in the world's first jet-fighter aircraft, the Messerschmitt Me 262. This engine was later used in the world's first jet bomber aircraft, the Arado Ar 234. Delays in engine availability caused these aircraft to arrive too late to decisively help Germany's position in the war.

After the War

After the war, the Allies carefully reviewed the German jet engine designs. German designs were generally more advanced aerodynamically than those of the British, but the simplicity and advanced British metallurgy meant that Whittle-derived engine designs were more reliable than their German counterparts.

By the late 1950s, jet engines powering U.S. and Russian fighters were using engines with compressors based on the axial flow design (as distinct from Whittle's centrifugal design). Almost all jet engines of fixed-wing aircraft today utilize aspects of the axial flow design.

British engine designs were widely licensed in the United States and were sent to the USSR in a technology exchange. The Rolls-Royce Nene centrifugal flow engine was reverse engineered in the USSR and used to power the Russian MIG-15. Apart from the General Electric J47 engine, which gave excellent service in the U.S. F-86 Sabre aircraft in the 1950s, axial flow types in general did not come into wide use until the 1960s. By the 1950s, the jet engine was used in almost all military aircraft. By then, many of the British jet engine designs were cleared for civilian use and were used in the de Havilland Comet and the Canadair Jetliner.

Improvements in bearing metallurgy and design enabled the shaft speed in the jet engine to be increased, thereby substantially reducing the diameter of the centrifugal compressor. The shorter engine length that this permitted was of benefit in helicopters. As centrifugal flow components became more robust, this engine enjoyed more utilization, since axial flow compressors are more subject to damage from foreign object (as from fan ingestion or the like).

Improvements in turboprop engines caused replacement of piston engines in spite of their generally lower efficiency. In the 1970s, the advent of high-bypass jet engines at high speeds and high altitudes produced high fuel efficiency.

In 1953, Whittle published *Jet* and in 1981 *Gas Turbine Aero-Thermodynamics*. Whittle developed health problems. He decided to live in America in 1976. He was a member of the faculty of the Naval Academy in Annapolis, Maryland.

His 1930 marriage to Dorothy Mary Lee was dissolved in 1976 (they had two sons), and thereafter he married Hazel "Tommy" Hall. In 1947, Dr. von Ohain came to the United States under the auspices of the American Operation Paperclip, which sought to harness the expertise of scientists and engineers who had worked under the Nazi regime. He became a research scientist at Wright-Patterson Air Force Base.

In 1963, he gained the position of chief scientist in the Air Force Aerospace Research Laboratories, responsible for almost all Air Force research in the physical and engineering sciences. In 1975, he became chief scientist of the Aero Propulsion Laboratory, with responsibility for maintaining the technical quality of Air Force research and development in air-breathing propulsion, power, and petrochemicals.

In all, von Ohain spent 32 years in U.S. government service, during which he achieved various inventions. He initiated and demonstrated the "jet wing" concept for cold air thrust augmentation for vertical and short takeoff and landing aircraft used in the Navy's experimental XFV-12A fighter. He demonstrated that the energy of fluid gases could be converted directly to electricity without using moving parts in an electrofluid dynamic generator to produce a practical power source. He also developed the economical retention of nuclear fuel in a gas core reactor, which made possible the investigation of its use in high thrust-to-weight space propulsion systems. Von Ohain published more than thirty technical papers and received nineteen U.S. patents; earlier, in his Heinkel employment, he had received fifty German patents.

Whittle and von Ohain had met briefly in 1966 at the American Institute of Aeronautics and Astronautics Goddard Award ceremony, but in early May 1978 they were again together on the occasion of the Air Force's commemoration of the seventy-fifth anniversary of the Wright brothers' first powered flight. Von Ohain was then the Air Force Research Laboratory's chief propulsion scientist. Scientists at the laboratory were given the opportunity to ask the two inventors questions in an open meeting at the Air Force Museum. There, although the two had served on opposite sides during World War II, they compared notes and gave their impressions of the difficulties faced in creating engines. The interview reportedly proceeded in a light and friendly manner.

Hans and Frank became good friends over the years after their earlier contacts. At the time of their meeting in 1978, Hans and his wife Hanny had Sir Frank (as they called him) to dinner. They observed that Sir Frank seemed to be frail and he told them that he had been in poor health for several years. Hans was then in good health.

In January 1979, von Ohain retired from U.S. government service, but he soon became associated with the University of Dayton Research Institute, which apparently then operated as a contract proposal firm. The University of Dayton asked von Ohain to teach graduate classes in thermodynamics and propulsion. Before long, the von Ohains bought a house in Florida.

During their respective careers, von Ohain and Whittle each received many awards. In 1991, von Ohain and Whittle were jointly awarded the Charles Stark Draper Prize for their work on turbojet engines.

Whittle died of lung cancer at his home in Columbia, Maryland, on August 8, 1996, and was survived by his wife, Tommy. Von Ohain died in retirement at his home in Melbourne, Florida, on March 13, 1998. He was survived by his wife, Hanny, and four children.

FURTHER RESOURCES

Margaret Connor. *Hans von Ohain: Elegance in Flight*. Reston, VA: American Institute of Aeronautics and Astronautics, 2002.

Hans Peter Diedrich. *German Jet Aircraft 1939–1945*. Atgien, PA: P. A. Schiffer Publishing, 1998.

John Golley. *Genesis of the Jet: Frank Whittle and the Invention of the Jet Engine*. Ramsbury Marlborough, UK: The Crowood Press, 1996.

Glynn Jones. *The Jet Pioneers*. London: Methuen, 1989.

Jacob Neufeld, George M. Watson, Jr., and David Chenoweth. *Technology and the Air Force: A Retrospective Assessment*. Darby, PA: Diane Publishing, 1997.

Sterling Michael Pavelec. *The Jet Race and the Second World War*. Westport, CT: Praeger Security International, 2007.

Royal Air Force History. Frank Whittle and the Jet Age. http:www.raf.mod.uk/history_old/e281.html.

Sir Frank Whittle. *Jet: The Story of a Pioneer*. Frederick Muller Ltd., 1953.

AP Photo

Nylon and Wallace Carothers

Nylon, a generic name for a thermoplastic, crystalline, polyamide polymer that is extrudable into monofilament, was the first plastic to be synthesized directly from starting monomers and was created in research on the chemistry of polymerization carried out at the du Pont Company under the leadership of Wallace H. Carothers in 1935. Carothers's work on polymerization was brilliant, productive, creative, and original. His invention of nylon led to a massive development effort by du Pont that resulted in the commercial production of nylon and the display of nylon stockings at the Golden Gate Exposition in San Francisco, California, in February 1939. During World War II, nylon production went for various military uses. After the war, besides satisfying enormous demand for hosiery, nylon usage in other fields expanded. Worldwide use of plastics expanded greatly.

HISTORY OF PLASTICS

A *plastic* is generally a material that is capable of being molded or shaped with or without the application of heat, such as, for example, soft waxes and moist clay. Here, the word refers to a material that is organic (that is, incorporates reacted carbon), and that contains one or more types of polymers (that is, substances comprising a large number of chemically interconnected individual units, usually monomers). Such a plastic as available commercially is usually compounded with other components, such as colorants, fillers, reinforcing agents, plasticizers, curatives, and so forth, and it may be thermoplastic (heat softenable) or thermoset (irreversibly set when heated).

The long history of plastics probably indicates humans' need. From early history, humans have used naturally occurring plastics and polymers, including shaped wood, animal hides, and natural fibers. Organic natural fibers, including for example, cotton, silk, wool, flax, and various other plant and animal fibers, have been found at sites of ancient civilizations all over the globe.

The oldest natural fiber for use in textiles was apparently flax, dating back to about 5000 BC. Cotton was first used about 3000 to 5000 BC, and wool about 3000 BC. Use of silk in China may be earlier than recorded history, based on archaeological sites along the lower Yangtze River. The main organic bulk-type plastic material used early on was wood, which was manually cut, shaped, or carved to produce particular objects for use. A few early instances of formed materials classifiable as plastics are known, where a particulate material, such as flour, sawdust, or clay, was mixed with a fluid, such as water, a tree-derived resin, or animal blood; the resultant paste or mixture was then molded and dried to a solid form. Adhesives, sealants, and coatings from natural vegetable or animal sources have been known and used for thousands of years. For example, a flour and water paste and beeswax were used as primitive adhesives, and tar or pitch was used for caulking, sealing, and adhering. Early glues derived from animal residues were variously used, for instance, to bond a veneer to planks of sycamore, or the like. For thousands of years, the only plastics known were composed only of the naturally occurring polymers that were available to humans.

As knowledge accumulated and accelerated, things began to change. In 1665, Robert Hooke (1635–1703) an infant prodigy who became an English physicist, who suggested some theories, and who drove the genius Isaac Newton to a nervous breakdown, described an idea for producing artificial silk from a gelatinous mass (which he never reduced to practice).

A brief history of cellulose work illustrates much effort in the field of plastics. John Wesley Hyatt (1837–1920), an American inventor, received several hundred patents. Seeking a substitute for ivory with which to make billiard balls, he experimented with Parkesine, a partially nitrated cellulose substance created and named by English chemist Alexander Parkes (1813–1890) in the early 1850s. Hyatt achieved in 1869 the first commercially practical plastic, which he called Celluloid (a solid solution of cellulose nitrate, camphor, and a flame-retardant such as ammonium phosphate; the term was once a trademark, but it is believed now to have fallen into common use).

Cellulose itself is a natural high–molecular weight polymer (a polysaccharide) having repeating glucose (anhydroglucose) units joined by an oxygen

Celluloid and John Wesley Hyatt

John Wesley Hyatt (1837–1920), an American inventor, received several hundred patents. At first, he worked as a printer and then established a factory in Albany, New York, where he manufactured checkers and dominoes.

Along with many others, in the early 1860s, he sought to win the prize of $10,000 offered for a substitute for ivory in billiard balls offered by Phelan and Collender, a billiard maker in Waterford, New York. Hyatt had heard of a new English method of molding pyroxylin (a low-nitrogen form of nitrocellulose, soluble in collodion) that had been created and named Parkesine by English chemist Alexander Parkes (1813–1890) in the early 1850s. Hyatt improved the method, and in 1869, he patented a method of making billiard balls using the material, but for some reason he did not win the prize.

He named the material "Celluloid," and it was the first synthetic plastic. It was essentially a solid solution of partially nitrated cellulose and camphor (or other plasticizer) plus a flame retardant such as ammonium phosphate. It enjoyed usage in shirt collars, baby rattles, photographic film, and the like, but it was quite flammable.

Hyatt made many other inventions, but none the equal of Celluloid. He was awarded the Perkin Medal in 1914.

linkage to form long, essentially linear chains. Cellulose is the fundamental component of all vegetable tissues and is the most abundant organic material in the world. Since fully nitrated cellulose is explosive, when cellulose is used in plastics, it is not fully nitrated; however, it is still quite flammable.

The first artificial silk from nitrated cellulose appeared in 1855. Though initially commercially impractical, a practical process for making was developed in France in the early 1880s by the chemist Comte de Louis Marie Hilaire Bernigaud Chardonnet. He extruded nitrated cellulose through little holes to form threads, as Joseph Swan in England had previously taught in an 1883 patent to form threads used in an unsuccessful process for carbonizing threads to make light bulb filaments. Presented at the Paris Exposition in 1891 as "Chardonnet silk," the fabric was a sensation. In 1894, the artificial (but not wholly synthetic, since it incorporates the natural occurring cellulose polymer) fiber became known as "viscose," and in 1924 it came to be called "rayon." Rayon is conveniently defined as a semisynthetic fiber comprised of regenerated cellulose, wherein hydrogen of component hydroxyl groups is replaced chemically by not more than 15% of other substituents; it is not a wholly synthetic polymer, such as is produced by chemically combining monomers.

Using various chemicals and process techniques, various types of rayon were developed, including regular (viscose) rayon, high-wet modulus rayon, high-tenacity rayon, and cuprammonium rayon. Paul Schutzenberger discovered that cellulose can be reacted with acetic acid anhydride to produce a cellulose triacetate. The discovery that the triacetate could be hydrolyzed and had solubility in economical solvents reduced cost and made cellulose acetate fibers cheap and efficient. Eduard Schweizer discovered that tetraamine copper hydroxide could dissolve cellulose, and process improvements in 1899 and 1901 led to an artificial cuprammonium silk product.

In 1893, Arthur D. Little of Boston invented what became known as acetate, a cellulose derivative that he developed as a film. By 1910, Camille and Henry Dreyfus were making acetate motion picture film and articles in Basel, Switzerland. In World War I, the brothers Dreyfus built plants in England and in the United States to make cellulose acetate dope for airplane wings and other products. Acetate in fiber form was developed in 1924 by the Celanese Company.

The reaction product of carbon disulfide and cellulose under basic conditions produced a highly viscous solution of zanthate, which was named viscose in 1894. This solution came to be used to make both rayon and cellophane (regenerated cellulose). Wood, which contains both cellulose and lignin, could be used as a raw source material, which made the system cheaper than methods needing lignin-free cellulose as the raw source.

Cellophane was invented by Jacques Brandenberger in 1900. While seeking to make an inexpensive tablecloth that would resist staining, he found that if he added rayon to cotton, the resulting fabric was too stiff and

brittle. He then proceeded to make a machine that would produce rayon sheets he called cellophane. Eventually he produced the clear, waterproof packaging film cellophane, usable for virtually any product.

Before Carothers's arrival, du Pont had acquired technology for manufacturing viscose rayon and acetate fibers and cellophane film and sheeting, and the company was commercially manufacturing and selling such products.

During the early 1900s, several attempts were made to produce "artificial silk" in the United States, but none appears to have been successful until Samuel Courtalds and Co., Ltd., formed the American Viscose Company and started making rayon in 1910.

Leo Baekeland (Belgium American chemist; 1863–1944) in the early 1900s in his own research created phenol-formaldehyde polymer reaction products, which he called "Bakelite"; in cured (cross-linked) solid forms, these products were water resistant, solvent resistant, weather resistant, fire resistant, heat resistant, electrically resistant, strong, chemically stable, and shatterproof. Bakelite was the first commercially practical thermosetting plastic (one that, once heat set, would not soften under heat). The material came into wide and continuing use and is considered to have stimulated the development of plastics. In 1944, when Baekeland died, the worldwide annual production of phenolic resins had risen to about 175,000 tons.

Bakelite and Leo Hendrik Baekeland

Leo Hendrik Baekeland (1863–1944) graduated from high school in Belgium at age 16 as the youngest and brightest, attended the University of Ghent on a scholarship, graduated in 1882, and in 1884 received his doctorate *maxima cum laude* at age 21. He received a professorial appointment in 1887 at the University of Bruges but won a 3-year traveling scholarship in 1889 and came to the United States.

He immediately got a good job and opened his own consulting office in 1891. In 1893, he invented a new photographic print paper he called Velox that was exposable and developable under artificial light. George Eastman of Eastman Kodak Company bought this paper invention for a fee between $750,000 and $1,000,000.

For several years, Baekeland undertook electrolytic research. After a brief visit to Europe in 1900, he started to create a substitute for shellac in his home laboratory in Yonkers, New York, attempting to make a reaction product of phenol and formaldehyde. At first, he could produce a tar-like residue but could find no solvent. The reaction itself had been investigated by Bayer in 1872, but Baekeland learned how to control the reaction to yield dependable results on a commercial scale. After continuing to experiment for several years, in 1909 he announced his new thermosetting plastic. The Bakelite

(continued)

company started in 1910, and the term "Bakelite" was trademarked in 1911. Later the company became a division of Union Carbide.

A thermosetting polymer (of which Bakelite is the first known) is produced by reacting phenol with aqueous 37–50% formaldehyde at 120–212°F with a basic catalyst. The polymerization is of the condensation type and proceeds through three stages. The first stage, called the A-stage or resole, comprises partially condensed phenol alcohols, which are fully soluble in common solvents, such as alcohols or ketones, and is fusible (can be thermoset) below 300°F. The A-stage is a constituent of most commercial laminating varnishes and is also used in special molding powders. Further heating converts (partially reacts) the A-stage to the B-stage, which is not fully fusible; this stage has limited commercial use. Further heating produces a fully thermoset (cross-linked) C-stage resin that is insoluble in all solvents.

However, thermoplastic condensates of phenol and formaldehyde called novolaks are produced with an acid catalyst and excess phenol. These resins can be reacted with hexamethylenetetramine, paraformaldehyde, and so forth for conversion to cured, thermoset, cross-linked structures by heating to 200–400°F.

Baekeland served as president of the American Chemical Society in 1924.

Some modern plastics were discovered by accident. Waldo Semon, a B. F. Goodrich chemist, developed polyvinyl chloride. He was attempting to adhere rubber to metal when he discovered accidentally that polyvinyl chloride was durable, easily molded, and inexpensive. In 1933, Ralph Wiley, a Dow Chemical employee, accidentally discovered polyvinylidene chloride, which Dow called "Saran." Because it would cling to almost anything, was economical, and sold well, Saran Wrap became a common household film. E. W. Fawcett and R. O. Gibson reacted ethylene and benzaldehyde under very high pressure, but the test tube lost pressure and the experiment failed. However, while cleaning up, they found a plastic substance that was waxy. The plastic was found to be useful and economically producible. Roy Plunkett, a du Pont chemist, pumped Freon gas into a cylinder and placed the cylinder into cold storage overnight in 1938. The gas turned into a white powder that was found to be impervious to acids and extreme temperatures. Du Pont named the solid "Teflon." Through subsequent commercialization, it became a common household coating.

Synthetic organic polymers for use in adhesives, sealants, and coatings (including paints) are typically thought to have originated in the early 1900s, generally concurrently with the origin of some other synthetic organic polymers. For example, phenol-formaldehyde adhesives for the plywood industry first appeared about 1910; various others arrived later. The very high strength-to-weight ratio associated with some synthetic organic polymers in adhesives in effect promoted the development of synthetic organic adhesives particularly suitable for usage in the aircraft and aerospace industries, sometimes as replacements for mechanical fastening means.

As the above brief review indicates, commercially practical, wholly synthetic, thermoplastic (that is, heat-softenable) polymers produced from combined monomers had not yet arrived when Carothers came to du Pont in 1927. In 1938, Paul Schlack of I. G. Farben Company in Germany polymerized caprolactam and created a new, different polyamide known as nylon 6.

DU PONT'S DECISION

E. I. du Pont de Nemours & Co. of Wilmington, Delaware (du Pont), had a central research department, sometimes called the chemical department, which in the 1920s was involved with applied research. Dr. Charles Stine, the man in charge of this department, presented to the company's executive committee on December 18, 1926, a brief memorandum titled "Pure Science Work." He proposed that the company enter fundamental research (involving new science discovery) as distinct from the already existing applied research (which sought new and sometimes improved products based on previously known scientific information). He pointed out that while fundamental research was undertaken in universities, it had already been undertaken successfully in the German chemical industry and in the General Electric Company regarding new products. He identified what he thought were reasons supporting his proposal.

The committee requested a more detailed proposal, which Stine produced in March 1927, and the committee promptly approved his proposal effective April 1927. The approval was accompanied by what was then a generous monthly budget plus funds to build a new fundamental research laboratory.

Stine began looking for twenty-five scientists. His preference was to seek men of proven ability and recognized standing in particular fields. Such men existed, were established in academia, and had developed particular areas of research. Acquiring them would not be easy. Alternatively, he would do what General Electric and Bell Laboratories had already done: hire men of exceptional promise but having no established reputation. Of course, the nature of all the hirees' work was to be largely determined by du Pont management.

By the end of 1927, he had employed eight men to work in certain areas of fundamental research, but only one of these men originated from academia: Wallace Hume Carothers. He was a 31-year-old Ph.D. instructor from Harvard University.

Carothers lasted 9 years at du Pont, during which time he made various significant contributions to polymer science and led research efforts that resulted in neoprene rubber and nylon fibers. These were the world's first thermoplastic, commercially producible and successful, totally synthetic polymers. The company management chose to spend much time, money, and effort to develop these polymers into commercially practical, large-scale plastics from their original, laboratory-prepared, very small amounts.

WALLACE CAROTHERS (1896–1937)

Early Years

Wallace was born April 27, 1896, in Burlington, Iowa, to Wallace and Helen Carothers. He was the oldest of four children, the others being a brother and two sisters. As a boy, he was most interested in tools and mechanical things, and he spent hours experimenting. He became interested in chemistry in the Des Moines public high school where he was a student. His father rose to be vice president of a small commercial college, the Capital City Commercial College, in Des Moines. Wallace entered this school, and after completing the accountancy and secretarial program in 1915, he entered Tarkio College in Missouri, where he could study science. There he did very well in chemistry. However, this was a small institution. During World War I, the only chemistry teacher, Arthur M. Pardee, left for the University of South Dakota, and replacing him proved impossible. Wallace agreed to teach chemistry though he was still a student in the college.

After graduation in 1920 from Tarkio at age 24, he entered the University of Illinois, where he obtained his M.A. degree in 1921 and his Ph.D. degree in 1924. Between these degrees, Wallace joined Pardee in South Dakota and taught courses in chemistry to gain enough money to support himself for his Ph.D. work at Illinois. He subsequently tried teaching chemistry at the University of Illinois and then in 1926 at Harvard University. He found, though, that it was not teaching but research that interested him.

In 1928, when the du Pont Company was initiating a program in basic research, Carothers was hired to run organic polymer research. Stine recruited him. Wallace hesitated to leave Harvard and at first refused. He disclosed in a letter to Stine that he suffered "from neurotic spells of diminished capacity that might constitute a much more serious handicap than here." Stine nevertheless wanted Carothers and sent an assistant, Dr. Hamilton Bradshaw, to visit him. Bradshaw raised Carothers's salary offer to $6,000 a year, talked chemistry with Carothers, and indicated that du Pont would provide support for Carothers's projected polymer research. Ten days later, Carothers accepted the du Pont position and started working in suburban Wilmington at the du Pont Experimental Station, a campus-like area for research, on February 6, 1928.

Carothers was very intelligent and original in his thinking. James B. Conant, a chemist who became president of Harvard College in 1933, noted, "Dr. Carothers showed even at this time the high degree of originality which marked his later work."

Du Pont wanted polymer research, and Wallace was interested in polymers. Before leaving Harvard, he set down his thinking about polymers and their structure. He thought that practical synthetic polymers, by analogy to naturally occurring polymers, probably could somehow be made from organic compounds (that is, compounds containing carbon atoms). He believed that the views about polymers of Hermann Staudinger (1881–1965), the German chemist, were correct, that is, that polymers were held together by the same chemical forces as small molecules and that in a polymer simple component units were usually connected linearly. Carothers devised a way to prove that these views were correct. His way involved producing polyesters (molecules containing a plurality of ester groups) from selected polymerizable monomers.

Neoprene

By the middle of 1929, Carothers and his staff had not produced a polymer with a molecular weight of much over 4,000, although Carothers had initially sought to reach polymers with a molecular weight of more than 4,200, a weight previously achieved by synthesis by Dr. Emil Fischer (1852–1919), the German chemist. However, his objective still had not been achieved by 1930.

In January 1930, Dr. Elmer K. Bolton of du Pont was appointed assistant chemical director of what was called the chemical department, under Stine but over Carothers. Bolton wanted realistic results in 1930, and he got them in two areas.

In one area, Bolton asked Carothers to review the chemistry of an acetylene polymer, such as had been achieved by Julius Nieuwland (1878–1936), a Belgian American chemist teaching at the University of Notre Dame in Indiana, who was evidently a du Pont consultant. Acetylenic polymers had previously been looked at by du Pont without any product success. Acetylene, a compound containing two carbon atoms, could combine with itself to form a four-carbon molecule and a six-carbon molecule. These larger molecules could add, by polymerization, a plurality of two carbon acetylene molecules, whereupon a giant molecule with some rubber-like properties resulted. Bolton hoped that a synthetic rubber could be created.

Under Carothers's leadership, it was found that if a chlorine was added at the four- or six-carbon stage, the properties of the resulting polymer product were quite similar to those of rubber. In April 1930, one of Carothers's chemists, Dr. Arnold M. Collins, isolated chloroprene (2-chlorobutadiene-1,3) and found that it was a liquid that polymerized to produce a solid, rubber-like material. Carothers undertook with his group to study divinylacetylene, a short polymer of three acetylene molecules. With purified divinylacetylene, prepared by distillation, Collins, under Carothers, produced, using chloroprene, clear films instead of the yellow ones that resulted from the impure material. This polymerized product was found to be a useful synthetic rubber and was named neoprene (polychloroprene).

Another du Pont laboratory was soon assigned the work of development and commercialization of neoprene. Du Pont began manufacturing and commercializing neoprene in 1931. Neither Nieuwland nor Carothers survived long after the invention of neoprene. Later, after Pearl Harbor, Japan cut off the supply of natural rubber. It was neoprene that kept the essential portions of the American economy rolling.

Polyesters

In the second area in which Bolton got results, by the summer of 1930, Carothers and his group were busy with other polymers. Dr. Julian Hill of the Carothers group had restarted work on producing a polymer with a molecular weight above about 4,000. He continued to pursue reacting monomeric carbon-containing materials that could react together to form polymeric products, as Carothers had initially proposed. This was a reaction of an alcohol monomer having a hydroxyl group (–OH) at each end of its molecule (called a diol) with a carboxylic acid monomer having a carboxyl group (–COOH) at each end of its molecule (called a dicarboxylate) to form ester polymeric products in which a repeating unit was an ester group formed by the reaction of a hydroxyl group with a carboxyl group. In each such alcohol monomer molecule and acid monomer molecule, it was known that their two terminal functional groups (the hydroxyl group and carboxylic acid group) were separated by some carbon atoms, which were themselves interconnected directly and adjacently together. In the resulting polymer, as had first been theoretically supposed, there was a chain of repeating but linked alternating alcohols and carboxylic acids, and in the polymer. each alcohol was linked to its adjacent carboxylic acid through an ester linkage. In the polymerization reaction, each ester linkage was produced when one hydroxyl group of an alcohol molecule reacted with one carboxyl group of a carboxylic acid molecule in a so-called condensation reaction, of which a water molecule was a by-product.

However, Julian Hill could not get the reaction product polymer molecular weights to be above about 4,000 (roughly corresponding to about twenty-five alcohol-acid ester groups per polymer). Carothers decided that the reason why higher molecular weights were not obtained was that the water, formed as a by-product during the alcohol-acid esterification reaction, was producing an equilibrium condition that effectively stabilized the esterification reaction at a molecular weight of about 4,000. To remove the water in regions adjacent to the esterification reaction, Carothers and Hill constructed an apparatus called a molecular still that was adapted to separate the water produced. Also, they decided that a long chain di-acid molecule (a dicarboxylate) reacted with a short chain di-alcohol molecule (a diol) would be expected to promote formation of longer chain polymer molecules, and so they selected a sixteen-carbon dicarboxylate acid molecule and a three-carbon alcohol diol molecule as starting monomers.

From the reaction of these monomers, and while using the molecular still, Hill found, as he removed some polymer product from the still, that the molten product could be drawn into fibers and that, unexpectedly, the cooled product could be stretched ("cold drawn") to achieve very strong fibers. The product was found to have a molecular weight of over 12,000. Carothers initial proposal to produce polymers above about 4,200 molecular weight had been achieved, and the resulting polymeric products had some promising properties. These products, polyesters, were apparently the first all-synthetic, high-molecular-weight plastic polymers with fiber-forming capacity.

The 3-16 polyester (as it was called for brevity) and related products from other new combinations were found to be not usable as practical textile fibers, though, because they all melted below 212°F, were at least partially soluble in dry-cleaning solvents, and were not stable in water.

In June 1930, Bolton became chemical director, and Stine advanced into the corporate executive committee. Bolton had opposed Stine's fundamental research program. Bolton wanted fundamental research managed so as to achieve new products that would give competitive advantages to du Pont.

Carothers wanted to publish these new synthetic fiber results, but Bolton's new assistant, Ernest Benger, refused, saying that the results needed to be first protected by patent. Carothers agreed to wait until patent applications were filed, but felt Benger had unilaterally changed company policy.

Bolton and Benger were excited by the work of Carothers and his group because their experimental polyester fibers retained most of their strength when initially wet and had an elasticity similar to that of silk, even though their melting points were unacceptably low. Particularly since simple amides (the reaction products of a carboxyl group with an amine group) usually had higher melting temperatures than the corresponding esters, Carothers and Hill decided to try to make useful polyamide polymers from polymerizing monomeric dicarboxylic acids and monomeric diamines. Such starting diamines would have a so-called amine group ($-NH_2$) at each end of their molecules instead of a hydroxyl group as previously employed. Unfortunately, in 1930, Carothers and Hill could not produce satisfactory polyamides. Nevertheless, evidently to prepare for further work by Carothers's group that might result in practical polymers, du Pont filed a broad patent application for synthetic fibers, including polyamides, in 1930.

Research: 1931–1934

In 1931, Carothers moved into a house in Wilmington that had acquired the name Whiskey Acres and that he occupied with three other du Pont chemists. His housemates he found to be "socially inclined," but, during his depression periods, as he advised a friend, he could not enjoy their activities. From about this period, it is believed Carothers always hung a capsule of cyanide poison on his watch chain.

Carothers strove to maintain a high professional profile. He undertook public speaking. He gave a speech at an organic symposium during the 1932 Christmas holidays, but confided in a letter to a close friend that "the prospect ... [had] ruined the preceding weeks and it was necessary to resort to considerable amounts of alcohol to quiet my nerves."

By 1932, Carothers felt he had used up his ideas for fibers and considered that the work had departed from the basic exploration he liked. Besides, he was uncertain about what Bolton expected from him. The Great Depression was worsening, and Bolton cut the money available for fundamental research. He encouraged Stine's original groups to pursue applied subjects and to keep in mind where profit could be gained. In 1932, Bolton effectively modified the agreement under which Carothers had been hired. There was to be "a closer relationship between the ultimate objectives of our work and the interests of the company." There was a shift from pure research to practical research. Carothers did not see himself as an applied researcher on commercial subject matter.

With the reputation he had previously gained, Carothers decided, apparently as early as late 1931, that he could pursue his basic research interests in polymerization mechanisms despite changes initiated by Bolton. He moved his research from linear compounds to cyclic compounds of 8 to 20 carbon atom rings that were synthesizable in the molecular still. He found that one of the compounds he synthesized emitted an intriguing aroma. It was the first synthetic musk and later was marketed under the name Astrotone.

He had a busy personal life during this period. He was having an affair with a married woman, Sylvia Moore. She filed for divorce in 1933. Concerned about the financial problems of his parents, he bought a house in Arden, about 10 miles from the Experimental Station, and, with his parents, moved into it. There was soon tension in the household. Though he was then 37, his parents did not approve of his seeing Sylvia Moore, who was now single. The parents returned to Des Moines in the spring of 1934.

Nylon

Carothers felt that he had completed his large ring research by late 1933 and looked about for something else. By mid-1933, Carothers had stopped synthetic fiber research in his group because he felt that the sought properties of high melting point and low volatility made fiber spinning impossible. Bolton, aware of Carothers's research situation, tried to interest him in further work on synthetic fibers. Responding to Bolton's suggestion to continue the polymer research, early in 1934, Carothers began pursuing a new approach. To avoid the previously demonstrated unreliability of the molecular still, he decided to use a carefully purified starting amino acid ester instead of the di-acid itself. To obtain a lower melting point, he proposed using a long-chain starting monomer.

In March 1934, Carothers suggested to one of his men, Donald Coffman, that he try to make a fiber from a polymerized aminononoanoic ester (that is, a monomer of nine carbon atoms with an amino group at one end and a preformed ester group at the opposite end). Coffman took five weeks to prepare this starting compound but then found that it could be quickly polymerized. Coffman felt he had made a high-molecular-weight polymer because, during cooling, it had seized the walls of the flask and had shattered them. The next day, in May 1934, Coffman drew a fiber from the few grams of polymer he had prepared by heating the polymer to just above its melting point and manually drawing a fiber filament using a glass rod. The properties of the filament were attractive (lustrous, not brittle, fairly tough, and cold-drawable).

Another experimental polymer was prepared and also displayed favorable properties. These were the first linear polyamide fibers. These results indicated that a practical synthetic polyamide fiber was technically feasible. Further research was pursued. It was a different polyamide that was eventually commercialized. Later, this class of polyamide fibers was sometimes called nylons. The polyamide was made by condensation polymerization, as distinct, for example, from radical polymerization, and the basic principles of condensation polymerization and the structure of condensation polymers were soon worked out in Carothers's group. In a condensation polymerization, two or more molecules combine with the separation of a simple substance, such as water or alcohol. The distinction from an addition reaction is not sharp, because some addition reaction reactions can be either stopped after the addition or carried a step beyond the addition.

In the summer of 1934, during this productive research but prior to the eventual development of a commercial polyamide, Carothers disappeared. He was found at a small psychiatric clinic, Phipps Clinic, which was associated with Johns Hopkins University in Baltimore, Maryland. Evidently, he had become so depressed that he drove to Baltimore to consult a psychiatrist, who put him in the clinic.

After several months, Carothers was released from the clinic, and he returned to his work. Bolton instructed Carothers to pursue polyamides with the aim of achieving a viable fiber. Chemists, apparently in Carothers's group, got the assignment of making polyamides, and apparently testing them preliminarily, from various combinations of diamines and dicarboxylic acids. Sometimes the monomeric starting materials were not available had to be synthesized first.

On February 28, 1935, a polyamide called "6-6" (from the number of carbon atoms in the diamine and the dicarboxylic acid) was first produced by

Dr. Gerard J. Berchet. Laboratory tests were favorable. During this effort, five polymer options resulted, but to Bolton the polymer 6-6 polyamide seemed to be a most viable nylon fiber.

Though difficult to work because of its high melting point, it was this polymer that Bolton chose to develop and that became the first commercial nylon fiber polymer. Bolton made polyamide development the major project in Carothers's group.

Bolton selected Dr. George Graves to work, apparently, at first as a sort of assistant group leader with Carothers on the polyamide project. Eventually, Graves replaced Carothers as the polyamide project leader. Dozens of du Pont engineers and chemists were eventually assigned to develop polyamide 6-6 into a commercial product, starting with something that was virtually just a laboratory curiosity of a few grams. Much development work had to be successfully accomplished. There was little prior technology to build upon. The project became a crash program that succeeded in bringing nylon to the marketplace in less than 5 years.

Deterioration of Carothers

Julian Hill, a member of Carothers's group, observed one time, probably in 1930, that Carothers carried with him a capsule of cyanide. Hill recalled that Carothers could name all the famous chemists who had committed suicide.

After his severe bout with depression in 1934, subsequent periods of depression became more frequent and severe. On February 21, 1936, he married Helen Sweetman, whom he had been dating since about 1934. She worked for du Pont, had a bachelor's degree in chemistry, and was involved with the preparation of patent applications. On April 30, 1936, Carothers was elected to the National Academy of Sciences. This was a very high honor. He was the first industrial chemist to receive this honor. In June 1936, he experienced a major breakdown and was involuntarily admitted to the Philadelphia Institute of the Pennsylvania Hospital. His psychiatrist was Dr. Kenneth Appel. One month later, he was released to go hiking in the Tyrolean Alps with friends. After the friends left, Carothers stayed on and hiked by himself, but did not send word to anyone.

On September 14, he suddenly appeared at his wife's desk at the Experimental Station. The company did not expect him to perform any further work, but he often went in to visit. At the request of his wife, who felt she was not emotionally strong enough to handle his problems, he began living at Whiskey Acres again.

On January 8, 1937, Carothers's favorite sister died of pneumonia. Carothers with his wife traveled to Chicago for her funeral and then to Des Moines for her burial. It is likely that her death added to the depth of his bouts of depression.

Back in Wilmington, he traveled periodically to Philadelphia to visit his psychiatrist, Dr. Appel, who advised a friend of Carothers that he thought suicide was the likely outcome of Carothers's case.

On April 28, 1937, Carothers went to the Experiment Station with intent to work. The next day, he traveled to Philadelphia and registered in a hotel. There, in his room, he dissolved his cyanide capsule in lemon juice and committed suicide. He left no note. He had just turned 41 two days previously. His daughter was born after his death. The significance of nylon was just then beginning to be realized by du Pont.

Carothers left a large harvest of technology and inventions. He had been a hard worker. His work with his team had resulted in more than twenty papers and patents in the field of acetylene chemicals (including isoprene, the molecular unit of natural rubber). In the field of polyamides (nylons), he and his team produced at least thirty-one papers and patents, which included the commercial nylon polymer and general theories about polymers (including natural polymers) and about how polymers form, and he even developed standardized terminology relating to polymers.

Even today, people still wonder about Carothers's depression attacks. Few facts are available; few questions are answerable. Did he have adequate treatment? A basic environmental circumstance he faced was the interplay between applied research and fundamental research. If Carothers had been left to himself to manage and operate his group, would he ever have achieved nylon? Did he have a treatable, controllable bipolar disorder that was characterized by periods of depression, or was his condition untreatable and irrevocably deteriorating? How did his environmental circumstances affect his mental status? Had he become obsessed, for some reason, with the thought that he had been a failure as a scientist or inventor? Did he think that his creativity in science had been adversely affected by his mental condition? Could that last visit to the laboratory have triggered a bout of depression? Did he then perhaps have some fresh insights to provide, but was turned away? In his weakened state, could he have then seen something causing depression? Perhaps it was the scene he viewed with his management-appointed successor situated and operating in his old office serving as leader. Perhaps his successor suggested to him that he should just go away.

DEVELOPMENT OF NYLON

New Problems

In April 1937, at the time of Carothers's death, the involved du Pont people had already advanced deep into nylon development work, and there were significant problems being faced. How much of this work had been known to Carothers or involved him is not known.

To make nylon 6-6, the starting adipic acid could be made by existing and probably available German technology, but the starting hexamethylene diamine was difficult to synthesize and had not been previously made in quantity. Could this diamine be made in large quantities? How was the polymer product molecular weight to be controlled? Could practical melt spinning of molten polymer into fibers be achieved? A sizing composition was necessary, but what sizing composition could be used on the filaments? These problems and others were met and solved.

Du Pont had spinning technology from its rayon and acetate fibers, but those fibers were spun from solution and did not melt. Could spinning of molten nylon polymer be carried out? Though this option was attractive, it was vastly different from solution spinning, but it was pursued. In 1936, the first melt spinning of nylon lasted about 10 minutes, but by May 1937, continuous melt spinning times extended from about 10 to 82 hours.

Du Pont management made various early decisions. One was to tackle nylon development by focusing on one major problem at a time. Another one was to first build a pilot plant and overcome the production problems before building a full-scale large plant. Another one was to first focus first on the full-fashioned hosiery market; other uses and markets could wait. These decisions had the effect of avoiding dilution of employee effort. Du Pont felt it had a controlling patent position. Du Pont elected to build a pilot plant before a large commercial plant to utilize personnel, to cut risks, and to solve problems. Manufacturing costs of nylon were found to be reasonable, and to make money it was not necessary to have the best process, but only a workable one, to make fiber and yarn.

High-quality yarn that comprised uniform strands of twisted fine filaments was necessary for making full-fashioned hosiery. When such yarn became available, then sufficiently large samples of yarn were needed for submission to textile companies so they could knit hose under existing standard commercial conditions. Du Pont had to make the filaments and the yarn because the technology was new and unknown to people such as the old silk throwers. Du Pont chose not to enter knitting and hose manufacture not only because of the machinery and know-how required, but also because established hose manufacturers had trained staff and capacity accomplish marketing to the trade.

Experimental stockings had to be made, sampled, and tested. Problems such as wrinkling arose and had to be solved. The polyamide development by du Pont from laboratory quantities to commercial product quantities had to be undertaken and successfully accomplished within realistic parameters. Data finally began accumulating that suggested that the du Pont nylon hosiery might be better than the prior silk product.

Selection of "nylon" as the commercial product name was a sort of evolutionary development from various starting efforts and was not settled until 1938. The loss of the company's trademark Cellophane in 1937 (due to generic usage) resulted in a negative position by management regarding registering nylon as a trademark. Instead, nylon began life as a generic word available for use by anyone.

Nylon Stockings

Du Pont, through Charles Stein, now du Pont's vice president, first publicly announced the achievement of nylon in October 1938 at the World's Fair Site. It was still a year and a half before the first nationwide sales of nylon stockings. The next day, the *New York Times* ran two articles on nylon. In one article, the phrase "coal, air, and water" was associated with nylon, and this association continued later, though these substances had nothing to do with the composition of nylon and may have been used in du Pont publicity to suggest the power or magic of science.

Sample nylon stockings became available and were used for consumer testing and comment purposes. Du Pont was pleased to see that durability was a typical response. Du Pont began commercial-scale production of nylon in 1939 and displayed nylon stockings at the San Francisco Golden Gate Exposition in 1939. Believing that nylon might prove to be superior to silk, the price of nylons was set initially about 10% higher than silks. The stockings were first offered in New York stores on May 15, 1940. More than 780,000 pairs were sold on the first day, and in the first year 64 million pairs were sold throughout the United States. "Nylons" soon became the name for hosiery.

After the United States entered World War II, the War Production Board on February 11, 1942, allocated all nylon production for military use, and all civilian production of nylon was stopped, so nylons departed, but in less than 2 years, nylon had displaced over 30% of the whole full-fashioned hosiery market. Nylon hose, which sold for $1.25 a pair before the war, sold on the black market for $10. Movie stars such as Betty Grable in war bond drives auctioned nylon hose for up to $40,000 a pair.

After the war, limited available supplies of nylon stockings became advertised and were available in September 1945. Women lined up to buy, and fights broke out involving disappointed potential customers. Several years passed before the demand could be met; in fact, it has been said that the supply did not satisfy the demand until the 1960s. Nylons at first covered only about two-thirds of a woman's leg, from foot to mid-thigh, and were fastened with garters and a belt. They were knitted on very complex machines.

The basic knitting machine had been invented by Reverend William Lee in 1589, and for centuries stockings had been woven from wool, cotton, or silk. Stockings or hose were originally mainly worn by men. The development of the power loom in 1884 enabled significant improvements and variations in fabrics.

Nylons could be purchased either as "full-fashioned," with seams on the back, or seamless. The full-fashioned (sometimes "fully fashioned") and seamless stockings were made to the shape of the leg, as nylon at first did not stretch. Nylon hosiery at first was sold in a large number of sizes. Seamed nylons were knitted flat and then the two sides were joined together manually in a small seam up the back.

In the late 1950s, yarn manufacturers discovered that stretch could be added to nylon by crimping the material under heat. In 1959, du Pont inventors created, and du Pont commercialized, a stretchable fiber called Spandex, which reportedly could expand at least 100% and recover its original shape. Hosiery could fit better and be more comfortable with this fiber.

In the 1960s, the miniskirt arrived, and tights came in as an alternative to stockings. Within 2 years, tights had 70% of the market. In the same period, hosiery companies provided hosiery knitted as a tube, with the consequence that nylons with seams almost vanished. "Tights" originally were a tight-fitting garment for the lower portion of the torso and legs worn by acrobats, dancers, etc. Hosiery companies managed to make tights as one-piece knitted panty hose that were sold widely. Although stockings have increasingly fallen out of fashion, hosiery technology has developed a variety of new products, including body-toning, control top tights and thigh-high, massaging, and moisturizing tights.

Toothbrush Bristles and Other Offerings

In 1938, when du Pont had a limited capacity to melt spin the 6-6 polyamide, its Plastics Department began marketing nylon bristles under the trademark Exton in Dr. West's toothbrushes. This was a beginning market entry for nylon and was attractive to du Pont because imperfect polymer from the pilot plant was sold as toothbrush bristles. Du Pont did not disclose publicly the nature or origin of these bristles.

During the war, nylon was used as a replacement for Asian silk in parachutes, tires, tents, ropes, ponchos, and various other military supplies. It was used even in high-quality paper for U.S. currency. Although cotton made up more than 80% of all fibers used at the start of the war in 1941, at the end of the war in 1945, cotton made up only 75%, and manufactured fibers stood at 15%.

AFTERMATH

After the war, nylon was used in many ways not only as a fiber but also as a solid plastic. Toughness, strength, and lightness were featured. Nylon and nylon products came into worldwide common usage, including apparel, carpeting, home furnishings, and industrial and commercial products. (In 1969, Buzz Aldrin planted a nylon flag on the moon, which presumably is still there.)

Nylon became by far the largest money-maker in du Pont's history. One result was that du Pont put money into further research seeking more nylon successes.

Wallace Carothers in early research at du Pont had experimented with polyesters without uncovering a product. However, later the Calico Printers Association in Great Britain with J. T. Dickson and J. R. Whinfield made commercially acceptable polyester fibers by condensation polymerization of ethylene glycol and terephthalic acid. Subsequently, du Pont acquired the U.S. patent rights, and Imperial Chemical Industries acquired the patent rights for the rest of the world. In 1953, du Pont began selling commercial polyester fiber. Soon, other fiber producers became involved.

Polyester commercialization in 1953 was accompanied by the introduction of triacetate, a cellulose derivative derived by directly acetylating

Engineering Plastics

Plastics with superior mechanical and thermal properties in a wide range of conditions apart from those of commonly used so-called common or commodity plastics have been developed and are usually classified as engineering plastics. A typical listing of such plastics includes the following.

Acrylonitrile-butadiene-styrene
Polycarbonates
Polyamides
Polybutylene terephthalate
Polyethylene terephthalate
Polyphenylene oxide
Polysulfone
Polyetherketone
Polyetheretherketone
Polyimides

Examples of commodity plastics that are used in large quantities include polystyrene, polyethylene, polypropylene, and polyvinyl chloride.

Typically, engineering plastics are sold in lower quantities and are more expensive (per unit weight) than commodity plastics. Engineering plastics tend to be used in applications where wood or metal can be, or need to be, replaced, or where high performance is sought, such as high heat resistance, high mechanical strength, rigidity, high chemical stability, or flame retardancy.

cellulose, whose production was discontinued in 1965. Orlon (a trademark for a copolymer containing 85% acrylonitrile) and Dacron (a trademark for a polyester fiber made from polyethylene terephthalate) from du Pont appeared in the late 1940s and 1950s. Acrylic fiber (a generic name) involving a polymer based on polyacrylonitrile, a wool-like product, appeared in the 1950s from du Pont, but du Pont production ceased in 1991. Beginning in the 1960s, consumers proceeded to buy increasing amounts of polyester clothing.

Though it had been an explosives maker in its first 100 years, du Pont now became mainly a plastics and fibers producing company, along with other diversified businesses. Synthetic textile fibers had in effect revolutionized the company and probably the world. Polyolefins, especially polypropylene, in 1966 became the only Nobel Prize–winning fiber.

Innovation is characteristic of the plastics industry. Many different polymers and plastics have come into extensive worldwide use. Whole development and business groups have been assembled about polymers such as polyurethanes, expandable polystyrene, polycarbonate, and other polymers.

However, du Pont regarded itself as basically being a producer and marketer of new products. In general, as older products matured in market, new and younger products appeared as replacements. The chemical industry generally had a long period during which a few new products occurred. Over the years, the areas of sales growth for du Pont slowly changed. By the 1990s, though it was by then a $45 billion a year oil and chemicals-based firm, du Pont management decided that the best future for the firm lay in the direction of being a science-based, market-driven firm in a resource-constrained world with businesses in specialty chemicals, biomaterials, crop protection, and seeds. For about a decade, chief executive officer (CEO) Charles O. Holliday Jr. reshaped the firm, so that by about September 2008 the firm had been downsized to about $30 billion a year. In 2004, the company sold its fibers business to Koch Industries for $4.4 billion.

Then, in late September 2008, du Pont announced that Holliday, at 60, was yielding the CEO position to Ellen J. Kullman, 52, who will be the first woman to lead a major public U.S. chemical firm. Since 2006, she had been an executive vice president with responsibility for four of the five businesses of du Pont. She has announced that sales growth will be du Pont's top priority. Kullman has a degree in mechanical engineering from Tufts University in Boston, Massachusetts, and a master's degree from Northwestern University.

FURTHER RESOURCES

Roger Adams. *A Biography, in High Polymers: A Series of Monographs on the Chemistry, Physics and Technology of High Polymeric Substances*. Vol. 1.

Collected Papers of W. H. Carothers on High Polymeric Substances. New York: Interscience Publishers, Inc., 1940.

Yasu Furukawa. *Inventing Polymer Science: Staudinger, Carothers, and the Emergence of Macromolecular Chemistry.* Philadelphia: University of Pennsylvania Press, 1998.

Ann Gaines. *Wallace Carothers and the Story of du Pont Nylon (Unlocking the Secrets of Science).* Hockessin, DE: Mitchell Lane Publishers, 2001.

Matthew Hermes. *Enough for One Lifetime, Wallace Carothers the Inventor of Nylon.* Philadelphia: Chemical Heritage Foundation, 1996.

News of the Week. *Chemical and Engineering News*, September 29, 2008, p. 7.

AP Photo

Nuclear Reactor and Enrico Fermi

CONTEXT OF THE ATOMIC BOMB

To learn whether uranium could be fissioned to release energy in a chain reaction, Enrico Fermi undertook the building of a self-sustaining, power-producing reactor using uranium and achieved a successful result. This was the world's first atomic reactor. It not only set the stage for the development of the atomic bomb and nuclear power, but also began the age of atomic energy.

PERSPECTIVE

Fear that Nazi Germany might achieve an atomic bomb of enormous power led to interest in America making the bomb first. Basic questions had to be answered: Was it even possible to make such a bomb? Could uranium be fissioned in a chain reaction and thereby release great energy? Affirmative answers were achieved when an operating, self-sustaining, power-producing nuclear reactor based on uranium was built by a team of scientists led by Enrico Fermi. The successful result ushered in the atomic age. Every nuclear power plant utilizes a nuclear reactor, but that one in late 1942 was the first ever attempted.

Though the atomic bomb could seem to be an invention that appeared suddenly in the 1940s, it actually was not a single invention but a pinnacle of developments and inventions undertaken in the midst of a global conflict. It arrived as a result of a massive secret undertaking that comprised applied, known technologies plus a plurality of new technological discoveries and advances.

The historical circumstances leading to the atomic bomb and to sustained power from nuclear energy are distinctly different from those of other great innovations. The atomic bomb, which took about 5 years to accomplish, from initiation of development to realization, is an example of a substantial breakthrough realized through governmental funding and control, leadership by gifted people, and talents of outstanding scientific and technological personnel under the structure of the secret Manhattan Project.

Although an atomic bomb was successfully tested in mid-1945 and utilized to win World War II, substantial development of nuclear power using nuclear reactors to generate electric power in privately funded central stations did not occur until the mid-1960s.

Taken in the context of the rapidly unfolding events occurring in the Manhattan Project in the earlier 1940s, one can get a glimpse of how a major innovation for humankind, with its associated developments and offshoots, fits into a larger picture of unfolding events in civilization. From the standpoint of history, in the more usual situation involving a major technological innovation and its development and offshoots, commonly a lengthy period of time would elapse. Of course, the present circumstances cannot be taken as typical since they may not have occurred but for the combined circumstances of the pressure of war, the availability of funds, and the existence of willing and able much talent.

TERMINOLOGY

Atomic physics and associated engineering has developed various specialized terminology, which grew out of certain chemical and physical fields. It was found that the present synopsis needed to use such terminology for reasons of accuracy, brevity and clarity of description, and characterization. These specialized terms are defined for the reader. A reader with technical background may overlook the following definitions and concepts, referring to them only if needed for an understanding of subsequent text portions.

An *element* is one of a class of substances (catalogued in the *Periodic Table*), each of which comprises an infinitesimal physical structure or atom having electrons located outside a nucleus.

An *atom* is a portion of matter that is too small to be viewed with the unaided human eye but includes a nucleus and at least one electron and that individually comprises an element.

A *nucleus* is the positively charged central mass of an atom of any element. It contains almost the total mass of an atom and is in the form of *protons* and *neutrons*. A proton and a neutron are considered to be fundamental units of matter (although they are in turn made up of certain subatomic components). Each is assigned (arbitrarily, for convenience) a mass number of 1 (equivalent to about 1.67×10^{-24} grams). A proton has a positive charge; a neutron, no charge.

An *electron* has a negative charge and a mass of only 1/1,837th that of a proton. The number of electrons normally in an atom of any element is the same as the number of protons in the atom's nucleus. In an atom, the total charge of the electrons is equal to the total positive charge of the protons. Electrons that are associated with an atom are arranged in from one to seven *shells* (also called paths or orbits), in which the electrons revolve around the nucleus. The number of electrons in each shell is determined by the laws of physics. Electrons can be removed, for example, from the atoms of metals and some other elements by heat, light, electric energy, bombardment with particles, or chemical reaction; when so removed, they are totally free from atomic orbit, and their energy can be used and directed by means of a conductor, semiconductor (such as a transistor), vacuum tube, or the like, in an electrical circuit.

All *chemical* reactions involve only electrons associated with atoms. During a chemical reaction, one or more electrons in orbit about a nucleus rupture, or change position, from their respective initial shells.

All *nuclear* reactions involve a change in the nucleus of an atom. A nuclear reaction usually results in the release of a tremendous amount of energy, while a chemical reaction usually releases a relatively small amount of energy. There are three types of nuclear reaction, *transmutation*, *fission*, and *fusion*.

The actual mass of a nucleus is always slightly less than the sum of the masses of its protons and protons, the difference being the mass equivalent of the energy of formation (binding energy) of the nucleus. This difference is called the *mass defect* and accounts for the high energy release from nuclear fission.

Transmutation refers to the transformation of atoms of one element into atoms of another element as a result of a nuclear reaction. In transformation, two nuclei may interact or a nucleus may react with an elementary particle, such as a proton or neutron. For example, radioactive decay of uranium can be regarded as a type of transmutation.

Fission indicates the splitting of an atom's nucleus by bombardment with neutrons from an external source and commonly propagated by the neutrons so released. Fission of a nucleus releases about 200 million electron volts (eV) compared with about 5 eV for rupture of a chemical bond. In the case of fission of U-235 or Pu-239 (which have fissionable (unstable) nuclei), the following events can occur: (1) the nucleus disintegrates into the nuclei of two or several other elements, called fission products or fragments, all of which are radioactive and have high kinetic energy; (2) the disrupted original nucleus emits an average of (secondary) 2.5 neutrons (U-235) or 3 neutrons (Pu-239), which in turn may enter and split other nuclei of the fissionable material in a self-perpetuating chain reaction; and (3) the nucleus absorbing a neutron may also emit the energy equivalent of the mass defect of the nucleus which is usually about 200 MeV per nucleus, some of which is in the form of gamma rays. Fission is used for electric power generation and for making radioactive isotopes.

In *fusion*, the nuclei of light atoms, especially the hydrogen isotopes deuterium and tritium, unite or fuse to form helium. Fusion yields large amounts of energy without radioactive hazard. Uncontrolled fusion was achieved years ago in the hydrogen bomb, where the required very high initiating temperature was supplied by a fission reaction.

The number of protons (and also the number of negatively charged electrons in the shells) in an atom of an element is called the *atomic number* of the element. Each element has an atomic number based on the number of protons in the nucleus. This number indicates the location of the associated element in the Periodic Table. Atomic numbers of naturally occurring elements range from 1 for hydrogen, upward consecutively to 92 for uranium. As the atomic number increases, the *atomic weight* and *mass* increase. Human-made (or synthetic) elements above 92 have higher atomic numbers that likewise are derived from the number of protons in the element's atom. The atomic numbers can be arranged in the Periodic Table, which conforms to the natural periodic law. When the elements are so arranged and thus aligned in increasing order, they constitute a succession of periods. The corresponding parts of the various periods thus indicated are adjacent. Similar elements are thus shown together in groups in the Periodic Table.

An element having a particular atomic number can exist in two or more forms, each having a different mass. The difference in mass is caused by a different number of neutrons in the element's nucleus.

Each element of the same atomic number but having a different mass is called an *isotope*. For example, "regular" hydrogen has an atomic number of 1 and a mass of 1 (caused by one proton with one associated electron); it is one of three isotopes of hydrogen. Of the other two isotopes, each has an atomic number of 1, but one of these isotopes has in its nucleus both a proton and a neutron (and is sometimes called deuterium) and so has a mass of 2, while the other isotope has a nucleus composed of a proton and two neutrons (and is sometimes called tritium) and so has a mass of 3.

The atomic mass of an element as it naturally occurs is the average weight percentage of all of its natural isotopes (determined by extensive analyses of naturally occurring material). Among naturally occurring elements, twenty-one have no isotopes. Isotopes can be nonradioactive or radioactive (either as they occur in nature or from artificial neutron bombardment). For example, in the case of uranium, its atomic number is 92, its atomic weight is 238.029, and it has three natural isotopes: U-238 (99%), U-235 (0.7%), and U-234 (0.006%). Typically, because of their relatively low frequency of occurrence, the heavier isotopes occur rarely in the mass naturally making up an element.

Radioactivity occurs in natural or artificial nuclear transformation. The energy emitted is classified as *alpha*, *beta*, or *gamma* rays (or particles). Alpha particles consist of helium nuclei. Beta particles consist of either electrons (negatively charged) or *positrons* (which have the same mass and spin as electrons but are positively charged). Gamma radiation has extremely short wavelengths and high energy; it is commonly known in the form of X-rays. Radioactivity is not affected by physical state or chemical combination. Radioactivity can be measured by *half life*, which is the time for the activity to decrease to one-half of the original level, and by the *curie* (and its metric fractions, the milli- and microcurie), which is the rate of radioactive decay, defined as 3.73×10^{10} disintegrations per second. A common unit is millicuries per millimole.

Cross-section (or *capture cross-section*) is a measure of the probability that a nuclear reaction will occur. It is the apparent or effective area presented by a target nucleus (or particle) to an oncoming particle or other nuclear radiation, such as a neutron, a photon, or a gamma radiation.

A nuclear *chain reaction* is a reaction yielding energy or products that cause further reactions of the same class, and so such a reaction becomes self-sustaining. In, for example, the fission of a uranium atom by a neutron, a neutron enters a uranium nucleus. As a result, more neutrons are released from that nucleus, and at least one or more so released neutrons (sometimes termed secondary neutrons) enter other uranium nuclei, cause further fissions, and so forth.

Critical mass is the minimum mass of a fissionable material (for example, U-235 or Pu-239) that will permit an uncontrolled chain reaction to occur, as in an atomic bomb or a nuclear reactor. The critical mass of the fissionable uranium isotope U-235 is about 33 pounds; that for Pu-239, about 10 pounds. A nuclear explosion will not occur until a critical mass of fissionable material is attained, that is, the smallest amount capable of sustaining a chain reaction. Also, a nuclear reactor will not produce power until its assembly achieves a critical activity. Critical activity occurs when the neutrons entering a reactor are very slightly in excess of those lost to it, and at this point, measurable power is generated.

Prior to about 1938, concepts such as critical mass and chain reaction were unknown.

A *nuclear reactor* is an assembly of fissionable material (U-235 or Pu-239) designed to produce a sustained and controllable chain reaction. A nuclear reactor for the generation of power, ultimately usually electric power, is a *thermal reactor.* A nuclear reactor includes (1) enough fissionable material to maintain a chain reaction at a desired or necessary power level; for example, at least about 50 tons of uranium may be required; (2) a moderator to reduce the energy of fast neutrons for more efficient fission (commonly using slow or thermal neutrons, such as graphite, beryllium, heavy water, or even "light," natural water); and (3) a control system. A source of neutrons may be needed or used. Also, adequate shielding, remote control equipment, cooling, and appropriate instrumentation are needed. The rate of energy production from neutrons entering a reactor rises exponentially. Control rods composed of cadmium absorb neutrons readily and thereby permit a reactor to function at an exact, predetermined level of activity.

A *moderator* is a substance of relatively low atomic weight, commonly beryllium, carbon (graphite), deuterium oxide (heavy water), or light (ordinary) water, which has little neutron absorption. Neutrons lose speed when they collide with the nuclei of the moderator. Moderators are used since slow neutrons are more likely to produce fission. For example, a graphite-moderated reactor may contain 50 tons of uranium for 472 tons of graphite and be cooled with light water.

A *breeder* is a particularly efficient nuclear reactor in which plutonium is produced from U-238 from neutron flux. The "fuel" used is a mixture of nonfissionable U-238 and fissionable plutonium-239 sealed in long, thin, hexagonal metal tubes that are in turn sealed in "cans" (commonly aluminum). Breeding cannot be accomplished in (conventional) thermal reactors, which are fueled with uranium or plutonium, because such fuels become depleted and must be replaced. Owing to the manner of construction, a breeder reactor has about one hundred times more fissionable material available, and all of the energy potential in U-238 can be released. A breeder reactor uses fast neutrons, typically uses liquid sodium as the coolant (because it has no retarding effect on neutrons), and produces more fuel than it consumes. A breeder produces about 2.9 neutrons per fission; a water-moderated thermal reactor produces only about 2.4. It is the excess of neutrons that makes it possible for a fast breeder to produce more fuel than it consumes.

HISTORY

Motivation and Actions

Dating from the late 1930s, and extending into the 1940s and World War II, compelling but incomplete evidence accumulated that indicated to knowledgeable people in the free world that Nazi Germany or Japan could and, it was presumed, probably would develop a very powerful atomic bomb. If achieved, the results would have been disastrous to the free world. Some scientists managed to draw the matter to the attention of the highest levels of American government. After receipt of a (secret) British report, and information from qualified Americans that a large and expensive effort would be required, President Franklin Delano Roosevelt initiated on an urgent basis a federally financed and federally directed secret atomic bomb research and development effort that actually began in 1941, before the Japanese attack on Pearl Harbor on December 7, 1941. Thereafter, the project was greatly augmented and became the vast, enormously expensive Manhattan Project.

New technology, various breakthroughs, large industrial efforts, and many inventions, supported and achieved by the best available scientific and engineering talents of that time period, resulted in successful achievement, as demonstrated by a carefully planned explosion of a test atomic bomb in New Mexico on July 16, 1945. Though Germany had already surrendered on May 6, 1945, subsequently, other embodiments of the bomb were released over Hiroshima and Nagasaki, Japan, on August 6 and August 9, 1945, respectively, which resulted in more than 200,000 deaths and induced the Japanese to surrender on August 15, 1945.

The Manhattan Project was without precedent as a secret, very large, government-controlled and liberally financed research and development effort. It took place under wartime conditions and was stimulated by fears that enemy development efforts might succeed first. The size, cost, scope, intensity, urgency, and control of this project suggest that it could not have been undertaken or achieved in an equivalent time period by individuals or private industry. Also, the variety of scientific and technical knowledge needed and utilized to accomplish the complex results suggest that the project could not have been similarly accomplished at an earlier period in history.

During the project, many critical technical problems arose that had to be solved. Inventions had to be achieved, and developments had to occur, all in an environment of urgency, stress, and secrecy, before success could be achieved. Many gifted people made substantial contributions. Perhaps owing mainly to the governmental tendency to seek results and to use results regardless of origin or even ownership (such as patent rights), little regard seems to have been paid during the project to matters of origin and inventorship of particular creations, developments, and advances. Today, approaching 65 years after the success of the project, it is difficult, perhaps almost impossible, to identify clearly, from records, all the particular individuals and advances involved in the project. Objectively and rationally identifying who created the inventions utilized, and all who were responsible for the breakthroughs, developments, and results that occurred along the way to the success, may not be possible, but we can consider now what appear to be the main events and people.

In particular, a synopsis of the historical circumstances that comprise the creation, achievement, and impact of the first nuclear reactor helps one to grasp what happened and how the stage was set for the further developments, and thereby perhaps to enable further study and analysis.

Background: Atomic Energy History

The history of atomic energy is perceived to be an important precursor to the Manhattan Project and to the first nuclear reactor and accordingly is briefly reviewed. Then follows a brief account of the Manhattan Project, which encompassed that reactor and culminated in successful atomic bombs. Finally, as a matter of general interest, the development of atomic energy for peaceful electric power purposes is briefly reviewed. Those mainly curious about the first nuclear reactor can skip other parts of the history.

The field of nuclear physics can be thought of as having begun with the discovery of X-rays by Wilhelm Roentgen in 1895 at the University of Wurzburg in Bavaria. Within months, study of X-rays led to the discovery of radioactivity by Antoine H. Becquerel at the Museum of Natural History in Paris. Within a year, more than a thousand papers on X-rays had been published.

In 1897, at the Cavendish Laboratory at Cambridge University in England, J. J. Thomson identified the electron. Ernest Rutherford, Thomson's successor, had by 1900 identified and named alpha particles (helium nuclei), beta rays (energetic electrons), and gamma rays (electromagnetic waves). By 1911, Rutherford had identified the nucleus and noted its diminutive size relative to the size of the nucleus, taken with its surrounding electron(s) in an atom. By 1914, he had identified and named protons (hydrogen nuclei), and in 1919 he was the first to carry out a transmutation (formation of hydrogen and an oxygen isotope by alpha particle bombardment of nitrogen).

Madame Marie Curie and her husband, Pierre, laboriously but successfully isolated from uranium-bearing ore the naturally radioactive elements polonium (July 1898) and later radium (identified December 1898, isolated in 1902). Their daughter Irene and her husband, Frederic Joliot, known as the Joliot-Curies, in 1934 discovered that aluminum impacted by alpha particles produced a radioactive isotope of phosphorous not known in nature, the first artificial radioactive isotope (more than a thousand others have since been prepared).

Beginning in the 1920s, German chemist Otto Hahn and Austrian physicist Lise Meitner, working as a team at the Kaiser Wilhelm Institute in Berlin, Germany, advanced the field of atomic knowledge through discovery of isotopes. In 1932, James Chadwick, working in the Cavendish Laboratory, identified the neutron. The availability of the neutron in particular stimulated research in atomic physics because of the ease with which it facilitated nuclear reactions. Owing to its neutral character, the neutron enabled penetration of the positively charged atomic nucleus and the occurrence of changes in the nucleus that could be measured. Werner Heisenberg soon suggested that the nucleus was composed of protons and neutrons. Hideki Yukawa of Osaka University in 1935 is credited with initiating the theory of the strong force operating in the nucleus as the force holding the nucleus together, the other three fundamental forces now being identified as the weak nuclear force (first identified by Fermi in the early 1930s), gravity, and the electromagnetic force (which is involved in the transfer of protons).

In general, research efforts during the 1920s and early 1930s were successful only in achieving transmutation. During the 1920s, only alpha particles (i.e., helium nuclei) emitted by naturally occurring radioactive elements were available for bombarding nuclei, and significant usage of alpha particles was achieved. Devices for subatomic particle acceleration to provide projectiles for nuclear penetration were needed and developed in the early 1930s, such as Cockcroft and Walton's voltage multiplier (1932, under Rutherford), and Van de Graaff's electrostatic particle accelerator (1931) at the Massachusetts Institute of Technology (MIT). The cyclotron, first embodied in 1930 by Earnest Lawrence of the California Institute of Technology, was particularly useful. It accelerated charged particles using high-voltage alternating current.

During the 1930s, Hitler's antagonism toward Jews caused many nuclear and other scientists to leave Germany. Numbers of Jewish scientists also fled Hungary and Italy. Some went to England, but many came to the United States. Many subsequently became involved in the American Manhattan Project (see below). Also in this period, Hitler, with German military force, proceeded to take over various European countries. These significant events have been extensively considered elsewhere; space does not permit detailing them here, but their occurrence ultimately stimulated American efforts and the Manhattan Project.

In the 1930s, after Chadwick's identification, Enrico Fermi had become interested in the neutron and in its utility in irradiation. Nuclei capture neutrons of certain energies more frequently than neutrons of other energies. As Fermi found, while working in Rome, Italy, when he initiated nuclear reactions with neutrons, slow neutrons, having less energy than fast neutrons, could be more effective for nuclear reactions. He initiated nuclear reactions with neutrons after first passing the neutrons through water or paraffin, where some of the starting "fast neutrons" characteristically had some of their rather high associated energy absorbed through nuclear collisions. The resulting neutrons moved only at the normal speeds of molecules at room temperature. Such so-called thermal neutrons (slow neutrons) stayed in the vicinity of a nucleus a fraction of a second longer and were thus more easily absorbed than fast neutrons. Fermi tried to form a so-called trans-uranium (a transmuted element) from uranium by bombarding uranium with slow (thermal) neutrons.

In 1938, Otto Hahn and Fritz Strassmann, working at the Kaiser Wilhelm Institute in Berlin, reported carefully conducted experimental work that indicated that perhaps the uranium nucleus could be split by slow neutrons. Hahn himself, perhaps then the world's best radiochemist, was at first uncertain that fission had occurred. Lisa Meitner, Hahn's former associate, herself an excellent theoretical physicist of Jewish descent, had by then escaped from Germany to Sweden. After review of Hahn's data, she published the first, and confirming, report, that this work by Hahn and Strassmann had truly produced uranium fission. Further experimentation by various other physicists in the United States, England, and France very soon confirmed that Hahn and Strassmann had indeed fissioned uranium.

Fermi's Early Biography

Enrico Fermi was born September 29, 1901, in Rome, the son of Alberto Fermi and Ida de Gattis Fermi. Enrico attended a local grammar school and displayed an early aptitude for mathematics and physics, which was recognized and encouraged by his father (who was Chief Inspector of the Ministry of Communications) and his father's friends. In 1918, he won a fellowship at the Scuola Normale Superiore in Pisa, and later spent 4 years at the University of Pisa, where he gained a doctor's degree in physics in 1922. Awarded a scholarship in 1923 from the Italian government, he spent months in Göttingen, Germany, with Max Born, and in 1924 worked in Leyden, the Netherlands, with Paul Ehrenfest, having received a Rockefeller fellowship. Then for 2 years he served as a Lecturer in Mathematical Physics and Mechanics at the University of Florence. In 1926 he worked out what became known as Fermi-Dirac statistics, a system for mathematically analyzing certain subatomic particles that became known as fermions.

Appointed a full professor in Rome at age 24, he there formulated the theory of nuclear beta decay. In experiments, he found that slow neutrons are efficient agents of nuclear transformations. He received the 1938 Nobel Prize in physics for his work with thermal neutron nuclear bombardment, though subsequent work by Hahn in Berlin and by others showed that the portion of Fermi's neutron work with uranium actually mainly involved uranium fission.

Fascist Italy had in 1938 established antisemitic legislation. Even though Fermi himself was not Jewish, his wife was, and their continued life in Italy for them became uncertain and dangerous. When in 1938, Fermi traveled to Sweden with his wife and children to receive the Nobel Prize, he used the prize money to escape directly from Sweden with his wife and family by ship to the United States. There he joined the faculty of Columbia University.

Fermi's Involvement with Physicists in the United States

When Niels Bohr, the preeminent Danish nuclear physicist, visited the United States in early 1939, Fermi learned of the discovery of uranium fission, as shown by reported work of Germans Hahn and Strassmann and Meintner's confirmation. Fermi immediately concluded that a chain reaction with uranium was possible. The fissionability of uranium much stimulated development of nuclear theory and research.

Various questions promptly arose: Would uranium sustain a chain reaction? Was the rate of neutron generation sufficient in a mass of uranium to sustain a chain reaction? If so, how much uranium was needed? When uranium fissions from neutrons, what secondary neutrons does uranium produce? How many secondary neutrons must each uranium nuclear fission produce for a chain reaction to occur? Would a chain reaction, once started, continue and even destroy the whole world? How could a chain reaction be made self-sustaining, moderated, and controlled? What other elements might be fissionable? Some of these questions took years to answer. Some physicists promptly theorized that fission could be used to produce power (energy), such as electricity or a bomb of enormous destructive force, but no data were available for supporting or disproving such ideas.

From about 1938 through 1941, internationally, advances and experiments in nuclear physics occurred with seemingly increasing frequency. Physicists worldwide mainly continued to adhere to a policy of publishing results as soon as possible. Scientists in all countries could keep abreast of developments.

Published results suggested that among naturally occurring elements, only uranium fissioned by slow neutrons, though certain naturally mildly radioactive elements, such as natural thorium (Th-90, no stable isotopes) and natural uranium (which is actually composed of three natural radioactive isotopes, U-234 [0.006%], U-235 [0.7%], and U-238 [99%]) fissioned with fast neutrons. If the addition of neutrons to uranium nuclei resulted in the production of other (secondary) neutrons, then a uranium mass might support a self-sustaining (chain) reaction that was initiated possibly by even a single neutron.

Further research confirmed the Niels Bohr's insightful supposition in 1939 that the uranium isotope U-235, even though present in natural uranium only to about 0.7%, was responsible for the observed fission of uranium by slow neutrons. Although the remainder of natural uranium comprised mainly the slightly heavier uranium isotope U-238, U-235 had an inherent energetic advantage over U-238 after absorbing a neutron: U-235 accrued energy toward fission simply by virtue of its slight increase in mass through the acquiring of a neutron. Any neutron at all potentially would fission U-235, but the neutron cross-section in U-235 for fast neutrons was much less than for slow neutrons.

No one, though, had demonstrated the occurrence of a chain reaction. Though U-235 was clearly more desirable than U-238 as a fissionable material potentially useful in a chain reaction, separation of the U-235 isotope from natural U-238 was everywhere (among nuclear physicists) perceived to be a daunting problem. The problem of separating U-235 from U-238 had at that time never previously been considered or solved. Experts such as Bohr in 1939 even opined, based on then-known technology, that sufficient practical (commercial) quantities of U-235 could not be made. Some, like Fermi, hoped that natural uranium without isotope separation could somehow be used for fission and for chain reactions.

In England back in 1934, before uranium fission was known but after Chadwick's identification of the neutron, Leo Szilard, a physicist from Hungary with Jewish origins who had received his Ph.D. in Germany but who had fled from Germany to England, had an important conception. He conceived of the hypothetical idea of a nuclear chain reaction using neutrons to induce an atomic breakdown. He even filed a speculative British patent application on this idea, proposing the breakdown of beryllium to helium, but this in fact did not constitute a practical chain reaction. Szilard never achieved in Europe scientific accomplishments of a scope comparable to those of Fermi.

Szilard had moved to the United States in 1937, and the news about uranium fissioning was appreciated by him (apparently among others) to suggest that this element had potential for a bomb. For example, in 1939, young Ph.D. William Shockley (who later was an inventor of the transistor) while working in research at Bell Laboratories (first in New York City, later in New Jersey) was assigned (in the light of recently reported European discoveries) with a friend and associate, James Fisk, to consider fission as an energy source. Shockley thought that if uranium was placed in separate chunks, the neutrons could be slowed down and then be enabled to hit the U-235. In about two months, he and Fisk had designed an early nuclear reactor. Their written report was promptly sent to Washington, where the government immediately gave it a secret classification and subsequently opposed attempts to patent it. After the war, Manhattan Project scientists learned about this reactor, but they had meanwhile invented independently the same technology.

In 1937, Szilard became fearful about German usage of the published information on these early fission experiments, although the newspapers and their reporters were without any knowledge of what was potentially indicated. Szilard asked that his pending British application continue to be kept secret, and also he got American physicists, including Fermi, to avoid providing any useful publications. He also got the U.S. government involved.

Advent of U.S. Government Involvement

During the late 1930s, in the United States, a small group of informally associated nuclear physicists, both European and American, perceived not only the potential military significance of recent published discoveries regarding uranium fission, but also the fearful potential consequences if the potential weaponry was produced and came under the control of Germany or Japan. They realized that the federal government needed to be informed, made aware of the situation and the potential, and involved.

A first contact attempt between nuclear physicists and the U.S. government is reported to have taken place in February or March 1939. Another attempt appears to have occurred beginning in July 1939.

As the summer advanced, Szilard in particular thought the matter urgent. He managed to get Albert Einstein to sign a letter that he (Szilard) had mostly written and to get an intermediate to deliver the letter with attachments to President Roosevelt (FDR). However, it was not until October 11, 1939, that, through the intermediate, the matter was brought to President Roosevelt's attention. FDR immediately set up a Uranium Committee consisting of the director of the Bureau of Standards (Lyman Briggs), an Army representative, and a Navy representative. A subsequent meeting of the committee with certain of the involved nuclear physicists resulted in a report that may have been read by Roosevelt, but that report and further committee work remained dormant well into 1940.

Demonstration of a Chain Reaction

In this period, Leo Szilard and Enrico Fermi were both at Columbia University, each working to answer the question about basing a chain reaction on uranium.

However, in hindsight, it appears that Szilard's speculations about atomic physics were not equal to those of Fermi, who was gifted with capacities in both experimental and theoretical atomic physics. The strong-mindedness of Szilard conflicted with the easy self-assuredness of Fermi.

Fermi, in 1940, chose to pursue and evaluate with neutrons if a chain reaction in natural uranium, rather than in separated isotopes, was possible.

For one thing, he was not yet convinced that U-235 needed to be separated. With Herbert Anderson and Leo Szilard, Fermi, in an experiment at Columbia, determined that when natural uranium was exposed to slow neutrons, a slight excess of about 1.2 neutrons was produced per thermal (slow) neutron in natural uranium. Thus, in about March 1940, experimental confirmation was obtained showing that uranium fission produced more than one secondary neutron per fissioned nucleus, so theoretically a chain reaction with uranium was possible.

Fermi of course envisioned a uranium "pile" (Fermi's word for a nuclear reactor; this word was also useful for reasons of secrecy), but a pile needed a moderator to increase the number of available secondary neutrons so that a chain reaction could occur. In a pile, uranium material would be carefully and systematically associated with a moderator. Although water was a possibility as a moderator, the hydrogen present not only slowed neutrons but also absorbed some slow neutrons, thereby actually reducing the number of neutrons available for fission.

Without a moderator, a chain reaction in a critical mass could occur substantially immediately upon the addition of a neutron, producing a bomb. With a moderator, a chain reaction could occur uniformly over time and produce power.

Fermi and Szilard each thought of using carbon as a moderator. The question arose, would a pile of graphite and uranium produce sufficient neutrons to permit a chain reaction to occur? If a chain reaction of uranium would work in graphite (carbon), then a bomb was possible. With a $6,000 grant from the U.S. government, Fermi and Anderson measured the neutron cross-section of carbon and found it to be usefully small: 3×10^{-27} cm^2.

Particularly in response to pleas from Szilard, Fermi agreed not to publish this important measurement, which the Germans had found earlier to be almost twice that found by Fermi. Apparently, their high neutron cross-section carbon measurement led the Germans to cease pursing carbon as a cheap, effective moderator and to seek to use only heavy water (deuterium oxide) for a moderator. Fermi's agreement not to publish apparently marks the start of a voluntary censorship among American physicists involving the publication of some (but not all) nuclear research.

Through 1941, Fermi and his coworkers at Columbia continued to develop technology for achieving a workable pile based on natural uranium. Since Szilard and Fermi could not work together without friction, Szilard directed his efforts toward obtaining highly purified materials, such as graphite that was not contaminated with trace impurities such as boron. Boron had a very large capacity to absorb neutrons, so in effect even small amounts could poison a carbon moderator's desired effect in a pile. In November 1940, $40,000 was received from the National Defense Research Council (NDRC) for physical constant measurements. Before an operable pile could be constructed, the critical mass had to be determined for a volume of uranium and moderator (graphite) sufficient to sustain a minimum neutron multiplication and to offset the neutrons inherently lost from outer surfaces. To make this determination, Fermi advanced through a series of experimental piles that yielded information needed regarding quantities, arrangements, and methods of control.

Beginning with his first experimental and test pile, Fermi defined a single basic magnitude, *k*, for assessing a chain reaction. *k* could be measured accurately and was the average number of secondary neutrons produced by one original neutron in a pile lattice of infinite size. For a pile (nuclear reactor) to achieve a self-sustaining chain reaction, *k* had to equal at least 1. By using uranium (which gradually became available in increasing purity), and also purified uranium oxide and carbon (as purified graphite), which gradually became available in increasing purity from American suppliers, and by improving stacking configurations in a pile of components relative to one another, *k* was gradually increased in experimental test piles, beginning at about 0.87. By late 1941, Fermi and his team thought they were in a position to build a complete, functioning but controllable (i.e., self-sustaining with a critical mass) uranium pile.

U-235 Alternative

Various experimental work by others was undertaken to identify a process for separating the inherently very small amount of U-235 from U-238. Otto Frisch, who had moved to Liverpool and worked for the British, discovered there in 1941 that thermal diffusion using gaseous uranium hexafluoride was evidently inoperable. A simple or direct process for separating U-235 did not appear to be available. However, British research work seeking a workable gaseous barrier diffusion process proceeded.

An alternative or substitute for U-235 was a desirable possibility to avoid undertaking the separating of U-235 from U-238. One promising idea was to produce artificial elements heavier than 92 (uranium, U-238). Such elements could theoretically be made by bombarding uranium with neutrons. One or more of these synthesized elements might be fissionable by slow neutrons. Thus, indirectly, U-238 could be put to work by transmuting it into possibly fissionable elements.

Element 93 (which became known as neptunium), produced by Americans in 1940 in microscopic quantities, did not appear to have usable fissionable characteristics.

The isotope of element 94 with a mass of 239, in late 1940, was thought by Fermi and Emilio Segre (who was visiting Columbia from the University of California, Berkeley), and evidently others, to be possibly a useful, slow neutron fissioner if this isotope could be obtained (produced). If this isotope were produced using U-238 and isolated and had such properties, then this material could substitute for U-235. Furthermore, this element 94 could theoretically be produced in a nuclear reactor fueled with ordinary uranium, but such a reactor did not then exist. Chemical separation could theoretically be used to isolate element 94 (once it was produced). Such an element would appear to eliminate or avoid the difficult problem of separating U-235 from U-238. These two physicists then met with Lawrence (who happened to be then in New York from Berkeley) and Dean Pegram at Columbia University, and they planned to use a cyclotron radiation of natural uranium to produce a sufficient quantity of element 94 for testing and evaluation.

However, Glenn T. Seaborg at Berkeley was already working with collaborators to identify and isolate element 94. Upon the return of Lawrence and Segre to Berkeley, Segre and Seaborg determined by experiments that they could make element 94 by cyclotron bombardment. When chemical separations and analysis showed on about February 24 that element 94 (later named plutonium) had been produced, a 1.2-kilogram sample of uranyl nitrate hexahydrate (UNH) was prepared, loaded into glass tubes, and subjected to extensive bombardment in the 60-inch cyclotron available at Berkeley. From the resulting irradiated UNH, less than a millionth of a gram of element 94 as isotope 239 was obtained after separation and decay of radiation. On March 28, 1941, tests disclosed that this isotope of element 94, which was relatively only mildly radioactive, fissioned with slow neutrons and so was a candidate as a substitute for U-235.

Meanwhile, the NDRC under Vannevar Bush had been authorized by FDR, and it had begun in June 1939. The NDRC had absorbed the previously established Uranium Committee under Briggs. Bush had immediately enlisted James Bryant Conant, the president of Harvard, to be chairman of the chemistry and explosives division of the NDRC. Briggs, who had demonstrated a proclivity toward inactivity, was to report to Conant.

When Lawrence in early 1940 learned that Seaborg and Segre had isolated element 94 (Pu-239) and had measured its neutron emission and estimated its cross-section at 1.7 times that of U-235, he became quite excited and telephoned Arthur Compton at the University of Chicago. Compton immediately relayed this news to Bush, pointing out that the consequence was that the fission potential was thus increased by over 100 times and urging Bush to push the uranium/graphite work at Columbia.

Evidently by June 1940, Briggs had received a copy of the April minutes of the British MAUD technical subcommittee regarding the feasibility of an atomic bomb. The MAUD subcommittee was set up and chaired by George Thomson; neither the initials nor the name meant anything. The designation was a cover. Though this report confirmed the feasibility of a fast neutron bomb, Briggs remained only interested in a (possible) slow neutron chain reaction for power production. A second National Academy of Sciences (NAS) report at this time was similar to Briggs's position.

The U.S. government at this time, operating in effect through both Bush and Conant, was very close to dropping fission studies from the war program, notwithstanding Lawrence's proposal to use plutonium to make a bomb and Compton's support. What saved the U.S. fission studies then was the news that a group of physicists in England in a so-called secret MAUD report had concluded that the construction of a bomb made out of U-235 was entirely feasible. This MAUD report had been approved on July 15 in Britain, but was not officially transmitted to the U.S. government until October, although Bush had been given a draft of the report in the interim. Concurrently in that summer, the fact that the German war machine had opened the eastern front stimulated a favorable decision regarding the MAUD report by Bush and Conant. It then became clear to both Bush and Conant that a push was needed.

Bush asked the NAS for an opinion involving the British position because he felt he was not an atomic scientist. In April, an NAS committee under Arthur H. Compton was formed, and by May this committee rendered a report favorable to the British position.

Recognizing that to pursue engineering development with government funding, a new umbrella agency was needed beyond just the research capacity of the NDRC, in May 1941, Bush created and FDR approved the new Office of Scientific Research and Development (OSRD), which he would take over as director, with the NDRC under Conant as chairman remaining under the OSRD.

Conant visited England in the winter of 1941 primarily for the purpose of opening a liaison office between the NDRC and the British government. During that visit, Conant, apparently through Frederick Lindemann (Lord Cherwell, scientific advisor to Churchill), learned of the definite possibility of constructing a bomb of enormous power: the atomic bomb. Lindemann described to Conant how two separate portions of U-235, when brought together suddenly, would result in a mass that would spontaneously undergo a self-sustaining reaction—that is, a violent explosion. Conant, however, reportedly thought that if Bush wished to be informed about such British ideas, then the matter should pass through channels involving Briggs. Back in the United States, at a subsequent meeting in March 1941 at MIT, Lawrence, the Californian inventor of the cyclotron, vigorously presented a case for urgently pursuing research and development of such a bomb. It appeared that Lawrence had been informed by the British and that the British were convinced of the feasibility of a bomb based on U-235.

During July, Bush conferred with Vice President Henry A. Wallace about spending large sums of government money on the uranium program, but Bush apparently chose to wait until after the MAUD report was officially received. In August, the British scientist Mark Oliphant visited the United States and helped promote a U.S. program by visiting and conferring with various scientists working in the United States. (Oliphant had, in 1934, done work on hydrogen isotope interactions that led to work on hydrogen fusion and later to the development of the hydrogen fusion bomb.)

On October 9, Bush carried the MAUD report to FDR and told the president and vice president of the British conclusions (and the U.S. confirmations) that a bomb core of perhaps 25 pounds might explode with a force of about 1,800 tons of TNT. That began governmental support for the U.S. atomic energy program, which was to be controlled by a small policy committee composed of FDR, Wallace, Secretary of War Stimson, Army Chief of Staff George C. Marshall, Bush, and Conant. FDR saw the long-term potential and reserved to himself policy regarding nuclear weapons. Bush in a memorandum to Conant of the same date wrote that "after-war control" was discussed "at some length" with FDR. The October 9 decision related not to the building of an atomic bomb but to the exploration of whether of not an atomic bomb could be built. This decision was decided by FDR without consulting Congress or courts, apparently as a sort of military decision on the basis that he was the Commander-in-Chief.

In effect, under this decision, program scientists were not given a voice in deciding military and political uses of the weapons to be built. Though participation of the necessary qualified physicists and scientists was then undertaken, and though patriotism was a factor in motivating the participation by physicists and scientists in the subsequent program, fear of a possible German triumph was probably a stronger motive for their participation.

Bush and Conant asked Compton for a third NAS review, which seems to have resulted in a theoretical calculation of the critical mass of U-235 and estimates of how the inherently small amounts of U-235 could be separated from U-238. From information acquired from experts, Compton called for a meeting of his committee on October 21, and the report was finalized before November 1. This third report verified the British findings again and functioned to commit the American physicists to the program. The gaseous diffusion and centrifuge research programs for isotope separation were noted to be approaching a level where practical testing might occur.

When the Japanese bombed Pearl Harbor on Sunday, December 7, 1941, a weekend meeting begun on December 6 in Washington was under way by Bush with Conant and Compton dealing with the research program. The surprise attack on Pearl Harbor, which resulted in the entry of the United States into the war against Japan and also Germany and Italy, accelerated the already existing U.S. atomic bomb development. In effect, the original exploration decision then became a full program of development.

The Army Involvement

The U.S. Army became involved with the nuclear development in about June 1942 but at first was ineffective. In mid-September 1942, General Leslie Groves became the Army's project leader on what then was called the

Manhattan Project, and he immediately proceeded to build a reputation for rapidly solving problems and moving all work forward. During a visit to the Chicago operations on October 5, Groves said, "There is no objection to a wrong decision with quick results. If there is a choice between two methods, one of which is good and the other looks promising, then build both." This position seems to represent the manner in which the project was pursued.

FIRST NUCLEAR REACTOR

Compton was placed by Conant and Bush in charge of the nuclear project. Fermi's design and experimental work to demonstrate a chain reaction with a pile was accelerated, and Fermi was placed in charge of this pile project. Compton decided to move the pile work from Columbia in New York City to the University of Chicago, a central location, where experimental pile development work continued under Fermi beginning about April 1942.

Fermi began planning the building of a full-scale pile in Chicago in May 1942 after one of the experimental test piles his team had built under the west stands of Stagg Field at the University of Chicago had achieved a k value that, at infinity, would be 0.995. After extensive preliminary experiments, measurements, and developmental work, the actual pile would be constructed of layers of blocks of purified graphite, lumps of purified uranium and uranium oxide, and cadmium rods built with many and careful calculations and associated with good instrumentation. Actual construction began on November 16, 1942.

However, construction workers for the originally projected building site of pile installation (at a remote location relative to the University of Chicago) had gone on strike, so building's projected completion date of October 20 could not be met. Fermi proposed an alternative site: the doubles squash court under Stagg Field, the location where his group had built previously a series of experimental piles.

Compton then had to make a decision as to whether or not the pile should be, and safely could be, built under the Stagg Field stands. His answer depended upon how good the controls were. He consulted Fermi, who considered that the controls were sufficient. Compton, relying also on the discovery of the Carnegie Institution in 1939 that a small amount of the neutrons associated with fission took a few seconds after emission to emit, decided that since the pile would operate at a k slightly above 1, the controls had time to respond so a chain reaction could be controlled. Compton authorized Fermi to begin construction of the pile under the west stands. The pile was designated "Chicago Pile Number 1" (CP-1). Compton, though, decided not to inform the University's president, Robert Maynard Hutchins, deciding that Hutchins was a lawyer and not able to judge a matter of nuclear physics.

The original design for CP-1, as worked out by and under Fermi, called for construction of a seventy-six-layered sphere, but the quality (purity) of the supplied components, graphite, uranium, and uranium oxide, kept improving even as construction proceeded. This could have been a formidable engineering problem, but not to Fermi, who made careful calculations and adjusted design throughout construction. The final reactor was a flattened ellipsoid of fifty-six or fifty-seven layers. The graphite was in the form of smoothed bars or blocks 4¼ by 4¼ by 16½ inches, some of which were drilled and cut to receive uranium or uranium oxide pressed bodies and a few of which were slotted to define control rod channels. The uranium and uranium oxide bodies were stacked over or alongside the graphite blocks in a regular pattern. There were no detailed plans for the pile or its supporting wood framing. The control rods were cadmium sheets mounted over wood strips.

As the pile developed and grew, Fermi frequently took and plotted the measured raw data on pile neutrons being produced and calculated a countdown. As the pile approached its slow-neutron critical mass, the neutrons generated by spontaneous fission in the pile multiplied through increasing generations before being absorbed. Winter weather arrived, and the unheated stands became very cold. The laborers included young physicists and high school dropouts awaiting draft notices. As the supplied graphite, uranium, and uranium increasingly improved in quality, Fermi continuously adjusted the plans. He found that the countdown would approach 0 and $k = 1.0$ when layers fifty-six and fifty-seven were reached.

The pile was completed in 17 days and was successfully demonstrated on December 2, 1942. It included 771,000 pounds of graphite, 80,590 pounds of uranium oxide, and 12,400 pounds of (purified natural) uranium. Total cost was given as somewhat over $1,000,000. Fermi's last flux measurement fell exactly on the line he had previously extrapolated from his graph of his countdown numbers.

On December 2, 1942, the cadmium rods were gradually and systematically removed under Fermi's direction until finally, at 3:45 p.m., the chain reaction became self-sustaining and the first nuclear reactor in history was operating, albeit unshielded and not cooled. Although Fermi allowed the pile to operate for only about 4.5 minutes, and it produced only about one-half of a watt, the pile represented the pinnacle of years of experiment, discovery, and development. The entire project was led and supervised by Fermi, to whom the success of the project is attributed, but many competent physicists, chemists, and other technologists were involved.

The pile proved the reality of the chain reaction and demonstrated the first nuclear reactor. The immediate result was that the enormous effort at implementing a bomb proceeded. The successful pile demonstrated the feasibility of uranium for the atomic bomb and for power production. Though unshielded and uncooled, the reactor was also the forerunner of all subsequent reactors used for power generation and breeding. The now-proven chain reaction in effect permeated the huge work of bomb development, as described below.

BUILDING THE BOMB

Groves Chooses Oppenheimer

From his beginning on the project, Groves seems to have surmised or assumed that a chain reaction would occur and that an atomic bomb would be produced. Before the actual construction of CP-1 began, in mid-October 1942, Groves, after receiving reluctant approval from the Military Policy Committee, chose Robert Oppenheimer, despite communist political activity in his background, to be laboratory director of a new laboratory aimed at bomb design, development, construction, assembly, delivery, and firing.

During the summer of 1942, Robert Oppenheimer headed at Berkeley a small group of theoretical physicists whose purpose was to provide theoretical and calculated information for actual design of an atomic bomb. Since theory regarding the proposed fission bomb seemed to be well developed, the group turned its attention to a fusion bomb. The fear that a runaway explosion (one that would include igniting the atmosphere) could occur was shown to be unfounded; such an explosion was an impossibility. The large calculated TNT equivalent for the H-bomb (fusion bomb) apparently actually stimulated fission bomb development.

The new laboratory was sited and built at Los Alamos, New Mexico (known as the Hill to those involved). Later, during the project development, Oppenheimer became administrative leader at Los Alamos. Construction of Los Alamos under the Army proceeded to the point where in March and April 1943 staff began to arrive, live, and work at the site. About a hundred scientists were initially hired, but by the end of the war in Europe and Asia, about 10,000 people were reportedly at the site. The Los Alamos population reportedly doubled about every nine months till the end of the war.

In July 1944, the Fermis became American citizens, and in September 1944, they moved to Los Alamos. Enrico directed the new operation F (for Fermi) Division that was intended to take advantage of his theoretical and experimental versatility. Edward Teller's group was under him.

By those involved, the bomb was sometimes called "the gadget." The experimental and developmental work required to be done at Los Alamos was large, complex, and varied. It included, for example, such matters as core design, core tamper assembly design, electronic controls, initiators, material tests, exploration of the physical properties of uranium and plutonium, the mechanical-type working of solid uranium and plutonium, including casting, shaping and machining, investigation of control of core expansion, achieving a satisfactory three-dimensional implosion to achieve assembly in a delivered bomb of an efficient mass (which turned out to be more than a critical mass) rapidly, designing a cannon for incorporation into a bomb core for firing a plug into the core to complete assembly of an efficient mass, bomb engineering and development, and so forth. Neutron characteristics of various materials had to be measured for purposes of designing, assembling, and firing a bomb. Bomb efficiency was soon recognized as a problem because, even with a good tamper and an explosive core, the pressures generated in bomb ignition could quickly stop and prevent a chain reaction from occurring or continuing to occur. Premature detonation would reduce efficiency and the force generated upon ignition. Various experimental and testing arrangements had to be conceived, developed, produced, and then actually tested—all usually without actual fissionable material, which was at first unavailable, and the arrangements then had to be evaluated.

Fissionable Isotope Separation

At first, little or no fissionable material was available for the work at Los Alamos. While the work at Los Alamos proceeded, work to separate U-235 from U-238, and to produce and separate Pu-239, occurred mainly elsewhere.

Early in his administration, Groves acquired 59,000 acres along the Clinch River in eastern Tennessee, on which a new town, Oak Ridge, was built, along with what became a vast factory complex that was sometimes called the Clinton Engineer Works. Initially it was hoped that an electromagnetic isotope separation plant and a gaseous diffusion plant would be built at the Oak Ridge site.

Electromagnetic Separation

The basic idea for U-235 electromagnetic separation is commonly associated with Lawrence. In operation, a vaporized and ionized uranium compound is subjected to a magnetic field that causes the vapor to move in a semicircular path. The vapor moving along the path is separated into two beams, with the heavier U-238 moving in a larger curve than the lighter U-235. At a collecting site, the U-235 atoms are discharged, collected, and deposited. A single unit of the apparatus was derived from the 1918 mass spectrograph invention of Francis Aston at the Cavendish Laboratory and was adapted by Lawrence from his cyclotron. A unit of the apparatus was called the calutron. As constructed at Oak Ridge, a calutron had a racetrack-like configuration in which a plurality of semicircular mass-spectrograph tanks were separated from one another by an electromagnet with silver windings.

The calutron was inherently inefficient, particularly because of the small mass difference between U-235 and U-238. The 1942 summer group of theoretical physicists headed by Oppenheimer had estimated that an efficient bomb needed 30 kilograms (66 pounds) of U-235, more than just a critical mass. Lawrence, in the fall of 1942, estimated that about 2,000 calutrons could be needed to achieve enough enriched material to make one bomb core of U-235 every three hundred days.

Even with the relatively efficient calutron racetrack design developed by the Berkeley people, designated Alpha, only about 5 grams per day of enriched uranium would be produced per single calutron racetrack. However, time constraints caused Groves to authorize a contractor to start construction of a racetrack building in early 1943, before the Alpha design was completed. In March he authorized start of further construction on a Beta-stage facility that comprised half-size calutrons. In August, an experimental Alpha unit was successfully operated, and Lawrence urged Groves to double the original Alpha plant. Since a new estimate from Los Alamos had raised the amount of U-235 needed for an efficient bomb to 49 kilograms, Groves agreed.

However, in the facility constructed at Oak Ridge, in early operation, shorts occurred in the silver windings for the electromagnets. The operation was shut down, and investigation revealed faulty design that could only be overcome by remanufacture of forty-eight magnets. By the end of 1943, virtually no U-235 had been produced.

Gaseous Diffusion

A gaseous diffusion separation process had been demonstrated in November 1941 by John Dunning and Eugene Booth at Columbia University. Authorization for building a full-scale plant was given by Groves. The American system, in contrast to the separately developed British gaseous diffusion system, operated at high pressures with nearly conventional pumps and an interconnected, continuous cascade of about four thousand stages. The American system required a barrier with smaller pores than the British. Both systems were based on the idea that if a uranium gas (uranium hexafluoride) was pumped past (i.e., through) a porous barrier, the lighter U-235 molecules would pass through more readily than the heavier U-238 molecules. Owing to the slight difference in mass between the U-235 gas and the U-238 gas, a single stage of diffusion could inherently achieve only a small separation.

Under time pressures, Groves went ahead with construction of a gaseous diffusion plant at Oak Ridge, but a practical barrier material needed for plant operation had not yet been achieved. A promising barrier material of suitably fine-pored compressed nickel powder that was achieved in early 1943 led Groves to order a contractor to build a plant to fabricate this barrier. However, another barrier material, allegedly further improved, was proposed in the fall of 1943. Groves, after deliberation, in January 1944 decided in favor of the new barrier material, even though a considerable delay resulted while the improved barrier was developed and manufactured.

Centrifuging

Separation of enriched U-235 from uranium by centrifuge apparatus underwent research and development at the University of Virginia, but this separation process was not considered to have reached the level of practicality needed for a full-scale plant. This method was subsequently (after the war) found to be economically superior to the gaseous diffusion process.

Pu-239

Groves thought producing plutonium in Tennessee could be risky because of the possibility that an accident might release radioactivity that would be damaging to nearby populations. A site in the west along the Columbia River seemed better and safer. The large site of 500,000 acres was acquired and came to be known as the Hanford Engineer Works.

At the site, Fermi's pile technology was adapted and used. What would become known as breeder reactors were built using purified graphite and purified uranium metal. In such a reactor, neutron-irradiated uranium that was thereby converted into plutonium was produced in inherently small amounts. Thereafter, the plutonium was separated and purified chemically.

Since the heat generated was substantial, a coolant for the breeder reactor was required. Eugene Wigner made calculations that showed that water could be used as the coolant for the breeder reactors instead of helium (as originally proposed). The k was higher than that reached in Fermi's University of Chicago pile. Wigner's group placed a reactor in a large graphite cylinder through which transversely extended more than a thousand aluminum tubes. Du Pont became the contractor in charge of building and operating the Hanford Works and made improvements in reactor structure and operation.

Such a reactor was large. The reactor's tubes were charged with uranium slugs that were individually housed in aluminum cans. When a chain reaction was in process, the uranium produced so much heat that 75,000 gallons of water per minute circulating through aluminum tubes were needed for cooling. Neutrons from fission transmuted uranium to plutonium at the rate of about 250 parts per million of U-238 to 1 part of plutonium. After about one hundred days, a little more than 0.01% of the uranium metal that had been charged had been transformed into plutonium. The resulting canned slugs were pushed out the back side of the aluminum tubes, and fresh slugs displaced them. The discharged slugs, quite radioactive, fell into a deep pool of fresh water, in which the short-lived fission product radioactivity was confined for about 60 days, after which the slugs were separated out and subjected to chemical separation to retrieve the plutonium using separation technology generally attributed to Seaborg. Because of the level of radioactivity even in the separated and water-aged slugs, automated chemical separation was required. The slugs were dissolved and advanced progressively through successive separation cells, all located in a large, elongated concrete building. Finally, by the middle of December 1944, the various problems had been overcome and plutonium production in quantity was underway at the huge Hanford works.

Liquid Thermal Diffusion

In conducting research at the Naval Research Laboratory, beginning in 1941, to develop a nuclear reactor small enough and powerful enough to provide motive power for a submarine, physicist Phillip Abelson adapted a previously German-invented thermal diffusion process for isotope enrichment. Abelson's aim was to increase the amount of U-235 relative to U-238 in natural uranium. This process utilized the tendency of lighter isotopes, such as U-235, to diffuse toward a hotter region, while heavier isotopes, such as U-238, diffused toward a colder region. Abelson used 36-foot-tall columns, with each column containing three concentric pipes. The innermost pipe, with a diameter of 1.25 inches, conveyed high-pressure steam at about 400°F. Surrounding that pipe were a nickel pipe and a copper pipe. Uranium hexafluoride ("hex") flowed as a heated gas between these two pipes. Surrounding both pipes was a 4-inch galvanized iron pipe that carried water at about 130°F, which was just above the hex melting point but which cooled the hex. Abelson reported in early January 1943 that a single thermal diffusion column could enrich the percentage of U-235 from 0.7% to about 1% or more. With several thousand columns connected in series, he thought U-235 of 90% purity could be produced at a rate of about 1 kilogram per day; however, this estimate was found to be optimistic.

Abelson proposed to build a plant of three hundred 48-foot columns operating in parallel. The Navy authorized plant construction, which began in January 1944, and the first one hundred columns were operating as a unit in July 1944. The only moving parts were the pumps that circulated the water.

This development led the Manhattan Project leaders to regard the various enrichment and separation processes as separate projects that could be harnessed together instead of being competitors. As the gaseous diffusion process and the electromagnetic separation process gradually became operative, the U-235 separation and enrichment processes became effective in combination to produce quantities of U-235 useful for bomb purposes.

Developments at Los Alamos

Early in Los Alamos history, two basically different bomb designs were pursued. One, sometimes called the "thin man," incorporated a gun for explosively driving a pellet (or plug) into a receiving socket in a preformed mass to achieve a critical or effective uranium mass. The second, sometimes called the "fat man," utilized a spherically arranged group of explosives, which upon explosion compressed bomb components together to achieve a critical or effective mass. Each design involved formidable technical challenges to achieve a workable, practical bomb, but the problem of achieving a spherical explosive arrangement for implosion was perhaps the greatest. The assembly of each bomb, which was to be completed in an airborne delivering bomber, was greatly different for each. The assembly of the fat man was complex and involved.

A tiny but critical part of a bomb was its initiator, the source of neutrons that caused the start of the chain reaction when a critical mass was achieved. What was needed was a neutron source that would actuate and release neutrons at precisely the right instant. For a uranium bomb in which a gun fired a pellet that entered a cavity in a shaped mass and completed the formation of a critical mass, the design of the initiator seemed relatively simple: For example, alpha particles from a source such as polonium (which emits an abundance of alpha particles) might be located with the pellet, and a neutron generating source, such as beryllium, that produced neutrons when struck with alpha particles might be located with a shaped mass that received the fired pellet.

For a bomb actuated by implosion, though, separation and mixing of initiator components in a convenient arrangement that functioned to produce neutrons at the precise instant and perhaps the location required for neutron production was not possible.

The mixing of initiator components at the right instant in an initiator design was evidently the subject of many creative efforts during the winter of 1944–1945. It is known that an amount of polonium Po210 equivalent to 32 grams of radium, if thoroughly mixed with beryllium, would produce about 95 million neutrons per second. However, in an imploding plutonium bomb, not more than nine or ten neutrons would be useful to start the chain reaction in the brief 1/10,000,000th of a second when they would be desirable. The initiator designs have never been declassified. Regarding the implosion bomb, Richard Rhodes speculates in his work, *The Making of the Atomic Bomb*, that irregularities machined into the outer surface of the beryllium-induced turbulence in the imploding shock wave "did the job." After a series of experiments, a committee chaired by Hans Bethe at Los Alamos selected a design on May 1, 1945.

In the summer of 1944, Seaborg's warning a year previously that the Pu-240 isotope might form along with the desirable Pu-239 isotope generated a problem. The Pu-240 fissioned at a higher spontaneous rate. This meant that a gun could not be used to assemble a critical mass because, even at approach speeds of 3,000 feet per second, the plutonium bullet and the target would melt down and fizzle before the two components had time to join. The result was that only the implosion assembly method appeared to be a practical assembly procedure. This result produced an all-out effort to achieve a successful implosion based on the technology of explosive lenses attributed to John von Neumann and George Kistiakowsky at Los Alamos.

Work on a suitable explosive lens implosion device proceeded and was successful because of new technology, particularly diagnostics. The casting and machining of more than a hundred different pieces of explosives were necessary for making the spherical shell of a single implosion device. A single such device weighed about 5,000 pounds. After successful testing, Oppenheimer froze lens design in March 1945.

The possibility of a thermonuclear bomb of perhaps a million tons TNT equivalent was perceived by 1944, but such a bomb was not developed until after the war ended. Work meanwhile continued toward making weapons based on U-235 and Pu-240 with the intent of ending the war. The surrender of Germany on May 8, 1945, did not alter the Manhattan Project.

In March 1944, planning for a full-scale test of an implosion-initiated bomb began. Many things could not be definitely known from theories, guesswork, and limited experiments. If the bomb was a dud, Americans should know it first. Code named "Trinity" by Oppenheimer, the test was successfully carried out on July 16, 1945, at a site in New Mexico—the first human-made nuclear explosion. The explosion was substantially greater than expected.

Delivery

Delivery of a bomb was a separate problem. Developing and adapting the then new B-29, the first intercontinental bomber, for delivering the bomb began in early March 1943. The bomb support and release mechanism had to be worked out from test drops. Apparently, about eighteen B-29s were specially modified and equipped for bomb delivery. Paul Tibbets, an experienced and exceptional bomber pilot, was made group commander of over 225 officers and 1,542 enlisted men. Because of the expected power of a bomb, his crews practiced bombing runs at over 30,000 feet and practiced executing a sharp diving turn immediately after bomb release to increase airspeed and move 10 miles away from the site of a delayed explosion.

A "thin man" bomb (based on U-235) was dropped and exploded over Hiroshima on August 6, 1945, and a "fat man" bomb (based on Pu-239) was dropped and exploded over Nagasaki on August 9, 1945. The Japanese surrendered on August 14, 1945.

The Fermi Paradox

In 1950, with colleagues over a luncheon table, Enrico Fermi succinctly asked "Where are they?" He meant, if we assume a high probability of extraterrestrial civilizations, then why have we not had contact with them? This has been termed the Fermi paradox and has generated a great deal of comment and speculation, which continue today.

(*continued*)

In 1960, Frank Drake developed an equation in a systematized effort to evaluate the possibility of alien life. Though the equation includes at least four terms whose values are completely unknown, Drake assumed values for all terms and concluded that there were likely to be ten civilizations in our galaxy with whom we might hope to communicate. Some have assumed that technical civilizations have a short lifetime because they destroy themselves. Purporting to use "best current estimates" for values in Drake equation terms, this number of civilizations has recently been placed at 2.3. However, some people, such as science fiction writer Michael Crichton, regard the Drake equation as being "literally meaningless."

The SETI (Search for Extraterrestrial Intelligence) Institute endeavors to detect extraterrestrial life by detecting transmissions from outer space (presumably other planets). The spectral frequency and methodology of such transmissions are unknown, and SETI has made various assumptions. Beginning with Drake in 1960, various efforts using radio have been carried out or proposed. An Ohio State University radio telescope in August 1977 received and recorded a strong signal, now known as the "Wow" signal, but the signal has not since been detected. In 1979, the University of California, Berkeley (UC Berkeley), started its SETI project SERENDIP (Search for Extraterrestrial Radio Emissions from Nearby Developed Intelligent Populations), which receives and analyzes radio signals from deep space. Various upgrades in equipment have been made since then, but most of the observation time has been spent near the 1.42 GHz (21 cm) neutral hydrogen and 1.66 GHz hydroxyl transitions at frequencies of 400 MHz to 5 GHz. In May 1999, UC Berkeley started its volunteer computing project, which became popular. Anyone can download and run the SETI@home software package. Results are automatically reported back to Berkeley.

Some think that sending signals into space in the hope that they will be picked up by an alien intelligence involves real dangers. As the editor of the science journal *Nature* said in the October 2006 issue, "It is not obvious that all extraterrestrial civilizations will be benign, or that contact with even a benign one would not have serious repercussions" (p. 606).

AFTERMATH

Even during the research and development in the Manhattan Project at Los Alamos and elsewhere, experimental nuclear reactors that were the precursors to electric power generators underwent research and development. Generating electric power by atomic fission after the war became a major industry. The single biggest problem of such power generation is probably the disposal of the radioactive waste generated. Demand for electric power increases. Various limitations associated with fossil and other oxidizable fuels of possible use for generating electricity suggest that based on presently known technology, nuclear power for generating electricity will need to be utilized in greater amounts as the future unfolds.

Improvements and developments in bomb technology after the war occurred. The enormously powerful fusion or so-called hydrogen bomb became possible after breakthroughs.

Various nations other than the United States have achieved nuclear explosives. Various efforts to control and contain manufacture and (potential) usage of nuclear explosives were soon begun and have continued. Actual use of nuclear explosives, large scale, small size or yield, or otherwise, under battlefield conditions or otherwise has not yet occurred.

It can be argued that the urgently carried out research and development of the atomic bomb resulted in a substantial breakthrough that took place in a relatively short time interval. Compared with most previous substantial breakthroughs, this one, under "normal" circumstances, might have occurred only over an extended time period of unpredictable length.

After the war (in 1946), Fermi accepted a professorship at the Institute for Nuclear Studies at the University of Chicago. He held this position until his death from stomach cancer at age 53 in 1954. At the institute, he turned his efforts to high-energy physics and pursued investigations into the pion-nucleon interaction. A group of graduate students formed under him, including Owen Chamberlain, Murray Gell-Mann, Yuan T. Lee, and Chen Ning Yang. He considered the origin of cosmic rays and developed a theory involving a universal magnetic field that acted as a giant accelerator to account for the high energies of cosmic ray particles.

It is recognized that Fermi's cancer likely resulted from his exposure to radiation at or near the nuclear reactor. Two of his graduate students who similarly were with him near the pile also died of cancer. These men appreciated that their work carried risk, but they deemed the outcome from their work to be of vital importance. Element 100, discovered the year after his death, was named fermium in his honor.

Fermi wrote numerous papers in both theoretical and experimental physics. He was always sought as a lecturer, and he gave several courses at the University of Michigan, Ann Arbor, and Stanford University in California. He experienced the development by Teller and others of the enormously powerful fusion or H-bomb. He, along with Oppenheimer and others, opposed this development, though he did approve the use of the fission or atomic bomb on Japan.

His nuclear reactor was adapted for use at the works in Hanford, Washington, where uranium was converted into plutonium used in atomic bombs, but he did not live to see his reactor being used by Christopher Hinton in the first large-scale power station, Calder Hall, in England, which produced some electric power in 1954, or its use by Hyman Rickover to power naval vessels (first achieved in 1955 in the USS *Nautilus*, the first nuclear submarine). He was the first recipient of a special $50,000 award, which now bears his name, for his work in atomic physics.

NUCLEAR POWER

In 1954, the year that Fermi died, Congress passed a statute intended to promote nuclear power development. The 1954 Atomic Energy Act terminated the government's monopoly of the technology and made possible the commercial use of atomic energy. The act assigned to the Atomic Energy Commission (AEC) both the function of promoting nuclear power and the function of regulating its safety. The Manhattan Project was officially transferred to the AEC on January 1, 1947, and its military components eventually became the Defense Special Weapons Agency.

At first, utilities were reluctant to pursue nuclear power because of the abundance of conventional fuels and because of questions involving economic uncertainties and safety. Beginning in the mid-1960s, though, a sudden and unanticipated expansion in nuclear power development took place when it was found that large nuclear plants could be economical compared with coal fired plants. Also, the development of interconnected electric grids enhanced the value of large plants. Furthermore, air pollution from fossil fuel combustion became a rising concern.

Following rising public debate on the issue of nuclear power safety, Congress in the Energy Reorganization Act of 1974 abolished the AEC and created the Nuclear Regulatory Commission (NRC), whose scope of activity was limited to safety of nuclear power and civilian applications of nuclear power.

The accident at the Three Mile Island nuclear plant in March 1979 caused the NRC to devote increased attention to various issues that were involved in the accident. However, the accident undermined public support for the technology. After 1978, no new nuclear power plants were ordered, and many previous orders were canceled. However, by 2000, 103 nuclear power plants were operating in the United States, generating about 20% of the national generating capacity.

FURTHER RESOURCES

John Bankston. *Enrico Fermi and the Nuclear Reactor.* Hockessin, DE: Mitchell Lane Publishers, 2003.

James W. Behrens and Allen D. Carlson. *50 Years with Nuclear Fission.* La Grange Park, IL: American Nuclear Society, 1989.

Doris Faber. *Enrico Fermi, Atomic Pioneer.* Upper Saddle River, NJ: Prentice Hall, 1966.

Laura Fermi. *Atoms in the Family: My Life with Enrico Fermi.* Chicago: University of Chicago Press, 1954.

O. Hahn and F. Strassmann. Über den Nachweis und das Verhalten der bei der Bestrahlung des Urans mittels Neutronen entstehenden Erdalkalimetalle. (On the Detection and Characteristics of the Alkaline Earth Metals Formed by Irradiation of Uranium with Neutrons.) *Naturwissenschaften* 27:11–15, 1939.

Lise Meitner and O. R. Frisch. Disintegration of Uranium by Neutrons: A New Type of Nuclear Reaction. *Nature* 143:239–240, 1939.

Richard Rhodes. *The Making of the Atomic Bomb.* New York: Simon and Schuster, 1986.

Emilio Segre. *Enrico Fermi, Physicist.* Chicago: University of Chicago Press, 1970.

Erica Stux. *Enrico Fermi: Trailblazer in Nuclear Physics.* Nobel Prize Winning Scientists. Berkeley Heights, NJ: Enslow Publishers, 2004.

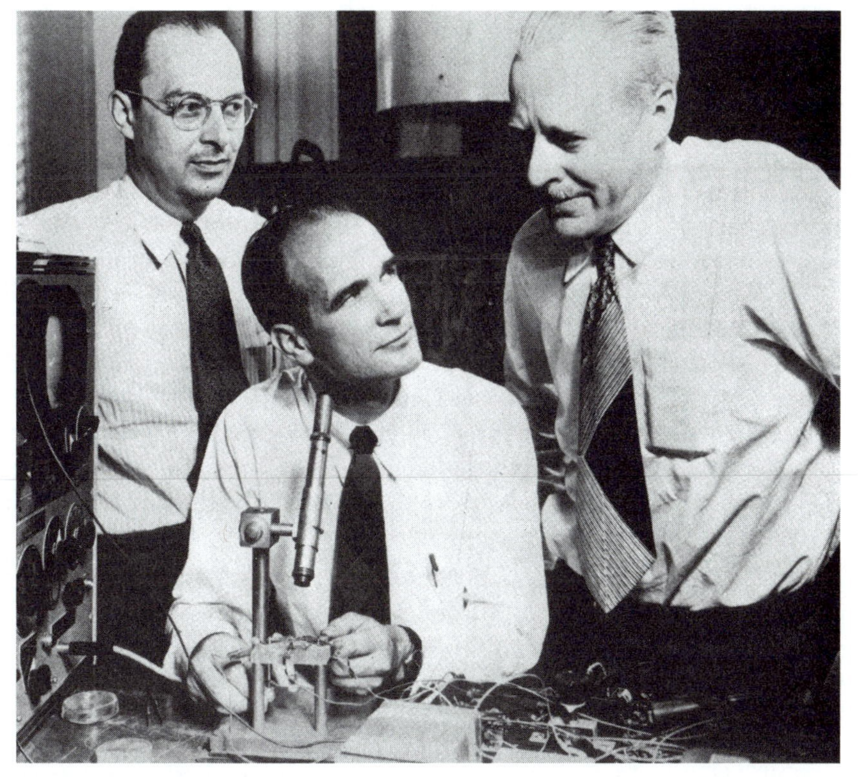

LSI Corp., National Canal Museum/AP Photo

Transistors: John Bardeen, William Shockley, and Walter Brattain

The transistor is a study of invention in a modern environment involving technically competent people. After over 2 years of disappointing research at the new solid state physics laboratory of Bell Laboratories, John Bardeen insightfully theorized in late 1947 that semiconductors inherently had a surface barrier layer of electrons. Walter Brittain soon thereafter made a point contact transistor using gold strips impressed upon germanium. William Shockley then created a junction transistor. Further research and development resulted in embodiments of increasing practicality. Considered generally as a small, solid-state device, the transistor was soon able to perform all the functions of the prior electronic vacuum tube. Further development occurred and applications were found. Today, the transistor is a basic component in electronic circuits that are ubiquitous and in universal usage. Though the transistor is considered by some to be the outstanding invention of the twentieth century, it is not really a single invention but rather an innovative developmental sequence.

HISTORY

The basic, pioneering, scientific electrical discoveries by James Clark Maxwell, Heinrich Hertz, Michael Faraday, and others in the eighteenth century; the successful production and usage of electric power in the nineteenth century by Thomas Edison and George Westinghouse; plus the devices and applications achieved by men such as Karl Ferdinand Braun, Gugliemo Marconi, Reginald Aubrey Fessenden, and others in the nineteenth and twentieth centuries led to the production and use of electrons, including the transmission and reception of electromagnetic (radio) waves.

One of the great natural barriers to development of radio devices was the fact that over distance, radio waves weakened and became difficult to detect. John Ambrose Fleming had improved the detection problem by his diode tube, which was later improved by the triode tube of Lee de Forest and improved in system by Edward Howard Armstrong. Vacuum tubes and applications for such developed, but vacuum tubes had problems, and people wondered whether somehow these problems could be overcome. A new solid-state device offered promise, but practical invention of such a device was not easily achieved.

One place to begin was reviewing the work of others prior to World War II. Germans Julius Edgar Lilienfeld in 1928 and Oskar Heil in 1934 had patented solid-state semiconductor devices using what they regarded as the field effect principle, but their work had largely been ignored.

Back in the 1930s, it had been discovered that materials with properties between those of conductors and insulators could be classified as semiconductors. A silicon crystal incorporated two distinct regions, a region that favored positive current flow, now called p, and a region that favored negative current flow, now called n. It was learned that impurities caused these regions and that they could be produced. The discovery of the p-n junction in semiconductors provided the basis for the subsequent development of the transistor and all other semiconductor devices.

After World War II, Mervin Kelly was director of research at Bell Laboratories (Bell Labs) in Murray Hill, New Jersey, then the research arm of the large telephone company, American Telephone and Telegraph Company (AT&T), which also owned Western Electric. Kelly decided to assemble a team dedicated to solid state physics research and development in an attempt to replace the fragile and cumbersome vacuum tube. He thought perhaps that some the advances in semiconductor technology achieved during the war involving radar might be considered. He chose as group leaders for this new team chemist Stanley Morgan and theoretician William Shockley. Shockley, before the war a Bell Labs employee, was a physicist and a Ph.D. who had previously proposed research on semiconductors and who, during the war, on leave from Bell Labs, was involved with radar research at Columbia University's Anti-Submarine Warfare Operations Group.

Kelly's new group, as assembled, incorporated various experts, including Walter Brattain, a Ph.D. physicist with a knack for constructing things, who during the war had been on leave from Bell Labs at Columbia University developing methods for submarine detection, and John Bardeen, a mathematical physicist and electrical engineer, who, after getting his Ph.D., had worked as an assistant professor at the University of Minnesota and during the war had worked at the Naval Ordnance Laboratory. Other experts in the group included physicist Gerald Pearson, chemist Robert Gibney, and electronics expert Hilbert Moore.

CREATING THE TRANSISTOR

In the spring of 1945, Shockley proposed a semiconductor amplifier he thought would utilize the field effect. His proposed device used a small cylinder coated with a thin layer of silicon. The cylinder was mounted adjacent to a small metal plate. Years later, Nick Holonyak of the University of Illinois called this "a crazy idea." Holonyak had worked at Bell Labs after Bardeen had left for the University of Illinois, had been Bardeen's first advanced student at the University of Illinois, and was a recognized expert in electronics and computer engineering.

When Shockley's device did not work, Shockley assigned Brattain and Bardeen to find out why. It is reported that Brattain and Bardeen worked unsupervised and that Shockley worked alone at home.

Bardeen and Brattain worked well together, with Bardeen suggesting experiments and analyzing results and Brittain building equipment and operating experiments. Various approaches were unsuccessful. Time sped by and they had little success. Shortly before Christmas 1947, Bardeen had the insight that contrary to the established idea that electrons traveled uniformly through all parts of the germanium solid they were working with, electrons in the surface area of germanium solid formed a barrier. Bardeen and Brittain, without telling Shockley, became enthusiastic about this idea. Seeking to evaluate Bardeen's idea, they immediately made changes in their investigation.

Brittain soon succeeded in making a point contact transistor using thin strips of gold foil on a plastic triangle that was pushed down into contact with a slab of germanium. Brittain noticed that two wires only two-thousandths of an inch apart were amplifying. It was a crude assembly of wires, insulators, and germanium, but it worked to amplify when demonstrated on December 23, 1947. They called Shockley to relate their invention. Shockley was pleased at the results but furious that he had not been asked to be personally involved.

Another Bell Labs engineer (believed to be John Pierce) suggested the contrived name "transistor" for the device. Bell Labs patent attorneys discovered that Shockley's field effect idea had been anticipated by prior art in 1930 by Julius Lilienfeld. A patent application on the point contact transistor was prepared by Bell Labs patent attorneys and was applied for on June 17, 1948, naming Bardeen and Brattain as coinventors. The attorneys at that time also filed a patent application directed to electrolyte-based transistors with Bardeen, Robert Gibney, and Brattain as coinventors plus two other patent applications. Gibney was a Bell Labs chemist who had suggested surrounding the point contacts between the semiconductor germanium and the conducting wires with electrolytes.

Shockley's name as inventor was not on any of these four patent applications. Shockley had sought to be named as a coinventor on these applications, or even as the sole inventor for the point contact transistor, but apparently the Bell Labs patent attorneys advised that from a legal standpoint, inventorship considerations did not justify his inclusion as inventor on any basis.

Thinking to preserve his status in Bell Labs, Shockley determined to achieve another, different transistor. With anger, and with a burst of creativity, he holed up mainly in a Chicago hotel room, and spent about four weeks with pen and paper. He created what he called a "sandwich transistor," which was subsequently called a junction transistor. Another patent application on that transistor, also prepared by Bell Labs patent attorneys, was applied for on June 26, 1948, only nine days later, naming Shockley as sole inventor of this junction transistor. Shockley had performed this work secretly, but he expected that this structure would likely be more valuable commercially than the point contact transistor because it would be easier to fabricate. It was several years, though, before Bell Labs people were able to fabricate a practical embodiment of Shockley's transistor.

Shockley continued to theorize and to develop semiconductor technology. His important book, *Electrons and Holes in Semiconductors*, was published as 558 pages in 1950. In this work, he worked out various concepts relating to drift, diffusion, and electron flow in solid-state crystals and provided differential equations, including his diode equation. Reportedly, the work was utilized by a generation of researchers and developers involved with semiconductors.

The first public demonstration of the transistor took place at a press conference in New York City on June 26, 1948, and involved a point contact transistor. Western Electric promptly placed the point contact transistor into preliminary production. Initially, news of the device seems to have been relegated to back pages of publications. Beginning in September 1948, a few electronics industry observers, such as Hugo Gernsback, began publishing speculative ideas about prospective uses for the transistor.

During the period of the research that culminated in success, mainly because of Shockley's personality, behavior, and inclinations, and even though he was the group leader, there was little coaction or esprit de corps among the three. Until their work nominally "under the same roof" resulted in a significant invention, theirs was not an impossible situation, but it seems to have been an environment with limited tolerability that vanished when the first invention appeared. Although Bell Labs consistently presented Shockley, Bardeen, and Brattain as a team, Shockley managed to take most of the credit publicly, which resulted in a deterioration in his relationships with Bardeen and Brattain. Bardeen began developing a theory for superconductivity and left Bell Labs in 1951. Brattain, at his request, was assigned to another group.

It seems that the first available, and allegedly commercially available, individual transistor was a point contact type identified as CK703. It did not immediately appear publicly until the January 1950 issue of the magazine *Radio-Electronics*, where it was accompanied by information about constructing a three-transistor radio. The CK703 transistor also appeared in the 1950–1951 issue of *Radio Master* catalog. However, it seems that this transistor was not produced in significant quantities and remained a market curiosity possibly because of problems of manufacture plus, possibly, its high retail price (about $18–$20 each).

Shockley's bipolar junction transistor was publicly announced by Bell Labs on July 4, 1951. It operates on a different principle from that of the point contact transistor. Various methods for making this device were worked out, but the method identified as "diffused base" was mostly used. The junction transistor proved to be manufacturable with consistent characteristics and soon became more frequently manufactured and used than the point contact transistor. It was early reported that junction transistors advantageously had lower noise and higher gain than the point contact transistors. Various technical problems with junction transistors were gradually overcome.

The laboratory and the divergent three researcher/operators were not nearly sufficient to allow a continuance of cooperative work and effort after

Shockley determined unilaterally that he would have to act fast and furiously to achieve a place in the sun. Yes, Shockley made a significant invention and prepared a significant treatise, but he destroyed his work environment.

TRANSISTOR DEVELOPMENT

The rate at which various different transistors developed and came into extensive use following their innovations is itself significant. A brief review follows.

Point contact transistors with consistent operating characteristics and with wires embedded directly into the semiconducting material were found to be difficult to manufacture consistently.

Various prototypes of all-transistor amplitude modulated (AM) radio receivers were demonstrated, but early transistors tended to be structurally unstable and suitable mainly only for low-power, low-frequency applications.

In 1952, Bell Labs reportedly began using point contact transistors in commercial Bell Telephone equipment and also began licensing its transistor technology. Reportedly the first licensees were General Electric, IBM, Raytheon, RCA, and a Japanese company that became Sony Electronics. The initial license price has variously been reported as ranging from $25,000 to $50,000. In early 1952, Western Electric was apparently the only manufacturer making junction transistors, at a rate of perhaps not more than 100 transistors per month.

In 1954, IBM reportedly made a prototype computer with no vacuum tubes using two thousand transistors. Until about October 1954, all-transistor AM radios reportedly were only prototypes and not commercially produced. The first successful, practical, significantly produced, all-transistor AM radio was apparently the Regency TR-1 first made in 1954 by the Regency Division of Industrial Development Engineering Associates (IDEA) of Indianapolis, Indiana. Sony, then a new firm, in 1955 produced its commercial TR-55 AM radio, and, in 1957, its much more advanced TR-7 model. As transistor technology developed, commonly, transistors were found to have long lives compared with vacuum tubes.

Transistor development by industry was effectively promoted by military perceptions, needs, and government contracts, beginning in the early 1950s, during the Cold War, particularly for applications using low power, small size, and long life. Usage examples included proximity fuses, missile guidance, underwater devices, and battlefield radio communications. Among consumer goods, hearing aids were an attractive area for transistors, providing a reduction in device size and a long battery life.

Prices of transistors soon dropped. For example, in early 1953, germanium junction transistors identified as CK722 were advertised to be available commercially from Raytheon for about $18–20 each, retail. The CK722 became a common "workhorse" for various applications. By December 1955, only 7 years after the origin of the transistor, the retail price of the CK722 was advertised for $0.99 each. Motorola's 2N2222 is evidently another example of a widely used transistor whose retail price dropped.

Germanium transistors were found to be limited because of sensitivity to temperature and humidity. Silicon, with a crystal structure identical to germanium but less sensitive to such conditions, was more reliable than other known semiconductor types. Research was expended looking for methodology that would enable silicon to be used, but at first high-performance silicon based transistors could not be made. In 1954, first Bell Labs personnel using growth rate fluctuations and npn junctions, and a little later Texas Instruments personnel using selected, sequential doping, produced desired silicon transistors. A significant advance occurred when improvements in fabrication technology enabled production and extensive use of silicon transistors. Silicon transistor types now predominate.

Various transistor types have been developed and have been extensively used. The bipolar junction transistor (BJT) was the first transistor to be produced in quantity. It conducts by using both majority and minority carriers. It has three terminals, the emitter, the base, and the collector. Inside a BJT, the two pn junctions, the base/emitter junction and the base/collector junction, permit currents at the emitter and the collector to be controlled by a relatively small base current. The BJT is useful in amplifiers.

The field effect transistor (FET) utilizes either electrons (n-channel FET) or holes (p-channel FET) and has four terminals, identified as source, gate, drain, and body (base or substrate). Usually, the body is connected to the source internally. Voltage between the gate and the source (body) controls current between the drain and the source. A conducting channel near the gate connects the drain to the source, and the drain/source current travels through the conducting channel. Channel conductivity is varied by an electric field that is generated by the voltage applied between the gate/source terminals. Thus, the current flowing between the drain and the source is controlled. FETs are classified into depletion-mode and enhancement-mode types, and also into junction FETs and insulated gate FETs (also known as metal oxide semiconductor FETs, or MOSFETs).

A large number of different transistors, usually made from silicon semiconductor, have become available, especially since the 1960s. Almost all modern electronic devices incorporate transistors in various ways for various purposes and functions.

Initially, transistors were made and used as individual components that were connected to other components, such as resistors, capacitors, inductors, and so forth. A manufacturing drive to reduce size of components incorporating transistors developed. Transistor-containing circuits could, for example, switch faster than tube-containing circuits. Circuit boards common in the 1950s with individual transistors and, in the case of computers, with magnetic memory cores, reached their limits and often came to be replaced with integrated circuits that apparently originated in the late 1950s from Texas Instruments (Jack Kilby, who received the Nobel Prize in physics in 2000 for his invention of a printed circuit) and Fairchild (Robert Noyce). Integrated circuits involved Kilby and are considered in a separate chapter. With integrated circuits, more complex components and circuits could be made concurrently with the same materials on single silicon chip, as taught by Kilby. Technology regarding integrated circuits with transistors greatly developed. It is said that the number of transistors per unit area has doubled every 1.5 years (attributed to Gordon Moore, an integrated circuit pioneer and a founder of Intel Corporation). An integrated circuit can involve a single little piece of silicon (or chip) holding millions of transistors together with other electronic components and wiring. Currently, for example, a personal computer (PC) chip may have 3.5 million or more transistors. In such a chip, the cost of an individual transistor is said to be practically nil.

BREAKUP OF THE INNOVATORS

As indicated above, following the invention of the point contact transistor, Shockley utilized technological information previously generated by Bardeen and Brittain. He placed himself in the foreground, to the disappointment and irritation of Bardeen and Brattain, though Bell Labs management consistently presented the men as a team. Shockley apparently deliberately and intentionally isolated Bardeen and Brattain and prevented them from working on his junction transistor. Bardeen and Brattain became motivated to withdraw from Shockley. Bardeen joined the engineering faculty at the University of Illinois at Urbana-Champaign in 1951. Brattain refused to work further with Shockley and was assigned to another group in Bell Labs. After the first year following the invention of their point contact transistor, neither Bardeen nor Brittain had much to do with developing the transistor.

NOBEL PRIZE

Bardeen, Brattain, and Shockley were awarded in 1956 the Nobel Prize in physics for their invention of the transistor in 1947.

NATURE OF THE THREE

Each of the three coinventors was a distinct, unique individual. Except for their Nobel prize, it appears that none received significant financial reward for achieving the transistor. A brief biography of each is relevant and instructive.

William Shockley (1910–1989)

Shockley was born in London on February 13, 1910, to American parents. His father was a mining engineer, and William (Bill) was raised in California. His childhood was not conducive to the development of socialization. He received his B.S. from the California Institute of Technology (Caltech) in 1932 and his Ph.D. from the Massachusetts Institute of Technology in 1936. After graduation, he joined Bell Labs.

Before the war, at Bell Labs, he had demonstrated intelligence, problem-solving ability, and the capacity to utilize both theory and experiment. With a coworker, he designed an early nuclear reactor in 1939 that was immediately placed on a secrecy status by the federal government and not divulged until after the war. During the war, apparently in May 1942, he left Bell Labs and took a research director position at Columbia University, where he was involved in antisubmarine work and helped improve the effectiveness of depth charges. In 1944, he became involved in a training program for B-29 bomber crews regarding use of the new radar bomb sight. In 1945, he prepared a report for the War Department regarding projected casualties from an invasion of Japan (it has been speculated that the government was then wondering whether or not to use the atomic bomb on Japan). He received the National Medal of Merit. After the war in 1945, he returned to Bell Labs, as indicated above.

Bell Labs management refused to give Shockley an executive position, apparently because of his abrasive style of management. It was also indicated by Bell Labs that Shockley was considered to be a greater asset as a theorist and research scientist. Shockley thought he deserved management and more money. He took a leave of absence from Bell Labs in 1953 and went to Caltech for an initial term of four months as a visiting professor.

A Caltech friend gave Shockley an opportunity to run his own company. In 1955, Shockley went with Beckman Instruments and became director of Beckman's newly capitalized Shockley Semiconductor Laboratory division in Mountain View, California. Shockley endeavored to hire some of his former staff people from Bell Labs for his new laboratory, but none would come. He then proceeded to seek the brightest graduates from universities and to try to build his company.

The records indicate that Shockley's management over time became more domineering and "increasingly paranoid." On one occasion, he demanded that the staff at his company take lie detector tests for the purpose of identifying who was responsible for the cut on a secretary's thumb. The cut, though, was found to be caused by a broken thumbtack on an office floor. The incident made his already hostile staff more so. His project for his staff to build a particular proposed new so-called three-state device proceeded slowly.

In 1954, he separated from and divorced his wife Jean, whom he had married while a student in August 1933. Their first child had been born in March

1934. On November 23, 1955, he married Emmy Lanning, a teacher of psychiatric nursing, and this marriage lasted until Shockley's death in 1989.

In 1957, eight of his employees resigned after Shockley decided not to continue research into silicon-based semiconductors. With capital from Fairchild Camera and Instrument Corporation, these people started Fairchild Semiconductor. Later, two of the eight left Fairchild and started Intel. The first integrated circuits, produced by Fairchild Semiconductor and Texas Instruments, made Shockley's proposed three-state device extraneous.

In 1961, he was in a serious car accident along with his wife, Emmy, and son Dick and required months to recover. His firm was sold to Clevite and never made a profit. He then joined Stanford University as a professor of engineering and applied science. He obtained his last patent, which involved a semiconductor device, in 1968.

In his later years, he exhibited great interest in race, intelligence, and eugenics. He proposed that individuals with IQs below 100 be paid to undergo voluntary sterilization. He donated sperm to a sperm bank catering to Nobel Prize–winning donors.

He liked to perform magic tricks at office parties and enjoyed elaborate practical jokes. With his wife present, he would endorse prostitution to solve marital boredom. He was an atheist and never attended church. He was estranged from his three children. An inventor in more than ninety U.S. patents, Shockley died of prostate cancer in 1989 at age 79.

John Bardeen (1903–1991)

John Bardeen was born May 23, 1903, in Madison, Wisconsin, as one of four children of Dr. Charles R. Bardeen and Althea Harmer Bardeen. His father was a professor at the University of Wisconsin.

John credited his seventh grade mathematics teacher with encouraging him to pursue advanced work and exciting his interest in mathematics. He graduated from high school at age 15; he could have done so earlier but was delayed partly because of taking courses at another high school and partly because of his mother's death. He received his B.S. and M.S. degrees in electrical engineering from the University of Wisconsin at Madison in 1928 and 1929, respectively.

He worked for Gulf Research Laboratories for 3 years in Pittsburgh, but the work failed to keep him interested. He was accepted for graduate work at Princeton University, where he pursued both mathematics and physics. He wrote his doctoral thesis under Nobel laureate Eugene Wigner and received his Ph.D. in 1936. After his father's death in 1935, he went to Harvard University on a 3-year postdoctoral fellowship.

He married Jane Maxwell, whom he had met in Pittsburgh, before he left Princeton. He began as an assistant professor at the University of Minnesota in the fall of 1938. As the war started in 1941, colleagues convinced him to work for Naval Ordnance Laboratory, where he remained for 4 years. A large offer from Bell Labs in 1945 was made because of his background in solid state physics, and he was induced to work there where he renewed an earlier friendship with Walter Brittain. He had met Shockley earlier, when Bardeen was a student at Princeton.

At the University of Illinois in 1951, his first Ph.D. student was Nick Holonyak (1954), who invented the light-emitting diode in 1962. Bardeen was a professor there until 1975 when he became professor emeritus.

He worked with his doctoral student J. Robert Schrieffer and with doctoral associate Leon N. Cooper to complete creating the so-called standard theory of superconductivity, which became known as the BCS theory, announced in 1957. Under the BCS theory, electrons (fermions) pair via phonon coupling and the resulting pairs (now bosons) condense into a single coherent ground state where the electrons move cooperatively through the crystal without scattering. Superconductivity is a state of matter first observed in 1911 in which materials lose their electrical resistance at low temperatures, and the theory is based on a model in which electrons form bound pairs. The theory explains fundamental processes in solid-state physics, nuclear physics, astrophysics, and particle physics, and for this work they received the Nobel Prize in physics in 1972. At the time, Bardeen was the first Nobel winner to receive a second prize in the same field.

Bardeen had a very unassuming personality. At the University, he was remembered by neighbors for throwing cookouts. He did not fit the stereotype of genius or "crazy scientist." He and his wife Jane had three children and six grandchildren at the time when he died. He died of a heart attack on January 30, 1991.

Walter Brattain (1902–1987)

Walter Brattain was born in Amboy, China, on February 10, 1902, to Ross R. Brattain and Ottilie Houser Brattain. His father had a teaching contact in China. He was raised by his parents on their cattle ranch in Washington state. He earned his B.S. in 1924 in physics and mathematics at Whitman College in Walla Walla, Washington; his M.A. from the University of Oregon in 1926; and his Ph.D. from the University of Minnesota in 1929.

After working at the National Bureau of Standards in Washington, D.C., he was hired in 1929 by Bell Labs. During World War II, he worked on methodology relating to submarine detection at Columbia University. After the war, he returned to Bell Labs.

After returning to Bell Labs, he became part of the newly formed solid state research group under William Shockley. Early in 1946, he started an investigation of semiconductors aimed at producing a practical solid state amplifier. Semiconductor rectifiers were known at the end of the war.

Shockley hoped to produce a new device that would have variable resistance and would be useful as an amplifier.

Brattain had a reputation for being able to conceptualize and build laboratory devices in research and make them work. His hands built the first transistor. In 1935, he married Dr. Karen Gilmore, and they had one son. After her death, he married Emma Jane Kirsch Miller in 1958. His professional career was spent at Bell Labs. He is believed to have lived a quiet, relatively uneventful adult life. On October 13, 1987, he died in Seattle, Washington.

IMPACT

Prior to the transistor, most great inventions had been achieved by one individual. With the rise of the industrial research laboratory, and its collection of diverse, expensive, state-of-the-art equipment, plus its collection of intelligent personnel with specialized training and experience, the stage was set for the making of important inventions such as forms of the transistor.

However, the first invention of the point contact transistor by Bardeen and Brattain proved over time to be of lesser importance than the second invention of the junction transistor by Shockley.

Though interesting and charged with possibility, these two inventions required an enormous amount of further research, development, further inventions, manufacturing, and marketing by others before their potentials could be approached.

FURTHER RESOURCES

Biography of Walter H. Brattain. Nobel Lectures, Physics, 1942–1962. Amsterdam, the Netherlands: Elsevier Publishing Company, 1964.

Lillian Hoddeson and Vicki Daitch. *True Genius: the Life and Science of John Bardeen*. Washington, DC: National Academy Press, 2002.

Paul Horowitz and Winfield Hill. *The Art of Electronics*. Cambridge: Cambridge University Press, 1989.

November 17–December 23, 1947: Invention of the First Transistor. *American Physical Society* 2000:9(10).

Michael Riordan and Lillian Hoddeson. *Crystal Fire: The Invention of the Transistor and the Birth of the Information Age*. New York: W. W. Norton & Company Ltd., 1998.

Joel N. Shurkin. *Broken Genius: The Rise and Fall of William Shockley, Creator of the Electronic Age*. New York: Palgrave Macmillan, 2006.

AP Photo

Integrated Circuits, Microprocessors, and Computers: Jack Kilby and Steve Wozniak

Integrated circuits, originated by Jack Kilby in 1958, soon became omnipresent in modern electronics because of features such as precision fabrication, miniaturization, practicality, cost, and performance. As soon as development permitted, these circuits became extensively used in diverse devices. Large, complex integrated circuits became microprocessors and entire central processing units (CPUs) for computers, thereby enabling the making of remarkably miniaturized computers, such as illustratively originated by Steve Wozniak. Some historians believe that integrated circuits have enabled the digital revolution, which is perhaps one of the most significant developments in human history.

INTRODUCTION

The rapid and significant development of electronics after the arrival of transistors is probably nowhere better illustrated than in the invention and development of small, compact integrated circuits incorporating multiple interconnected transistors and other solid state components. From relatively simple integrated circuits, complex, advanced microprocessors evolved. Many electronic and electromechanical devices incorporating microprocessors appeared. The development of central processing unit (CPU) microprocessors can be considered to have enabled the creation of small computers commonly called personal computers and laptop computers.

Without doubt, the invention and development of integrated circuits into microprocessors and the like have been overall progressive and truly iconic in significance. Many microprocessors have been developed, and many more will be. Microprocessors have been achieved, produced, and utilized in many places and applications, all in a relatively short time interval. Many innovators and developers are involved. At this point in time, when microprocessors have become commonplace, it does not seem to be possible or proper to assign invention of microprocessors as a class, their manufacture, or even their applications and use to a single inventive entity (solely or jointly).

The appearance of microprocessor CPUs and their utilization in the development of small computers that are widely used seems to be one of the best illustrations of microprocessor utilization. Also, this event appears to be localizable, with apparently identifiable innovators.

In earlier days, when iconic inventions occurred, it appears that they commonly tended to be spaced in time relative to one another and to their development and use. Today, relative to former times, if one were to generalize relative to the field of electronics, generally product development and usage seem to have quickened, perhaps because of civilization, changes in circumstances, and more rapid utilization of technological advances. Perhaps a mix of the following elements is involved:

- Scientific advances and rapid communication
- Competition (in some instances)
- Many educated and capable people
- A general desire for progress and improvement
- Enhanced manufacturing capability
- Customers able and willing to purchase advanced products
- The attraction of potentially large markets and profits from sale and use of products
- The pace of new inventions

An understanding and appreciation of the nearly contemporary circumstances associated with the presently considered iconic inventions in the field of electronics can perhaps be gained from the following related and sequential accounts, beginning with integrated circuits, progressing through microprocessors, and reaching a peak in computers.

INTEGRATED CIRCUITS AND JACK KILBY

Summary

Jack Kilby, at the time he was hired by Texas Instruments (TI) in 1958, was interested in miniaturization of electronic components. At TI, he soon conceived of directly fabricating little circuits on individual semiconductor wafers using known materials without first separately fabricating and then connecting individual components. His conceptual product was soon embodied and called an integrated circuit (IC; also called a chip or microcircuit).

The materials he used were known because prior exploration and research had identified them. Basically, they were obtained from among the elements of the Periodic Table and involved compounds and compositions thereof that were found suitable for use as solid-state electronic components.

Background

An IC is a miniature or microelectronic device that includes several interconnected active and/or passive components, such as transistors, diodes, resistors, dielectrics, and capacitors, with interconnectors, mounted on or in a localized site, usually a small, thin, semiconductor chip, such as a silicon chip. The chip itself is usually mounted on a substrate member, and the resulting device is usually packaged in a plastic or ceramic case or wrapper with external attaching pins or the like. A simple integrated circuit usually has one specific function or purpose, such as an amplifier, oscillator, timer, counter, memory, sensor, or data-processing unit.

The concept of an integrated circuit is believed to have been first proposed by a British radar expert, Geoffrey W. A. Dummer (1909–2002), who worked for the Royal Radar Establishment of the British Ministry of Defense and who published his idea in Washington, D.C., on May 7, 1952. In 1956, though, Dummer tried without success to build such a circuit.

Before the integrated circuit, to make a circuit, individual elements, composed of assembled, discrete, individually prefabricated components, were functionally and suitably interconnected together by discrete wires and solder and conveniently associated with a support. Transistors, after their creation, were developed and replaced the prior vacuum tubes. Transistors came to be used in solid-state circuits for most applications.

Vacuum tubes were bulky, required large amounts of power to operate, produced considerable amounts of heat, were relatively fragile and unreliable, and were costly to fabricate. They required labor-intensive efforts not only to fabricate into circuits that were incorporated into, or used in, devices, but also to maintain in circuits, especially in large, increasingly more complicated systems, such as early computers. In contrast, the individual new solid-state transistor structures were comparatively very small, tended to be relatively inexpensive, used relatively little power, produced less heat, and were relatively durable and reliable.

At first, a solid-state circuit contained only a few transistors and associated components, but advances in electronics coupled with the transistor effectively encouraged and stimulated designers and engineers to fabricate electronic circuits of increasing complexity and numbers of transistors and associated components. Very soon, the number of individual components in a manually fabricated circuit could range into thousands, even hundreds of thousands.

Prior to the IC, fabricating large transistorized circuits was a major problem. The prefabricated individual components had to be suitably individually located and interconnected to form the circuits, and the interconnections had to be formed by individual bits of wire that had to be hand soldered in place. One subproblem was that hand soldering each of thousands of wire bits and associated components was time consuming, tedious, and expensive. Another subproblem was that hand soldering was not completely reliable. Each soldered joint was potentially a source of trouble.

Another subproblem was circuit size. Particularly in a complicated circuit, or in a circuit with a high density of components, if some of the connecting wires were too long, or if some of the components were too large, then electric signals passing through did not travel rapidly enough. One consequence was that the circuit operated too slowly.

In the electronics industry, the general problem of fabricating circuits with many components came to be called the "tyranny of numbers," and the problem threatened to limit certain advances in microcircuitry. Engineers soon comprehended the facets of the problem and realized that fabricating increasingly larger and more complex circuits would be impossible by the then conventional techniques.

By the mid-1950s, significant efforts were being made to find and implement components and circuits that would permit alternative routes of construction and achieve reliability with associated low cost. One research and development program was sponsored by the U.S. Army Signal Corps and was called the Micromodule program. Apparently, Jack Kilby himself had proposed this program to the U.S. Army, and it looked promising in 1957. It involved creating small ceramic squares (called wafers), with each one containing a different circuit component or components. It was aimed at eliminating wires used in interconnections and at making the individual circuit components of uniform size and shape. The necessary interconnecting wiring was to be fabricated with these components so that two components, subcircuits, or modules could be brought together adjacently and perhaps connected simply by being snapped together. Texas Instruments was involved with the Micromodule program in July 1958 when Jack Kilby became an employee at TI's new Semiconductor Building in Dallas.

Kilby's Early Years

Born in Jefferson City, Missouri, on November 8, 1923, Jack spent much of his early life in Great Bend, Kansas. His father operated there a small electric supply company but worked his way up to become president of the Kansas Power Company. Jack developed an early interest in electronics and wanted to be an electrical engineer like his father. However, he failed to pass the Massachusetts Institute of Technology qualifying exam, so after he graduated from Great Bend high school, he attended the University of Illinois. However, after Pearl Harbor and the start of World War II, he entered the Army and as a Corporal was sent to India, where he repaired and improved army radios in northeastern India. After the war, he went back to the University of Illinois and in 1947 received a degree in electrical engineering.

He graduated 1 year before the transistor became common knowledge and available. He found that although his training in vacuum tube technology became obsolete, he now considered that he had "great opportunities" to use his physics-related studies. He took a job with an electronics parts manufacturer, Centralab Division of Globe Union, Inc., in Milwaukee, Wisconsin. From 1947 to 1958 he was there reportedly involved with the design and development of thick film integrated circuits. He was also involved with ceramic based silk screen circuit boards and transistor-based hearing aids and seems to have been named as an inventor in twelve U.S. patents. In Milwaukee, he took evening classes at the University of Wisconsin extension, working toward a master's degree in electrical engineering, which he received in

1950. Though he was 6 feet 6 inches tall, he was known to be a humble man who lived simply.

In 1958, he took a job with TI and in July moved with his family to Dallas, Texas. He later said, "It was the only company that agreed to let me work on electronic component miniaturization more or less full time."

Invention of Integrated Circuit

In July 1958, at TI, Kilby found himself in an almost deserted building. Most of the TI employees had left for their annual two-week vacation period. Kilby, as a new employee with no vacation time, was left to staff the laboratory.

Kilby, from his work at Centralab in Milwaukee, was familiar with the "tyranny of numbers" problem, and at the TI laboratory began work on the Micromodule program. He soon concluded that the Micromodule program could not be the best answer to this fabrication problem because the program did not deal adequately with the basic problem of large numbers of components in large, complex circuits. So Kilby began considering and searching for an alternative and better answer.

Soon, in July 1958, he began making entries, including sketches, in a laboratory notebook. It suddenly occurred to him that all parts of a circuit, not just each transistor, could be made out of silicon. He decided that about the only thing a semiconductor house such as TI could make cost effectively was a semiconductor. As he later explained in a 1976 article titled "Invention of the IC":

> Further thought led me to the conclusion that semiconductors were all that were really required—that resistors and capacitors [passive devices], in particular, could be made from the same material as the active devices [transistors]. I also realized that, since all of the components could be made of a single material, they could also be made *in situ* interconnected to form a complete circuit.

No one at that time was making capacitors and resistors out of semiconductors. If this was possible, then an entire circuit could be built out of or on a single crystal, and the circuit would be small and relatively easy to produce. His boss liked the idea and suggested that he try to build a prototype.

By September 1958, Kilby had achieved a prototype demonstration embodiment of a working IC. He had fabricated the embodiment on a slice of germanium measuring 7/16 by 1/16 inch. The slice had protruding wires that were glued to an adjacent glass slide. A transistor and other components were formed and suitably interconnected on the slice. When Kilby pressed a switch, an unending sine curve undulated across an associated oscilloscope screen. His invention worked, he had solved the problem, and he had made a breakthrough at age 35.

His demonstration on September 12, 1958, of his relatively simple device to a few coworkers in the TI semiconductor laboratory had also been witnessed by several TI executives, including Mark Shepherd, the former TI chairman. Although the U.S. Air Force showed some interest in Kilby's integrated circuit, the industry reacted skeptically. As Kilby related, the IC and its relative merits "provided much of the entertainment at major technical meetings over the next few years."

On February 6, 1959, TI filed a patent application for a "solid circuit" on a germanium chip. In March 1959, TI evidently publicly demonstrated its first "solid circuit," which was the size of a pencil point. Kilby proceeded to pioneer various commercial, industrial, and military applications at TI. He headed teams developing microchip technology; one developed the first military system, another the first computer incorporating integrated circuits.

Potential Competition

In California, in January 1959, about six months after Kilby's IC invention, Robert Noyce (1927–1990), an employee and a manager of the small startup company, Fairchild Semiconductor, achieved a somewhat similar conception of an entire circuit produced on a single chip. Apparently in contrast to Kilby, who some allege without evidence was then or at first still trying to make and use individual components by emplacement on a semiconductor surface, Noyce allegedly sought to produce the individual components on a semiconductor surface.

Later it became known that, apparently from the TI public demonstration in March 1959, Fairchild knew about the TI IC development and that TI had already filed a patent application. This accelerated development efforts at Fairchild. One of Fairchild's original founders, Jean Hoemi, invented a so-called "planar" method for making connections between components. His method used oxidation and heat diffusion to form a smooth insulating layer over the surface of a silicon chip. The technique enabled an insulating layer to be interposed between layers containing transistors and other components in silicon, and the insulating layer permitted adjacent layers to be electrically isolated, thereby eliminating the need to cut apart the layers and wire them back together. Fairchild filed a U.S. patent application covering the planar method on July 30, 1959, which included a highly detailed disclosure that evidently aimed to distinguish in a patentable manner Fairchild's work from that of Kilby and TI. The Fairchild application apparently included a chip of silicon with many transistors all etched into its working surface.

On April 25, 1961, the U.S. Patent Office issued a patent to Noyce and Fairchild for an integrated circuit on a silicon chip. Kilby's (and TI's) original U.S. patent application was still pending. A legal battle resulted lasting over a decade between TI and Fairchild, which terminally resulted in the U.S. Court of Claims holding that Kilby and Texas Instruments had first invented a working integrated circuit but that Fairchild had invented certain interconnection techniques. Apparently, eventually, the two companies agreed to a cross-licensing arrangement.

Kilby's U.S. patents on the IC that were subsequently issued reportedly included U.S. Patents 3,138,743, 3,138,747, 3,261,081, and 3,434,015. However, Noyce's device, known as the planar IC, reportedly dominated the market, at least at first.

IC Fabrication

At first, using the available technology, just a few transistors and associated components could be fabricated as an integrated circuit upon a single chip, such as silicon. However, intensive research and development efforts by individual technologists and teams at different companies, driven by commercial possibilities and competition, soon resulted in chips having more complex circuits including many transistors and other components.

Many people have worked to develop and bring IC (chip) fabrication and the associated technology up to its present levels. IC fabrication is now considered to be the most complicated fabrication process commercially used. Many inventors have contributed improvements and features.

Chip production evolved and improved over the years subsequent to the creation of the IC. Design of a new commercial IC embodiment, particularly with a complicated circuit, was found to be involved, time consuming and costly, but distributed over a large number of individual copies (chips) of a given embodiment, the cost of each embodiment was relatively small.

Chip fabrication illustratively, characteristically, and usually involves the use of photolithography. The progressive making of many superimposed layers on usually a silicon monocrystalline substrate is usual and involves what is usually a complex, multistep process. A single IC fabrication facility (sometimes known in the industry today as a semiconductor fab) in 2005 was estimated to cost over $1 billion, mainly because of the extent of automation.

Though detailed IC fabrication methodology of large-scale circuit devices is too involved to detail presently, briefly, a typical silicon wafer slice about 12 inches in diameter is coated at an early stage with a photolithographic photosensitive film layer. A controlled high-intensity ultraviolet (UV) light is directed on the coated slice through a mask that holds an image of selected component parts of the circuit so that light-struck portions of the coated slice become selectively imaged. Light in the visible spectrum is not used because individual mask sizes of individual components are usually very small, and the wavelengths of visible light are too large to create the desired sharp imaged patterns on a photosensitive layer; hence the preferential use of ultraviolet light for imaging. One example to illustrate the small sizes involved is to consider that it is theoretically possible for several thousand transistors to fit on an area about the size of a cross-section of a human hair.

After imaging, the resulting latently imaged, photosensitive film of a coated layer is developed, usually by chemicals, and, depending, upon the chemical compositions used, either the exposed portions or the unexposed portions in the light-struck layer are removed. The resulting surface thus has protected and unprotected areas corresponding to components, or component portions, positioned in a preselected relationship.

Characteristically, because of imperfections of various sorts, chip fabrication can involve a high rate of rejects, which ultimately adds to the cost of successfully made (marketable) chips. For example, one major cause of chip imperfections is caused by masks. All masks have flaws. A chip surface imaged through a mask may incorporate one or more flaws that are "expressed" (imaged) on the chip. At first, a chip producer simply threw away chips with mask-caused imperfections. Even in the late 1970s, this commonly meant throwing away at least 70% of chips. The engineers at MOS Technology learned the trick of fixing their masks after they were made, which enabled the engineers to correct major faults with a series of minor fixes so that masks could be produced and used with only a very low flaw rate. The result was that the number of chip rejects was reportedly reduced perhaps to less than 30%, and usable product chips were made faster and more economically.

To avoid environmental contamination of chips during manufacture, chip production is carried out in a so-called clean room environment, in which the presence of particulate foreign bodies is minimized. For example, air in the clean room is maintained at a slightly elevated pressure (relative to ambient pressure), is circulated, and is carefully filtered. Air flow in the room is relatively rapid, and workers wear special outfits ("bunny suits").

The unprotected areas, for example, on a chip undergo controlled processing to change their electrical properties. Multiple process steps can be involved. Many overlapping layers, and many subsequent successive coatings of various photolithographic photosensitive materials, each usually followed by development, can be used. Different colors may be used to identify particular layers. Selected dopants, for example, can be selectively diffused into the substrate in unprotected areas to produce diffusion layers as desired for p and n locations on semiconductors. Ions can be implanted to provide implant layers. Material to define insulators or conductors (such as polysilicon or metal, such as aluminum) can be deposited. Connections between, for example, conducting layers, can be defined to provide contact layers, called "via."

Individual components, which are typically very small, must be carefully structured and treated to obtain desired particular electrical characteristics. Resistances, for example, can comprise undulating stripes on a surface, but variables, such as stripe length, width, and resistivity define the actual characteristics of a particular resistance. Capacitances can comprise parallel conducting plates separated by insulative composition, but characteristically only very small capacitances can be defined in an IC.

To build on a chip a sequence of layers for a desired circuit and to form desired active and passive components of the circuit on the substrate, at least one additional layer of material is coated onto a resulting surface holding components. For each layer, a basic sequence of process steps is repeated. For example, a layer of conducting metal may be selectively deposited upon the previous coated layers of a substrate. To achieve that layer, a coating of photosensitive film is applied over it. The resulting surface is again exposed to intense ultraviolet light through a different mask that holds an image of individual wires that are to connect components formed on the surface portions. The resulting imaged film is developed, and exposed portions of the photosensitive metal are removed. The resulting exposed but unwanted metal portions may then be removed with a selected acid wash or the like, to leave individual components joined by wire-like connections defined by the remaining conducting metal.

With, for example, a 12-inch wafer, a process called "stepping" is commonly used. For example, the face of a wafer of silicon is coated with photosensitive material, and individual chips are defined systematically next to one another in a checkerboard-like pattern. During the subsequent light exposure procedure, for precision, the wafer is progressively moved laterally and stepwise under the mask relative to the axis of the light so that individual chips or circuits are successively exposed to ultraviolet light, thereby to "image" each chip or circuit in an identical manner. Thus, successive chips can be prepared using the same mask and UV exposure conditions.

It is possible for several weeks to elapse during progressive fabrication of chips on a wafer owing to the time-consuming character of individual process steps and the possibly of testing and debugging between steps. A resulting IC on each individual chip defined on a wafer is individually tested, usually with automatic equipment. To debug a fabrication process, it is essential for operators to use electron microscopes because the individual features on individual circuits are so small.

After successful testing, a wafer is cut into rectangular (or perhaps square), chip-holding blocks, each called a die. To form a device for commercialization, each die is associated with a package having aluminum or gold wires that are welded to the edge portions of a die at locations called pads. After packaging, devices are again tested using the same or similar test equipment. Depending on the device, testing (including labor and equipment) can add significantly to the cost of an IC product.

Usage and Growth

The IC, or chip, was a great improvement over prepared circuits composed of hand-assembled, discrete, interconnected components. The relative ease of fabrication, reliability, and relatively low unit cost promoted rapid and expanding IC usage.

Though development of integrated circuits on semiconductor chips proceeded after the invention of ICs, it took decades to work out methods for achieving significant details, such as creating semiconductor crystals without crystalline structural defects in fabricating integrated circuits.

Digital integrated circuits, as used, for example, in computers, microcontrollers, and other devices, can each contain from a few thousand to many million logic gates, multiplexers, flip-flops, and other circuits, all located in a few square millimeters. Digital ICs work with binary mathematics and use only 1 and 0 signals.

Analog integrated circuits, used, for example, in sensors, management circuits, operating amplifiers, and other devices, can contain similar numbers of circuits and can be used to perform functions such as amplification, active filtering, signal mixing, and so forth. Analog and digital circuits can be combined on a single chip, but they must be carefully designed to avoid problems such as signal interference.

The earliest ICs on chips contained a few (perhaps only about a dozen) transistors and were called small-scale integration. Such chips were used in the Apollo program (1961) and in the Minuteman missile program (1962), where lightweight digital computers that used multiple chips were needed for the inertial guidance systems. From 1960 to 1963, substantially all of the ICs available were used in these programs, and the demand served to fund production improvements and lower costs. The unit cost of ICs has been estimated to have reduced from perhaps about $1,000 in 1960 to about $25 in 1963. Use in consumer products evidently began soon, in the early 1960s.

In the late 1960s, ICs containing hundreds of transistors per chip appeared, called medium-scale integration; these ICs used smaller circuit boards. By the mid-1970s, economic considerations permitting, ICs containing tens of thousands of transistors per chip, called large-scale integration, became available. Starting in the 1980s, ICs with hundreds of thousands of transistors per chip arrived, called very large-scale integration. The growth has continued. In mid-2007, a single chip could hold several hundred million transistors.

Kilby reportedly built the first computer incorporating integrated circuits at Texas Instruments in 1961. In the early 1960s, as a demonstration project for accelerating widespread use of integrated circuits, Patrick E. Haggerty, former TI chairman, challenged Kilby to create a calculator as powerful as the large, electromechanical desktop models of the day, but small enough to fit in a coat pocket. Over about 6 years, as a group leader, Kilby invented and developed with a coinventor an electronic handheld pocket calculator incorporating an integrated chip circuit, which TI made and sold and successfully commercialized.

Kilby's IC chip invention had far-reaching utility. The chips came into increasing usage in many applications. Many products utilizing electronics would not have been developed without this invention. Chips have replaced mechanical controls in most machines.

The modern electronics industry (including the computer industry) is considered to have been made possible by the chip. It enabled, for example, the transforming of the previously room-sized computers into today's various compact mainframes, minicomputers, and personal computers.

Communications were restructured and developed by the chip, and new ways for people, businesses, and governments to exchange information were enabled. Space developments and operations were made possible by the chip. Chips are utilized in hearing aids and in most medical devices.

Many applications for the chip in education, manufacturing, transportation, and entertainment exist and are still developing. A new $150 million Kilby Center at TI aims to achieve new and improved IC products.

The integrated electronic circuit has enabled the whole electronics industry to grow and develop. The worldwide electronics market has reportedly grown annually from about $29 billion to over $1,400 billion. If the projections prove to be correct, electronics involving the IC will become the world's largest single industry. Various historians have observed that the revolution involving digital electronics was made possible by integrated circuits and constitutes a significant occurrence in history.

Awards and Personal Life: Kilby

It came as no surprise when the Nobel Prize in Physics was awarded in 2000 to Jack Kilby for his invention of integrated circuits. In 1969, Kilby was awarded the National Medal of Science, and in 1982, he was inducted into the National Inventors Hall of Fame. In 1990, he was awarded the National Medal of Technology, and in 1999, he received the Eta Kappa Nu Vladimir Karapetoff Award. He received an honorary degree of Doctor of Science in 1986 from Rochester Institute of Technology and another from Southern Methodist University. The Merchiston Campus of Napier University in Edinburgh has named its Jack Kilby Computer Center in his honor.

In 1970, Kilby took a leave of absence from TI to do some independent work. During his leave, one of his projects concerned how to apply silicon technology to help generate electrical power from sunlight. From 1978 to 1985, he was Distinguished Professor of Electrical Engineering at Texas A&M University.

Kilby is also the inventor of a thermal printer used in data terminals. In all, he achieved about sixty patents. In 1983, he retired from TI, and he died on June 20, 2005, at age 81, in Dallas, Texas, following a brief battle with cancer.

TI and the Jack Kilby family created the Historic TI archives and the Jack St. Clair Kilby Archives at Southern Methodist University, which are cataloged and stored in the University's DeGolyer Library. The collections include such firsts as the commercial transistor, the integrated circuit, the electronic calculator, and the single-chip microprocessor. Also included are early digital watches and early cell phone technologies.

Awards and Personal Life: Robert Noyce

Robert Noyce was awarded the IEEE Medal of Honor in 1978 for "his contributions to the silicon integrated circuit, a cornerstone in modern electronics." Born December 12, 1927, and the son of a preacher in Burlington, Iowa, he grew up in Grinnell, Iowa, and graduated from Grinnell College. He obtained a Ph.D. in physics from the Massachusetts Institute of Technology in 1953. After a brief time making transistors for Philco, he went to work for Shockley Semiconductor.

Noyce was one of a group of eight who left Shockley Semiconductor in 1957 and started Fairchild Semiconductor. Noyce was general manager of the new company, and he remained there until 1968, when he left with Gordon Moore to start Intel.

Noyce was pretty much always the leader of his associates and always projected an impression of substantial confidence. When the eight left Shockley, he was the leader and became general manager of their startup, Fairchild Semiconductor. At Fairchild, he apparently encouraged a casual working environment with much room for his employees to pursue whatever they wished. Later, at Intel, the culture was not as relaxed and informal. This firm reportedly promotes heavily from among its own personal. The headquarters building of Intel in Santa Clara, California, is named for Noyce, as are the Science Center at Grinnell College and the conference room at the Santa Fe Institute in New Mexico.

In July 1990, at age 62, at home in Austin, Texas, Noyce died unexpectedly of a heart attack in the midst of working to prevent the acquisition of a Silicon Valley materials supplier by a Japanese concern.

MICROPROCESSORS

Summary

Microprocessors are technically innovative and economically important, but it is not now known who could be considered the inventor entity.

A microprocessor is commonly composed of one or more integrated circuits, usually all on one semiconductor chip. It is commonly a programmable device that is adapted to perform one or more particular functions, such as performing calculations, remembering and updating information, processing programs, or directing or executing functions. It can incorporate from thousands to millions of transistors. The difference between a simple integrated circuit and a microprocessor seems to turn on the number of transistors involved, but no defined difference is recognized.

Background

Microprocessors tend to be complicated ICs. They can be considered to have developed from simple ICs. Though ICs at first were small and contained only a few transistors and associated components, circuit designing and fabrication technology advanced under the stimulus of competition and good market for products, even though microprocessors with multiple incorporated components, including circuits, were not easy or simple to fabricate. Over time, microprocessors came to have thousands, then hundreds of thousands, then millions of transistors. Though complex, in time, single microprocessors each constructed on a chip and having a complete CPU adaptable for computer-type operation were fabricated. Eventually, microprocessor CPUs replaced CPUs comprising discrete associated components, such as a group of interconnected integrated circuit chips. Even as early as the mid-1970s, it appears that the microprocessor CPU had replaced all other CPU types.

Once the IC had been invented, many people worked to achieve and produce commercially useful embodiments for various applications. Even though it would be technically difficult and would involve at least thousands of transistors and other components, and in some cases require years of development, innovators and developers at various companies promptly perceived that it would be desirable to produce ICs of various capabilities and to indicate (from a sales and marketing viewpoint) that they had produced microprocessors.

The increase in the processing capacity of increasingly more advanced and complicated integrated circuits has been known generally to follow what became known as Moore's Law. This was the observation made in 1965 by Dr. Gordon Moore, cofounder of Intel, that the number of transistors per square inch on a chip used in an integrated circuit had approximately doubled every year since the integrated circuit had been invented. Later, due to changes in IC fabrication, Moore altered his observation to a doubling every 2 years. Most experts, including Moore himself, expect the law to hold for at least another decade in spite of limiting factors, such as the minimal effective IC size, the maximum density of components such as transistors, and the maximum amount of permissible heat generated (apparently due to inherent current leakage).

Today, microprocessors are widely used in every type of control and computing device, from mainframes to the smallest handheld devices, including machine control systems. One set of market statistics suggests that in 2003 about $44 billion worth of microprocessors were made and sold. About half of this money was spent on CPUs used in desktop or laptop personal computers. However, apparently computer usage accounts for only about 0.2% of all CPUs sold.

Microprocessor History

In about 1970, at almost the same time, at least three concerns succeeded in producing a microprocessor: Texas Instruments with its TMS 1000, Intel with its Intel 4004, and Garrett AiResearch with its Central Air Data Computer (CADC). It is unknown which of these had the first microprocessor operating on the laboratory bench.

Garrett AiResearch

Designers Ray Holt and Steve Geller of Garrett AiResearch responded to an invitation to produce a digital computer to replace existing mechanical systems for the main flight computer of the U.S. Navy's new F-14 Tomcat fighter. Apparently they started in 1968, by 1970, they had completed and implemented a design that evidently used a metal oxide semiconductor-based chipset as the core CPU, identified as the CADC and also as the MP944 of Garrett AiResearch. The design evidently was only a fraction of the size of the prior mechanical systems it replaced and was much more reliable. It was used on all the early Tomcat fighters. The Navy considered the system to be so advanced that it refused to allow publication of the design until 1997.

Intel

Intel was founded in 1968 by two unhappy engineers who were working for Fairchild Semiconductor, Bob Noyce and Gordon Moore. They decided to quit and create their own startup company at a period when many Fairchild employees were leaving and attempting to do similarly. Noyce had prepared a one-page typed proposal regarding aims for their new company, and this was sufficient for San Francisco venture capitalist Art Rock to back their venture. In less than 2 days, Rock raised $2.5 million.

Their new company, Intel, produced as its first successful money-making product the 3101 Schottky bipolar sixty-four-bit static random-access memory chip. Intel has developed into the world's largest semiconductor company and producer of various microprocessors sold to computer manufacturers and others.

In late 1969, Busicom, a potential client from Japan, asked to have twelve different chips custom designed and fabricated for a proposed Busicom calculator. The separate chips would function for keyboard scanning, display control, printer control, and the like. Marcian (Ted) Hoff, who had joined Intel in 1968 upon graduation from Rensselaer as Intel's twelfth employee, determined that Intel could build one chip to accomplish the work of the twelve chips requested. Hoff was one of the earliest people to recognize that IC technology and the new silicon-gated technology could be used to fabricate a single-chip microprocessor CPU. Intel and Busicom agreed to fund the project of designing and building a new, programmable, general-purpose logic chip.

Federico Faggin headed the design and construction team for the new chip, which included Ted Hoff, Stanley Mazor, and Masatoshi Shima. They produced a silicon chip that contained a CPU that had four bits and 2,300 MOS (metal oxide semiconductor) transistors, operated at 740 kHz and 0.06 MIPS, and was built on a chip reportedly 1/8th inch by 1/16th inch having a surface area of 3–4 square millimeters. This chip had as much computing power as the earlier ENIAC computer, which had occupied 3,000 cubic feet and had 18,000 vacuum tubes. Intel designated the chip as its 4004. The chip had to be programmed from a set of instructions to execute specific work.

Intel appreciated that the chip had commercial value and decided to buy back the design and marketing rights from Busicom, which it did, after negotiation, for $60,000. Within a year, Busicom became bankrupt and never produced any items containing the 4004. Gordon Moore liked to call this microprocessor "one of the most revolutionary products in the history of mankind."

Intel introduced its 4004 chip into the market on November 15, 1971. With a good marketing plan, Intel had the 4004 widely used in a few months. However, Intel had other products, and the 4004 did not become a core of Intel's business until about the mid-1980s.

Intel typically gives itself credit, along with Texas Instruments, for the invention of the microprocessor (without clear evidence). TI and other concerns each developed a microprocessor at about the same time.

Current sixty-four-bit microprocessors use designs similar to that of the old 4004. The microprocessor remains the most complex mass-produced electronic product made. New microprocessors reportedly can contain about 1 billion transistors.

According to *A History of Modern Computing* (MIT Press, pp. 220–221), Intel entered into a contract with Computer Terminals Corporation (CTC; later called Datapoint) of San Antonio, Texas, for Intel to produce a microprocessor chip for a computer they were designing. CTC subsequently decided not to use the chip. Intel then improved and marketed this chip as its 8008 beginning in April 1972; this was the first eight-bit microprocessor. This chip was the core of the well-known Mark 8 computer kit advertised in the Radio Electronics magazine in 1974. The 8008 and its successor, the 1974 Intel 8080, which was used in the famous Altair computer kit, are commonly understood to have opened up the market for microprocessor chips.

Subsequently, Federico Fagan left Intel in 1974 after working on the 8080 and started Zilog (a California concern), which manufactured, beginning in 1976, the Zilog Z80, an eight-bit computer, and its later derivatives. The Z80 was apparently compatible with the Intel 8080. The Z80 emphasized low cost, used small packaging, had simple computer bus requirements, and had circuitry that avoided a separate chip. This helped the home computer market to expand in the early 1980s.

Texas Instruments

TI developed its four-bit TMS 1000 and a version identified as TMS1802NC, evidently released on September 17, 1971, as a "calculator on a chip." TI filed a patent application on this microprocessor, which resulted in the issue of U.S. Patent 3,757,306 on September 4, 1973, to Gary Boone, relating to single-chip microprocessor architecture.

In both 1971 and 1976, Intel and TI reportedly entered into broad field licensing agreements (details not disclosed). Intel is believed to have paid royalties to TI under the TI microprocessor patent.

U.S. Patent 4,074,351, for a "computer on a chip," which combines a microprocessor CPU, some memory, and input/output leads on a single chip, was awarded to Gary Boone and Michael J. Cochran, both of TI. In the industry, a microcomputer conventionally is sometimes considered to be a computer that uses one or more microprocessors as its CPU, but the patent indicates a concept for a microprocessor that is similar to a microcontroller.

Motorola

A competing microprocessor chip, the Motorola MC6800, was released in August 1974. The MC6800 CPU competed comparably with the Intel 8008, but was apparently easier to implement than the Intel 8080. The MC6800 needed only one operational voltage and no support chips. This microprocessor was sold mainly for peripheral and industrial controls.

In 1978, Motorola began selling the MC6809 eight-bit microprocessor, which was a complex, hard-wired logic design. It was very successful and long lived.

MOS Technology

Several engineer designers of the Motorola MC6800 microprocessor left Motorola in about 1975 and joined the small company called MOS Technology. A design team at MOS was headed by Chuck Peddle and included Bill Mensch. The team achieved a new 6501 CPU microprocessor that somewhat outperformed the Motorola 6800 microprocessor yet included several design simplifications. Its pin compatibility with the Motorola 6800 led to a lawsuit between the two companies. Subsequently, MOS replaced its 6501 with the 6502, which was not pin compatible with the Motorola MC6800. The MOS 6502 was reportedly used in the Apple computers.

The Digital Age of the 1980s

In the 1980s, microprocessors and uses for microprocessors seemed to explode. Various microprocessors became components of equipment that sold for many subsequent years. Various CPU clones appeared, such as the NEC V20 and V30, clones of Intel CPUs.

Microprocessors constituted a major development, as the above review of microprocessor developments suggests. However, study of available published information suggests that no single inventive entity can be said to have been responsible for the invention of the microprocessor. Basically, the invention of the integrated circuit by Kilby seems to have stimulated creation and development by others of various microprocessors. Various improvement-type inventions relating to particular types or forms of microprocessors have appeared.

SMALL COMPUTERS

Summary

Although microprocessor CPUs appeared on the market, at first no one seems to have made a prototype, manufactured and sold a small computer based thereon, although experts suggest that small prototype computers were made by amateurs from CPU microprocessors.

One particularly well-known example of a personal computer incorporating a microprocessor CPU involves Steve Wozniak, who found a commercially offered, inexpensive microprocessor CPU. Steve Wozniak was then a member of a California club of electronic enthusiasts. In the mid-1970s, he bought the CPU and created and built a little computer. His friend Steve Jobs thought there was a market for this computer, and, as the Apple I, hundreds were made and sold. Apple I was succeeded and replaced by the improved Apple II computer, and about 2 million were made and sold. Those computers utilized the microprocessor CPU and are today commonly regarded as the first and iconic embodiments of the personal computer fitted with a microprocessor CPU.

Background

A computer today is usually definable as a programmable electronic device that performs at least one function in response to a program. It contains at least one CPU and can be functionally associated with peripherals that can receive input data for the CPU and dispense output data from the CPU.

A CPU is a programmable electronic device that is incorporated into a computer that has display and keyboard capacity. It stores information, processes information, performs calculations, controls operational sequences, and initiates and executes commands. A CPU includes a control unit, a clock, and an arithmetic and logic unit, and necessary registers and links to or for information storage and to or for peripherals. A microprocessor can be fabricated to incorporate or comprise a CPU.

Brief History of Computers

To place the modern small computer and the microprocessor in the relevant contextual field, it is desirable to know the history of the computer. Originally, a "computer" was a person, usually a woman, who, seated at a "counting table," performed repetitive, boring calculations needed for navigation tables, planetary positions, tidal charts, taxes, retail prices, and so forth. For centuries, inventors sought mechanisms to perform such tasks.

Mechanical calculators arrived first. The abacus (a memory aid for calculations now attributed to the ancient Babylonians) was followed by John Napier's (1550–1617) "bones" in 1617. Napier, the Scottish mathematician, in 1594 invented logarithms, and in 1617 his logarithm sticks ("Napier's bones") marked lengthwise with logarithm values. William Oughtred (1574–1660), the English mathematician, invented the slide rule in 1622 in England, embodiments of which were treasured personal equipment of engineers for centuries. A sketch of a gear-driven calculating machine was found posthumously in one of Leonardo da Vinci's (1452–1519) notebook entries, but it was never embodied. The first known actual gear-driven machine calculator (sometimes at first called a "calculating clock" because of its resemblance to a clock mechanism) was independently invented by German professor Wilhelm Schickard in 1623. Blaise Pascal (1623–1662), a French child prodigy, invented a gear-driven adding machine called the Pascaline in 1642 to aid his father, a tax collector. A few years later, German Gottfried Leibnitz (1646–1716), coinventor of calculus with Isaac Newton, created his "stepped reckoner," a calculator that employed cooperating fluted drums, with ten flutes per drum, which could perform the four basic functions of addition, subtraction, multiplication, and division. Leibnitz also proposed the binary number system, which avoids problems with usual base 10 (decimal) systems and which system is common to modern computers.

Machine-readable punched cards were invented by Frenchman Joseph Jacquard (1752–1834) in 1801 for use in his automatic power loom, but it was Englishman Charles Babbage (1792–1871) who is credited with conceiving the first general purpose mechanical computer (which he eventually called the analytic engine) that was program-operated by punched cards and used gears as its CPU beginning as early as 1822. A program—an essential requirement for a true computer—is a conditional instructional statement that allows a computer to achieve a different result each time the program is run, depending upon the input. Although Babbage had invented the true computer, construction of an embodiment of his computer was never completed. Babbage's young friend, Ada Byron, daughter of the poet Lord Byron and later, by marriage, Countess Lady Lovelace, learned enough about Babbage's analytic engine to begin constructing programs or sequences of instructions for the machine. She invented the subroutine and identified the importance of looping in programming. Babbage is considered by experts to be the father of the modern computer.

In response to a prize offered by the U.S. Census Bureau for a device to help count the 1890 Census, Herman Hollerith entered the contest and won the prize by inventing the Hollerith desk. This invention comprised a card reader that sensed holes in a punched card, which enabled a gear-driven mechanism to count and to display the results of the count. With his cards, Hollerith incorporated a read/write technology (which, unknown to him, had already been invented by Babbage). His cards were successful. Hollerith formed a company, the Tabulating Machine Company, which later, after a small series of buyouts, became International Business Machines (IBM). As this business grew and its equipment became extensively used in accounting, the use of punched cards became universally present and known to consumers and to the public in things such as bills to homeowners, toll road charges, and ballots.

By World War II (WWII), the U.S. military needed better equipment for calculating the firing tables needed in estimating ballistic projectile trajectories and the like. These tables are based on complicated equations incorporating variables such as wind, drag, gravity, and muzzle velocity. Human computers were not adequate. The first programmable digital computer was constructed in 1944 and included switches, relays, clutches, and a 50-foot rotating shaft, as well as electronic components. Called the Mark 1, it used paper tape readers (instead of punched cards). Its main designer was Howard Aiken of Harvard University, and it was achieved by a partnership of IBM and Harvard. It ran nonstop for 15 years.

One of the Mark 1's primary programmers was the talented Grace Hopper, who once found a dead moth in the equipment. Ever since, Hopper's word "debugging" has been used to describe a process of eliminating program faults. Hopper innovated the world's first compiler, which is a necessity to translate a program language into the binary language of a computer.

Various attempts were made to construct all-electronic computers. In 1937, J. V. Atanasoff, a professor of physics and mathematics at Iowa State University, attempted to construct one such, and by 1941, with a graduate student, Clifford Berry, he succeeded. Though this machine could store data as a capacitor charge (as in some computers of today), it was not programmable (so, technically, it was a calculator). This machine was not pursued further until after WWII.

During WWII, the Colossus computer was secretly built in Britain for the purpose—in which it was successful—of breaking the cryptographic codes being used by Germany in its Enigma machine, which encrypted and decrypted messages intended to be secret. The Enigma machine was thought by Germany to produce unbreakable coded messages. With the Colossus, Britain was routinely able to "read" coded German radio transmissions. The Colossus, though, was not a general purpose, reprogrammable computer, and it employed some mechanical elements.

Between 1943 and 1945, the ENIAC (Electronic Numerical Integrator and Calculator), an all-electronic digital computer, was built at the University of Pennsylvania by professors J. Presper Eckert and John Mauchly, who were funded by the War Department because they said they would build a device what would replace all the women "computers" who were calculating firing tables for the army's big artillery. As built, ENIAC filled a 20 by 40 foot specially cooled room, weighed 30 tons, used more than eighteen thousand vacuum tubes, and used paper card readers from IBM. To reprogram it, many patch cords had to be rearranged between selected settings among more than three thousand switches. Many experts thought the machine would not work because of the unreliability of vacuum tubes, but Eckert solved this problem by very careful circuit design. The first program run computed whether or not it was possible to build a hydrogen bomb, a task which ENIAC completed in 20 seconds, answering affirmatively. The program used remains still classified today.

Eckert and Mauchly next teamed with John von Neumann (1903–1957), a professor at Princeton, to design EDVAC, which provided stored program capability and avoided the reprogramming problem associated with ENIAC. Nevertheless, ENIAC is considered by experts to be the origin of the U.S. commercial computer industry.

In 1959, IBM achieved its Stretch computer, which had a 33-foot length and held about 150,000 transistors, which replaced the vacuum tubes, but the transistors still were individual elements necessitating individual assembly.

In 1965, the work of German Konrad Zuse was first published in English, and many English and American computer experts found that Zuse had antedated some of their independent developments. In Nazi Germany, Zuse had produced a group of successively improved general purpose computers beginning in 1936. His first Z1 was constructed in the living room of his parent's home. His third machine, Z3, constructed in 1941, was clearly a general purpose, software-controlled (programmable) digital computer. Zuse had unknowingly reinvented Leibnitz's binary representation of numbers and Babbage's concept of programming. Although Z1, Z2, and Z3 were destroyed in Allied bombing, the Z4 survived because Zuse hauled it in a wagon up into the nearby mountains.

Until the invention of the integrated circuit, its development into the microprocessor, and the microprocessor into a CPU, computers were large, cumbersome, and expensive. Frequently, each was used by more than one person on a time-sharing basis.

The pace of technological development, particularly in the latter half of the twentieth century, is well exemplified by the swift progression of solid state electronics, as illustrated by the following sequence:

- The appearance of the transistor in 1947
- The invention of the integrated circuit in the 1950s

- The development of integrated circuit microprocessors complex enough to contain a CPU in the late 1960s
- The commercial realization of small, low-cost computers in the mid-1970s

The microprocessor CPU—the key computer innovation—had been made possible not only by the advent of practical, reproducible technology enabling the placing of large numbers of transistors and other components on a chip surface, but also by the associated reduced cost of processing capacity.

Large electronic computers had been around for about 20 years when the first commercially available microprocessor, the Intel 4004, was offered by Intel in 1971 (Intel did not invent the modern computer). Motorola in 1974 began commercially offering another microprocessor, the MC 8800, in 1974. However, it was not until 1975, when MOS Technology began offering still other, less expensive microprocessors, the MC 6501 and the MC 6502 (which replaced the MC6501), that the small computer, dubbed a personal computer, suitable for broad public utilization, became commercially possible, as described below.

As the preceding brief, even truncated, review of computer history shows, before the microprocessor and the capability of achieving a small CPU, computers were large, clumsy, and quite limited in functionality. Beginning about 1970, in view of developments in integrated circuit technology, a technically knowledgeable person might have asked, when will a small or personal computer appear?

Steve Wozniak

Childhood

Steve Gary Wozniak was born August 11, 1950, in San Jose, in suburban Santa Clara, California, in the area now known as Silicon Valley. His father was an engineer for Lockheed and his mother was president of a Republican women's club. Already a technological center, Silicon Valley growth expanded after the launch of the Soviet *Sputnik*. As a boy, Steve was deeply involved with electronics. He made many devices from kits and from purchased components.

Thought bright, school bored him. After graduating from high school, he was admitted to the University of Colorado but flunked out. He got a job at Hewlett-Packard in Palo Alto, California; there, while working on a mainframe computer, he was introduced to Steve Jobs by a mutual friend, Bill Fernandez, when Steve Wozniak was 21 and Steve Jobs was 16.

Creation of the Low-Cost Personal Computer

About 1975, when Steve Wozniak was about 25, he became a member of the Palo Alto Homebrew Computer Club, a group of electronics hobbyists.

Some members had assembled computers from the kits offered through *Popular Mechanics* magazine for the Altair 8800 (available from Micro Instrumentation and Telemetry Systems), which was based on the Intel 8080 CPU and the IMSAI (available from IMS Associates and also based on the Intel 8080 CPU). Steve Wozniak had a video teletype. He was motivated by these club friends and their assembled computer kits to associate his video teletype with a microprocessor and achieve a complete computer.

However, the only microprocessor CPUs available to him were the Intel 8080, which retailed for $179, and the Motorola 6800, which retailed for $170. Steve liked the 6800, but both microprocessors were out of his price range. He just watched, learned from other members, and designed some computers on paper. He waited for a day when he could afford to buy a microprocessor CPU.

He did not have long to wait. In 1976, MOS Technology offered its 6502 microprocessor for $20. Steve may not then have known it, but the 6502 was designed by the same engineers who had designed the 6800 for Motorola, but they had left Motorola and joined MOS Technology, a small company. At MOS, these engineers had discovered how to fix photolithographic mask flaws after the masks were made. (The masks were used in the manufacture of integrated circuits.) The consequence was that the number of successful chips they made increased from about 30% to about 70%, thus making manufacturing costs less and allowing the selling price to be reduced.

Steve wrote a version of BASIC for operating this MOS 6502 microprocessor and proceeded to design a computer for it to operate. Although the MOS 6502 was different from the Motorola 6800, he found that, to use the MOS 6502, he needed to make only minor changes in his earlier paper-designed computer that used the Motorola 6800. Steve built his computer and took it to Homebrew Computer Club meetings to demonstrate it and show it off.

At one meeting he attended with his new machine, his old friend Steve Jobs also attended. Jobs was interested in the commercial potential for small hobby machines and was immediately struck by Steve Wozniak's machine. Apparently, portable, small-size, low-cost computers with an associated keyboard and viewing screen, limited memory, and software that enabled it to be started, turned off, and used for some limited, programmed tasks seemed novel and appealing.

Steve Jobs

Early Years

Steve was born February 24, 1955, in San Francisco to (American) Joanne Carole Schieble, a graduate student, and (Syrian) Abdulfattah John Jandali, a graduate student who later became a political science professor. One week after birth, Steve was offered for adoption by his (unmarried) mother, and he was adopted by Paul and Clara Jobs of Mountain View, Santa Clara County, California, who named him Steven Paul Jobs.

His biological parents later married and had another child, Jobs's sister, who became the novelist Mona Simpson and whom Steve did not meet until they were adults. This marriage ended in divorce years later. Steve considers Paul and Clara Jobs to be his parents.

Jobs attended Cupertino Middle School and Homestead High School in Cupertino, California. He attended after school lectures at Hewlett-Packard in Palo Alto and was soon hired as a summer employee. He worked with Steve Wozniak.

After his high school graduation in 1972, Jobs enrolled in Reed College at Portland, Oregon, but dropped out after one semester. He continued to audit some courses at Reed, including one in calligraphy. In 1974, he returned to California and started attending meetings of the Homebrew Computer Club with Steve Wozniak. He also took a job as a technician at Atari, the manufacturer of video games, with the aim of saving money for a spiritual journey to India.

Jobs and Wozniak became interested in the earlier discovery in the 1960s that the promotional toy whistle in each box of Cap'n Crunch breakfast cereal could be partially taped over and then was able to reproduce the 2600 hertz tone used by the AT&T long distance telephone system. After a visit with John Draper, who popularized "phone phreakers," Jobs and Wozniak briefly went into business in 1974 building "blue boxes" that enabled illicit free long-distance calls.

Jobs then went to India and backpacked about with a Reed College friend, Daniel Kottke, seeking spiritual enlightenment. He returned with his head shaved and wearing traditional Indian clothing. He got his old job back at Atari and was assigned the job of making a circuit board for the game Breakout. Atari offered $100 payment for each chip that could be removed from the game machine. Since Jobs had little knowledge of circuit board design, he got Wozniak to work on the matter and offered to split the payment money equally with him. Wozniak reduced the number of chips by 50, but the resulting design supposedly could not be reproduced on an assembly line. Jobs told Wozniak that Atari had given him only $700, not $5,000, so that Wozniak's share was $350.

Building and Marketing the Small Computer

Jobs, now 21, decided Wozniak's computer, if manufactured, was marketable compared with the competition, which was, for example, the kit for the Altair (see the review of the Altair in the chapter on Bill Gates). He persuaded Wozniak to assist him in this effort.

The two pooled their financial resources: Jobs sold his Volkswagen van and Wozniak sold his Hewlett-Packard scientific calculator. The proceeds totaled $1,300. They assembled the first prototypes in Jobs's bedroom and later, for space, proceeded in the garage. Wozniak named the computer the Apple I; it incorporated a single circuit board with some ROM, a little RAM, a keyboard, and a monitor, so that altogether it could be considered to comprise a complete microcomputer.

Designing the Apple I was not as simple as it seemed, though this computer was apparently not the first of its type to use monitors and cassette storage with a keyboard. Wozniak's machine used a TV tube as a display screen. It displayed text at only the rate of 60 characters per second, but this was still faster than the contemporary teletypes. His machine was easy to start up (boot), with its bootstrap code on ROM. It had a cassette interface for loading and saving programs at the then rapid rate of 1,200 bits per second. Overall, his computer was quite simple, had good design, and used fewer parts than similar machines.

Jobs took an embodiment of Wozniak's machine and solicited the owner of a local computer store chain, The Byte Shop, to buy it. The owner, Paul Terrell, said he would order 50 and agreed to pay $666.66 each on delivery if the machines were fully assembled, and he gave Jobs a purchase order. Jobs took the purchase order to a national electronics parts distributor, Cramer Electronics, and ordered the parts needed for the computers. When the credit manager asked about payment, Jobs presented the purchase order and said he would pay in 30 days, in which time the computers would be built, delivered, and paid for. Then the credit manager called Terrell and verified the validity of the purchase order and payment. Terrell also said that upon delivery, he would pay for the computers, so that Jobs would have more than enough money to pay for the parts. The parts were given to Jobs by Cramer.

Working day and night, Jobs and Wozniak, along with their friend Ronald Wayne, built and tested the computers and delivered them on time to Terrell, who immediately paid. More than two hundred Apple I computers were eventually built and sold, some apparently in kit form.

Development

With incoming revenue from sales of Apple I, Wozniak now focused on fixing shortcomings of the Apple I and worked on developing a new, improved design for the Apple II. He retained simplicity and usability, but added high-resolution graphics, enabling pictures to be displayed. He commented, "I threw in high res. It was only two chips. I didn't know if people would use it." By 1978, he had designed an inexpensive floppy disk drive controller. With Randy Wigginton, he wrote a simple disk operating system and file system. For the disk operating system, a contract with Shepardson Microsystems was worked out to build a simple command line interface. In addition to selecting the hardware, Wozniak wrote most of the software initially furnished with an Apple. He wrote, among other things a programming language interpreter, a set of virtual 16-bit processor instructions known as SWEET 16, a game called Breakout (complete with added sound), and the code needed to control the disk drive. The TV interface was completely redesigned. Color was eventually added. Jobs pushed for an improved base and keyboard.

Influence of Jobs

Building units of the new Apple II machine would cost a lot of money. Wayne was hesitant to continue and eventually dropped out. Jobs looked for cash, but banks were reluctant to lend him money. He eventually found "Mike" Markula, who cosigned a note for a bank loan of $250,000. Markula, along with Jobs and Wozniak, effective April 1, 1976, formed Apple Computer Inc. Wozniak quit his job at Hewlett-Packard and became vice president in charge of research and development at Apple.

The Apple II was introduced at the West Coast Computer Faire on April 16, 1977. The new computer was successful and is apparently generally credited with creating the home computer market. More than two million Apple IIs were shipped at $970 each for the 4 KB model. Apple II for years was the principal source of profit for the company.

The Apple II was one of three personal computers that were introduced in 1977, which had to compete against each other. The other two were the Tandy Corporation TRS-80 and the Commodore International Commodore PET. All Three used the MOS 6502 microprocessor CPU. The company MOS Technology was then owned by Commodore. The Apple II was not the market leader in the late 1970s. The TRS-80 was retailed through Tandy's more than three thousand Radio Shack stores. It came with a monitor, was priced substantially under the Apple II, and had the largest software selection available in the microcomputer market. The Commodore PET was succeeded by the VIC-20 and then in 1982 by the C64, which sold very well.

The Apple II soon incorporated a 5¼-inch floppy disk drive and interface. Various model releases occurred, including the II+, IIe, IIc, and IIgs. Beginning in October 1979, the spreadsheet program Visi-Calc became available for the Apple II because its programmers, Daniel Brocklin and Bob Frankston, chose Apple II as the desktop platform. This was apparently the first spreadsheet program. It created a business market for the Apple II and helped sales for the Apple II even in the home market. By the end of the 1970s, Apple Computer had a production line and a staff of computer designers.

In December 1980, Apple Computer had an initial public offering (IPO) of its stock. This IPO generated a great deal of money and immediately produced more millionaires (said to be about three hundred) than any prior company. Jobs and Wozniak each became multimillionaires.

Jobs after 1980

In the early 1980s, Apple Computer found itself facing increasing competition, especially from the long-established IBM, which was then entering the field of desktop computing. The IBM PC (IBM model no. 5150) used the Intel 8088 microprocessor and Microsoft's DOS operating system. Apple Computer introduced the Apple III in May 1980 and the Lisa in 1983. Briefly, the Apple III failed allegedly because Jobs decided that the machine needed no cooling fan, as a result of which thousands of units failed or were recalled because of overheating. The Lisa (named after Jobs's daughter) incorporated a new graphical user interface (GUI) and mouse instead of the then-standard text-based interface. Lisa failed allegedly mainly because it was overpriced. Apple Computer continued to advance because of its sales of Apple II, particularly into the education (school) market.

In 1983, looking for an experienced executive, Jobs hired John Sculley from Pepsi-Cola to be Apple Computer's chief executive officer (CEO).

In January 1984, Jobs at the annual shareholders meeting announced Apple's new Macintosh computer, which later became the first successful computer with a GUI. However, near the end of 1984, a sales slump was blamed for causing a "deterioration" in the Sculley-Jobs working relationship. Following significant layoffs and an internal power struggle in which the board of directors backed Sculley, Sculley relieved Jobs of his position as head of the Macintosh division in 1985. Sculley allegedly feared Jobs would produce a product that, like the Lisa, would be unsuccessful.

Jobs left Apple Computer and proceeded to found a new computer company, NeXT Computer, in 1986, which soon developed the NeXT workstation. Though technologically advanced, this product never achieved sales of mainstream status, allegedly because of its high cost. Jobs marketed the product to the academic and scientific fields, emphasizing the innovative, experimental technologies it incorporated. Perhaps the most famous user was Tim Berners-Lee at CERN in Switzerland, who developed the World Wide Web using the product. By 1993, only 50,000 machines had been sold.

Meanwhile, at Apple Computer, things had deteriorated for various reasons. The Macintosh Portable introduced in 1989 failed, but it was followed by the more popular Macintosh Powerbook in 1991. Various desktop models of the Macintosh were made and introduced evidently with mediocre success in the late 1989s and the early 1990s. Computers with a GUI and based on the IBM PC seemed to dominate and control the market. The Apple II computers were discontinued in the early 1990s. In 1994, Apple began using IBM and Motorola hardware, such as the IBM PowerPC processor. Apple's sales continued to become a progressively smaller share of the computer market.

In 1996, Apple Computer agreed to buy NeXT for $402 million. This Apple/NeXT deal brought Jobs back into Apple Computer. In July 1997, Jobs became Apple's "interim CEO" when the then- CEO, Gil Amelio, was fired by the board of directors. Jobs at once terminated a number of projects and employees and infused major components of NeXT, especially software, into Apple products. Licensing of the Apple operating system to competitors was discontinued. The Mac OS X system was evolved, and the iMac computer and other new products were developed and introduced. Jobs emphasized appealing designs and strong branding. Sales increased significantly. In 1999, the Motorola PowerPC 7400 microprocessor CPU was used in the Apple Power Mac G4.

At the 2000 Macworld Expo, Jobs dropped the "interim" from his title and became the permanent CEO. He announced a 5-year deal with Microsoft under which Microsoft would develop software for Apple computers. A lawsuit between Microsoft and Apple was settled, and Microsoft bought $150 million of nonvoting stock in Apple. However, Apple's market share continued to decline and reached 3% by 2004.

To pursue a broader market, Apple opened retail stores in the United States and elsewhere. In June 6, 2005, Apple through Jobs announced that Macintosh computer would use Intel microprocessors, and in early 2006 Jobs announced that Apple would transition to Intel processors on all computer hardware.

In January 2007, Jobs shortened the company name to Apple, Inc., from Apple Computer, Inc., since the company was then selling the very successful iPod and TV receivers, not just computers. The iPod media player had been wildly successful for the company. Jobs also announced that Apple would enter the cell phone field in a partnership with Cingular Wireless (now part of AT&T). Since cash reserves increased substantially in 2006, Apple and Jobs created Braeburn Capital in April 2006 to manage its assets.

Wozniak after Apple II

Although Wozniak withdrew from the University of California in 1975, he later returned and gained a B.S. degree in electrical engineering and computer science in 1986 under an alias, Rocky Clark.

In February 1981, Steve Wozniak crashed his single-engine Beach Bonanza while attempting a takeoff from Santa Cruz Sky Park. The National Transportation Safety Board investigation and report showed that Wozniak was not rated for this aircraft, did not have a "high performance" endorsement on his pilot's license and so was unqualified to operate the airplane, and had a "lack of familiarity with [the] aircraft." The cause of the crash was a premature liftoff, followed by a stall and a "mush" into a 12-foot embankment.

From the crash, Steve suffered retrograde amnesia and temporary amnesia. He had no recollection of the accident and at first did not know he had had a crash. He at first could not remember things such as the day of the week or why he was in a particular room. Gradually, he pieced together some facts, but it was not until he asked his girlfriend, Candi Clark, if he had been in an accident and she told him about the event that his short-term memory was restored. He also attributes his restoration of memory to playing Apple II computer games.

After recovering, he did not return to Apple. He married Clark. In 1983, he returned to product development at Apple, but he wanted only a role as an engineer and to be a "motivational factor" for the Apple employees. In February 1987, after about 12 years with the company, Wozniak terminated his full-time employment with Apple, but he remains an employee and a stockholder and receives a paycheck. He maintains connections with Steve Jobs.

Also in 1987, he divorced from Candi. They had three children. He then renewed his friendship with Suzanne Mulkern, head cheerleader and homecoming queen of his high school years. They were married in 1990 but divorced in 2004.

After the Apple II and after Apple Computer's IPO, Wozniak involved himself with various projects:

- In 1982 and 1983, Wozniak sponsored a combined technology exposition and rock festival.
- In 1987, he started a venture called C.L. 9, which developed and marketed a universal TV remote control. Evidently, the venture was not a big success.
- In about 1987, he joined the Masons for the purpose of spending more time with his first wife Alice, (married in 1976, divorced 1977), who was a member of the corresponding women's group, the Eastern Star.
- About this time, he began teaching fifth grade students and undertaking charitable activities in technology programs of his local school district.
- In 1997, he became a benefactor of the Children's Discovery Museum of San Jose and a Fellow of the Computer History Museum.
- In 2001, he cofounded Wheels of Zeus (formed so as to have the acronym Woz, his nickname) to create wireless global positioning system technology. The venture was closed in 2006.
- In 2002, he joined the board of directors of Ripcord Newworks, Inc., a new telecommunications venture.
- Later in 2002, he joined the board of directors of Danger, Inc., the maker of Hip Top.
- In 2006, Wheels of Zeus was closed and, along with Apple alumni Ellen Hancock and Gil Amelio, he founded Acquicor Technology, a shell company for buying technology companies and developing them.
- In 2006, he published his autobiography (*iWoz: From Computer Geek to Cult Icon: How I Invented the Personal Computer, Co-Founded Apple, and Had Fun Doing It*), which was coauthored with Gina Smith.

He has received honorary doctorate degrees from the University of Colorado, North Carolina State University, Kettering University, and Nova Southeastern University. Also, he received the National Medal of Technology in

1985 from President Ronald Reagan and was inducted into the National Inventors Hall of Fame in 2000. He received the Telluride Tech Festival Award of Technology in about 2004.

Wozniak and Jobs after Apple II

After Apple II, Steve Wozniak was evidently only peripherally been involved with computers and computer development. Steve Jobs has continued as a developer and manager in the Apple computer business and has become involved in related fields. *Fortune Magazine* in November 2007 named him "the most powerful person in business."

FURTHER RESOURCES

Paul E. Ceruzzi. *A History of Modern Computing*. Cambridge, MA: MIT Press, 1998.

D. A. Hodges, H. G. Jackson, and R. Saleh. *Analysis and Design of Digital Integrated Circuits*. New York: McGraw-Hill, 2003.

Jack Kilby. IEEE Virtual Museum, 2008. http://www.ieeeghn.org.

T. H. Reid. *The Chip*. New York: Random House, 2001.

Steve Wozniak and G. Smith. *iWoz: From Computer Geek to Cult Icon: How I Invented the Personal Computer, Co-Founded Apple, and Had Fun Doing It*. New York: W. W. Norton & Company, 2006.

J. S. Young. *Steve Jobs: The Journey Is the Reward*. Glenview, IL: Scott Foresman & Co., 1987.

Jeffrey Zygmont. *Microchip: An Idea, Its Genesis, and the Revolution It Created*. Scranton, PA: Perseus Books Group, 2003.

AP Photo

Satellite Communications and John R. Pierce

John Pierce was, as far as is known, the first to describe unmanned communication satellites. After a successful demonstration of communication feasibility using a passive metalized sphere, a demonstration with an active satellite able to receive, amplify, and retransmit a signal convincingly showed feasibility for an operational system. Pierce's employment position was utilized to make the development advance, though he himself was not the actual inventor of any unmanned communication satellite or any portion of one. Unmanned communication satellites developed into global significance and became key portions of various commercial products and services.

BACKGROUND

Artificial satellites, that is, objects placed in orbit by human actions, began as hypotheticals: Edward E. Hale with his serialized short story "The Brick Moon" in *The Atlantic Monthly* beginning in 1869, Jules Verne's *The Begum's Millions* in 1879, Konstantin Tsiolkovsky's Russian work *The Exploration of Cosmic Space by Reaction Devices* in 1903, and Herman Potocnik (writing as Herman Noordung) in his book *The Problem of Space Travel—The Rocket Motor* in 1928. English science fiction writer Arthur C. Clarke, though, seems to have been the first to imagine communications satellites for global mass communications in his article in *Wireless World* in 1945.

The United States is reported to have been considering orbital satellites beginning in 1945, and in 1946 the U.S. Air Force Project RAND released its report "Preliminary Design of an Experimental World-Circling Spaceship" and indicated that a "satellite vehicle" could be "one of the most potent scientific tools." The White House on July 29, 1955, announced that the United States intended to launch a satellite by spring of 1958, and the USSR announced on July 31, 1955, that it would launch a satellite by fall of 1957. On October 4, 1957, the USSR launched the *Sputnik 1* satellite, and on January 31, 1958, the United States launched the *Explorer 1* satellite (using a Jupiter C rocket).

In the 40 years following the start of the space age in 1957, when the Soviets launched the unmanned *Sputnik*, the only commercial technology to emerge was satellite communication. From communications satellites by private corporations, the Communications Satellite Corporation (COMSTAT) was established by Congress. Later, COMSTAT became the American component of the emerging global system, the International Telecommunications Satellite Consortium (INTELSAT), in 1964. The INTELSAT system grew and now includes more member states than the United Nations. By the end of the century, orbiting satellites were producing billions of dollars annually in sales of products and services and were concerned with telephone and Internet communications, commercial broadcasting, and business, scientific, and individual exchanges worldwide. John Pierce is the reason we have satellite communication.

Since 1957, the U.S. Space Surveillance Network (SSN) has tracked space objects. It is reported that the SSN currently tracks more than 26,000 objects down to about 10 centimeters in diameter, of which more than 8,000 are human-made orbiting objects. About 7% of them are operating satellites (about 560 satellites), and the remainder are "space debris." It is estimated that there are about 2,465 artificial satellites orbiting the earth in geocentric orbits, that is, orbiting around the earth, as distinct from some other orbit, such as a heliocentric orbit (orbiting around the sun).

Operational satellites can be variously classified into several different types, depending upon matters such as size, function, orbit, and altitude. However, communications satellites seem to have had more effect upon humankind than any others. Also, satellite communications has the only truly commercial space technology. It generates billions of dollars annually in sales of products and services.

JOHN R. PIERCE

John R. Pierce is regarded as the innovator of the important communication satellite system that is used around the world, even though he is not the actual inventor of any technical part of the system, of a communication satellite, or of a rocket used to lift and position a satellite in orbit. His work, nevertheless, deserves recognition. His situation is unique among the iconic inventions considered in this work.

So far as is known, John Pierce was the first to propose unmanned satellites for use in global communication, such as are in fact always used. In addition, from what is known about his activities and position, he was the instigator, leader, and developer who brought communication satellites into reality. A search of available records seems to confirm and strengthen this conclusion.

Early Years

John Robinson Pierce was born in Des Moines, Iowa, on March 27, 1910, the son of a traveling salesman. Around 1929, the family moved to Long Beach, California, where John went to high school and graduated in 1929. It seems that the family then moved to Pasadena, where John attended the California Institute of Technology (Caltech) and received his B.S. in 1933, his M.S. in 1934, and his Ph.D. in 1936. He had an early interest in gliding and participated in the development of the Long Beach Glider Club in Los Angeles, California, which was one of the earliest glider clubs in the United States.

Shortly after leaving Caltech, he started work at Bell Telephone Laboratories (Bell Labs, later renamed AT&T Bell Laboratories) in New York City, where he was involved with problems in vacuum tube design.

During World War II, Bell Labs became involved in microwave research and supported the U.S. radar development effort. John helped develop an improved type of reflex klystron tube that was used in radar receivers.

In 1951, Rudolf Kompfner from Austria became a Bell Labs employee and began working on a traveling-wave tube, a new type of microwave amplifier. There was then practically no electronic device that could amplify microwaves in a broad range of frequencies. Pierce worked with Kompfner and, along with other Bell Labs associates, developed the tube into a practical microwave amplifier. In the tube, the waves were slowed down sufficiently to be associated with an electron beam, which transferred energy to the waves. The tube remains useful in the construction of microwave relay stations both on the ground and in orbiting communications satellites.

Echo

John recalled that at least from his teens he had read science fiction stories. When 20 and a student at Caltech, he had written a science fiction story that took second place in a contest. He continued to write science fiction stories and reported that he had published twenty-two such stories and also twenty-two factual articles in science fiction magazines under the pseudonym J. J. Coupling to avoid Bell Labs' release procedures that "had nothing to do with the sort of things [he] wrote." Early science fiction typically involved robots and space travel. The German V-2 rocket, used for bombarding London near the end of World War II, brought space travel closer to him. Apart from his work at Bell Labs, as a science fiction fan and expert, he gave lectures on space, the effects of zero gravity on astronauts, and other science and science fiction things.

In 1954, when asked to give a space talk to the Princeton section of the Institute of Radio Engineers (IRE) (now part of the Institute of Electrical and Electronics Engineers), he decided that a science fiction talk was not appropriate. Instead, he talked of the idea of communications satellites. He was not then familiar with Arthur Clarke's 1945 article on the use of manned synchronous satellites for communication. In a 1952 article, John had calculated the power necessary to transmit signals between earth and the moon, planets and stars, and large balloon-type satellites and earth. Such satellites, he found, would scatter intercepted radio signals and reflect back to an antenna on earth only about a billionth of a billionth of the original transmitted signal. He also made calculations for active satellites with radio receivers, amplifiers, and transmitters that were both at low altitude and at 20,000 miles above the earth's surface. He "was amazed and delighted" that his calculations, using then-available microwave equipment, showed that the power required on active satellites was "very small."

His IRE lecture was published in April 1955. Based on known information, Pierce was in fact the first to discuss unmanned communication satellites.

Nothing really came from his idea of unmanned communication satellites until after October 4, 1957, when the Soviet Union launched *Sputnik*. His reaction? "It's like a writer of detective stories going home and finding a body in his living room." For him, "space had suddenly changed from science fiction to a technology that could be put to some sensible use."

Following a stepwise developmental progression, John thought a satellite comprising a passive metalized sphere would demonstrate satellite communication feasibility. Such a sphere would reflect but would not generate or amplify radio or other electronic signals. He found that a National Aeronautics and Space Administration (NASA) employee, William J. O'Sullivan, had built a 100-foot-diameter, aluminum-coated Mylar balloon as a device to measure atmospheric density 1,000 miles above the earth, although the drag on the balloon orbiting as a satellite might be considered a problem. O'Sullivan had been unable to get NASA to launch it. The balloon was the same size that John had previously considered for evaluation of passive satellite communication. Its aluminum coating was found to be sufficient to reflect almost all the microwaves that struck it.

At first, John's idea that the balloon was a good way to get started was not accepted. Some wanted an operational military communication system immediately and proposed a satellite called Advent in 1960, but in 1962 that program was "reoriented" (abandoned and something else substituted) after an expenditure reportedly of $170 million.

By this time, John Pierce had gradually advanced a long way up into Bell Labs and AT&T management. He was well liked, had contacts and friends throughout the organization, and had demonstrated a capacity for management. He was able to influence and shape developments and projects.

Finally, NASA and Bell Labs in early 1959 reached an agreement about the passive balloon, then called Project Echo. NASA would supply the balloon and launch it. Bell Labs would build an east coast ground station and lease it to NASA. The Jet Propulsion Laboratory (JPL), then a part of NASA, would build a west coast terminal. The east coast station was built at Holmdel Laboratory in Crawford Hill, New Jersey, and the west coast station at JPL's Goldstone facility in California.

Finally, all was ready and the Echo balloon was launched into a polar orbit on August 12, 1960. A special message from President Dwight D. Eisenhower was transmitted from Goldstone to the satellite and reflected from the satellite back to earth. The project was successful and attracted attention.

For substantially longer than expected, the Echo balloon orbited. Meanwhile, John turned his mind and attentions to an active communication satellite, that is, one that could receive, amplify, and retransmit electronic signals.

It should be noted that the first active communication satellite was the U.S. government's Signal Communication by Orbiting Relay Equipment (Score) satellite that was launched December 18, 1958. Score functioned for 13 days until its batteries ran down. A further development of Score was the U.S. Signal Corp's Courier satellite project. Courier had not been launched until shortly after Echo. Courier operated for 17 days. Neither Score nor Courier received much public recognition. Each seemed to convince very few that communication satellites would have an important role in telecommunications or otherwise.

Telstar

It appears that John R. Pierce of AT&T's Bell Telephone Laboratories was the first person to evaluate carefully the options in satellite communication and to carefully evaluate the financial prospects of such. In a 1954 speech and in a 1955 article, he elaborated upon the utility of a communications "mirror" in space involving both a medium-orbit "repeater" and a 24-hour-orbit repeater. He compared the potential communications capacity of a satellite, which he estimated at a thousand simultaneous phone calls, with the communications capacity of the first transatlantic telephone cable (TAT-1), which could carry 36 simultaneous phone calls, at an estimated initial cost of $30–50 million. He estimated that such a satellite would be worth a billion dollars.

Because of congressional fears of duplication, NASA limited itself to experiments with mirrors (passive communications satellites), such as those in the satellite Echo, and the Department of Defense pursued repeaters (active communication satellites), which would amplify a received signal at the satellite and retransmit it and would provide much higher utility and quality communications than a passive satellite.

Under Pierce's direct and indirect leadership, AT&T developed active satellite communications technology, starting even before the launch of Echo. A design for an active satellite had been written by Roy Tillotson on August 24, 1959. The satellite, which came to be called Telstar, was to be at an altitude of 2,500 miles and would have a broad-band frequency modulation (FM) transmitter of 1 watt power and a 100-MHz bandwidth. There would be transmission capability for one TV channel or several hundred separate telephone channels. Tillotson's design turned out to be close to that of the actual Telstar satellite.

In 1960, AT&T petitioned the Federal Communications Commission for permission to launch an experimental active communications satellite. If it was successful and practical, then an operational system would be implemented. The petition was unexpected, and the government hastened to develop a policy regarding regulation of this new medium of communication. The John F. Kennedy administration opposed permitting AT&T to have a monopoly on satellite communications, which would, it was thought, extend the monopoly it had already effectively achieved on earth. Accordingly, in 1961, to offset AT&T's apparent lead in these new technological developments, NASA gave contracts to RCA and Hughes Aircraft to build communications satellites, which became identified as Relay and Syncom, respectively. Relay was a medium-orbit (about 4,000 miles high), active communication satellite. Syncom was a 20,000-mile-high satellite that was aimed at allowing twenty-four-hour use.

AT&T was building its own medium-orbit satellite, Telstar. About a year later, a military satellite proposal, called Advent (mentioned above), was canceled because of the complexity of the spacecraft, delay in launcher availability, and cost overruns. Telstar was launched on July 10, 1962, and was very successful. Among other accomplishments, it broadcast the first live television signals across the Atlantic.

Echo and Telstar seem to have convinced many that communication satellites would have an early and important role in communications. Although neither satellite could have been built and launched without NASA, or without the American missile programs (all launch vehicles prior to the space shuttle were adaptations of ballistic missiles), the leadership, expertise, and hard, dedicated work of Bell Labs and Pierce had been essential to their achievement. The factors that made these achievements possible were that (1) Pierce was the executive director of the relevant portion of the Bell Labs research department, which was crucial to these undertakings; (2) he was on good terms with his own bosses and with the people qualified to do the work; and (3) the technical people involved respected his technical ability and were willing to pursue the opportunity he pointed out.

Rockets and Robert Hutchings Goddard

A satellite was possible only because a powerful vehicle was available to carry it into space. The rocket is such a vehicle; it achieves thrust by emitting a rapidly moving fluid from an associated engine means.

Rocket history can be traced back at least to the thirteenth century AD. Rockets were used in military operations to deliver for inflammatory agents and explosives.

Robert Goddard (1882–1945) was a prominent rocket pioneer. He received a Ph.D. in physics from Clark University in Worcester, Massachusetts, in 1911 and returned there in 1914 to teach for almost 30 years. He had become interested in rocketry from reading H.G. Wells as a teenager. In 1914, he received two patents on rocket apparatus in 1914, and in 1919 he wrote a book, *A Method for Reaching Extreme Altitudes*. In 1923, he built and tested a liquid fueled rocket engine that used gasoline and liquid oxygen (probably

(continued)

the first such engine), and in 1926 he sent off his first rocket at his aunt's farm. His wife took a picture of him standing next to it just before launch, and it shows this rocket was about 4 feet high, about 6 inches in diameter, and supported on a frame. In 1929, he sent off a larger rocket holding some instruments, but the noise brought the police and he was ordered to stop rocket experiments in Massachusetts.

Many thought him a crackpot, but Charles Lindbergh grew interested in his work and got Charles Guggenheim to provide him with a $50,000 grant. Goddard set up an experimental station at an isolated spot near Roswell, New Mexico, where he built rockets and developed concepts that are now conventional in rocketry. Some rockets achieved speeds of 550 miles per hour and rose to 1.5 miles, some carried steering devices and gyroscopes, some were multistage. He obtained about 214 patents, but the U.S. government never grew interested in his work until World War II. One of his early inventions was developed into the bazooka.

After the war, when German rocket experts were brought to America, they were amazed by the American neglect of Goddard, for they acknowledged most of what they knew had been acquired from Goddard's teachings. Before the neglect could be remedied—many suggest that he could be credited with initiating the space age—he died of throat cancer. The government issued a grant of $1,000,000, of which one half went to the Guggenheim Foundation and the other half to Goddard's estate. The Goddard Space Flight Center in Maryland is named in his honor.

Rockets and Wernher von Braun

Wernher von Braun (1912–1977) was a key scientist in Germany's rocket development program, but after World War II he entered the United States through the then-secret Operation Paperclip.

In 1932, he started work for the German army to develop ballistic missiles. During this work, he received his Ph.D. in physics in July 1934. He led a team that during the war developed the liquid fueled V-2 ballistic missile rocket, which was about 46 feet long, weighed 27,000 pounds, flew in excess of 3,500 miles per hour, and delivered a 2,200-pound explosive head to a target 500 miles away. The rocket was developed at Peenemünde on the Baltic coast and manufactured with slave labor at a nearby factory called Mittelwerk. Scholars are still reviewing von Braun's role in these controversial activities.

Beginning in September 1944, the V-2 rocket began to be used, mainly against Great Britain, but by early 1945 von Braun realized that Germany

(*continued*)

would not win and began postwar plans. Preceding the Allied capture of the complex, von Braun led the surrender of five hundred of his top rocket scientists and engineers, together with plans and test equipment, to the Americans. With his team, he was moved first to Fort Bliss, Texas.

For the following 15 years, he worked with his team for the U.S. government in development of ballistic missiles. In 1950, they built the Army's Jupiter ballistic missile. In 1960, they were transferred from the Army to the new National Aeronautics and Space Administration (NASA) and undertook building the giant Saturn rockets. Von Braun became direct of NASA's Marshall Space Flight Center and was the chief designer of the Saturn V rocket superbooster, which lifted Americans to the moon.

Von Braun became probably the most prominent person advocating space exploration in the 1950s. His interest in space exploration was keen. In 1970, NASA asked him to move to Washington, D.C., to be head of NASA's strategic planning effort.

In 1972, he retired from NASA and began working for Fairchild Industries of Germantown, Maryland. In 1973, it was discovered that he had kidney cancer, and it was later found that the cancer could not be controlled by surgery. He continued to work to the extent possible.

When, in early 1977, the 1975 National Medal of Science was awarded to him, he was hospitalized and could not attend the White House ceremony. He died in June 1977 and was buried in Alexandria, Virginia. His wife, Maria, also died in 1977. They were survived by a son and two daughters, all born in the United States.

DEVELOPMENTS

On August 31, 1962, Congress passed and the president signed the Kennedy administration–sponsored legislation that created COMSTAT, in which ownership was divided equally between the public and the telecommunications companies, such as AT&T, RCA, Western Union, and ITT. Pierce felt that the legislation had taken him out of satellite work abruptly and finally and was disappointed.

Later, COMSTAT became the American part of a new global system called INTELSAT, which was formed in August 20, 1964. The INTELSAT organization would eventually take over ownership of the satellites and have responsibility for maintenance and management of the global system.

By 1964, two AT&T Telstars, two RCA Relays, and two Hughes Syncoms had been successfully launched and operated in space, and technological know-how had been transferred to companies other than AT&T. From

Tokyo in 1964, live television broadcasts of the Olympics served to provide a preliminary public exhibit of instantaneous global communications.

In April 6, 1965, COMSTAT's first satellite, Early Bird (or Intelsat 1), was launched from Cape Canaveral, marking the beginning of global satellite communications. Unlike Telstar, Early Bird was a geosynchronous satellite located 22,300 miles above one spot on earth: it rotated with the earth and stayed above the same point on the earth's surface with the satellite revolving around it in the same direction. Early Bird was built by Hughes Aircraft and the design was close to that of the Hughes Syncom 2, which had been the first synchronous communication satellite that NASA had launched July 26, 1963. Early Bird was a great success.

Syncom had a design that was ingenious and relatively simple and that included active station-keeping and attitude-control systems. It was lighter than Telstar, which did not include these systems. Together with some improvements in propulsion, the Thor Delta vehicle which had put Telstar in a lower orbit, was able to launch Syncom into a much higher synchronous orbit. Pierce attributed the Syncom design to Harold Rosen and his coworkers Don Williams and Tom Hudspeth.

Although initially the satellite technology was American in origin, by 1965 various countries, including France, Germany, Italy, Great Britain, Brazil, and Japan, had operating ground stations. From only a few members and a few hundred telephone circuits in 1965, the INTELSAT system expanded to include more members than the United Nations and to have sufficient technical capacity to provide service for millions of individual customers. Much of the early use of the system was to provide circuits for the NASA Communications Network. The final worldwide coverage involved the Indian Ocean and was completed only days before the *Apollo* landing on the moon on July 20, 1969, which was watched by an estimated half-billion people.

Cost to carriers per circuit and to individual customers declined very substantially with time. By the beginning of the new century, sales of products and services were generating billions of dollars annually. Besides telephone communication, scope of operations included commercial broadcasting, exchanges between businesses and sciences, and Internet communications.

Because of the inherent time delay in transmission caused by the distance between a land-based originating site and a receiving/transmitting satellite, and between this satellite and a land-based receiving site, telephone communication by means of satellite phone involves gaps of typically 2 to 4 seconds while an originating spoken communication travels to a satellite and then to a receiver. For television signal transmission, this time delay is commonly not critical because of the nature of such a transmission (i.e., the originating signal tends to be continuous and not interrupted by an oppositely moving signal).

Over the years since satellite communication became a reality, various land-based communication means have come into existence even for remote geographical locations. For telephone communication purposes, in remote locations served by such land-based communication means, little or no gap in transmission time occurs.

Domestic communications satellites have now become common. The 1972 Canadian TELSAT satellite, ANIK, was the first. The first domestic communications satellite was Western Union's WESTAR ON April 13, 1974. In early 1976, AT&T and COMSAT began the COMSTAR series. The 1962 legislation did not keep AT&T out of domestic satellite communications. These satellites, though initially used for voice and data, soon became used largely by television. By the end of 1976, 120 transponders were available over the United States, each capable of providing 1,500 telephone channels or one TV channel. An inexpensive method of distributing video involving satellite communications permitted large growth of cable TV.

Over the subsequent two decades, there have been some changes. Western Union is gone, Hughes is now a satellite operator and also a manufacturer, AT&T is no longer a partner with COMSAT but still is a satellite operator, and GTE is now a major satellite operator. Television is still the dominant factor in domestic satellite communications, but very small aperture terminals and small antennas are commonly used.

PIERCE'S LATER WORK

Pierce at Bell Labs became director of electronics research in 1952, research director of communications principles in 1958, and executive director of Bell Labs Communication Sciences Division. He reportedly oversaw work relating to mathematics, statistics, speech, hearing, behavioral science, electronics, radio waves, and microwaves, and he seems to have chiefly focused on electron devices, especially traveling-wave tubes. He invented the Pierce gun, a vacuum-operated device that transmits electrons and that is used in satellites and also, among other things, in the klystrons that power the Stanford Linear Accelerator. This gun is reportedly used in all linear-beam microwave tubes. His 1948 paper, "The Philosophy of PCM [Pulse Code Modulation]," with two others marked the beginning of the change from analog to digital signals, starting with the digitization of short-haul telephony by the Bell System in the 1960s.

After 35 years at Bell Telephone Laboratories, during which he had been involved in the design work of a large number of communications systems, including Echo and Telstar, he retired in 1971 at age 61. During this period, he wrote science fiction articles in various popular magazines, though none of these articles was in his own name. He used the pseudonyms John Roberts and J. J. Coupling.

After resigning from Bell Labs, Pierce involved himself in teaching, in doing research in communications, and in the acoustics of music, especially computer music and psychoacoustics, at Caltech and Stanford University. He took an engineering professorship at Caltech, and from 1979 to 1882 was chief technologist at the Jet Propulsion Laboratory.

He became associated with Stanford in 1983 and he served as visiting professor of music, emeritus at Stanford's Center for Computer Research in Music and Acoustics (CCRMA) for more than 12 years. He helped bring intellectual and financial support to the Center. He was interested in psychoacoustics and how people perceive sound, as well as computer music. A professor of music at CCRMA, Chris Chafe, has said he used Pierce's book *The Science of Musical Sound* as a textbook for many courses he taught at Stanford.

Pierce authored or coauthored about twenty books and more than three hundred papers and book sections. He was the inventor or coinventor in more than 90 patents. About two dozen of his science fiction works were published. He knew science writers Isaac Asimov, Ray Bradbury, and Arthur C. Clarke. Many of his technical books were written in a manner intended to permit understanding by a semitechnical audience of modern technical subjects.

Though some the origin lies with others, he was famous for quips and aphorisms, such as:

- "Nature abhors a vacuum tube."
- "Artificial intelligence is real stupidity."
- "I thought of it the first time I saw it."
- "After growing wildly for years, the field of computing appears to be reaching its infancy."
- "Rudy Kompfner invented the traveling-wave tube, but I discovered it."

He shared with Harold Rosen the Charles Stark Draper Prize for communications satellites, and they split the $400,000 award. He received dozens of honorary degrees, medals, and awards.

In 1948, he supervised the team at Bell Labs that invented the transistor, and he created the term "transistor" for the new device, apparently at the request of Walter Brattain. He died April 2, 2002, at age 92, in a rest home at Sunnyvale, California. He was survived by his wife; a son, John Jeremy Pierce; and a daughter, Elizabeth Anne Pierce.

FURTHER RESOURCES

Andrew J. Butrica, ed. *Beyond the Ionosphere: Fifty Years of Satellite Communication.* NASA SP 2417, 1997.

John R. Pierce and Ed C. Posner. *Introduction to Communications Science and Systems.* New York: Springer, 1980.

M. Richharia. *Satellite Communications Systems: Design Principles.* 2nd ed. New York: McGraw-Hill, 1999.

David J. Whalen. *The Origins of Satellite Communications, 1945–1965.* Washington, DC: Smithsonian Institution Press, 2002.

Jim Whiting. *John R. Pierce: Pioneer in Satellite Communication.* Hockessin, DE: Mitchell Lane Publishers, 2003.

AP Photo

Software and Bill Gates

The advent of small computers generated a need for program technology (software) to operate them. Bill Gates essentially arrived at the beginning of the age of small computer hardware structures and was able to produce and sell various types of software of wide utilization with different computers. Perhaps his greatest achievement was the creation, with his staff, of standardized software operating platforms for computers, which enabled large-scale and rapid adoption of computer technology. The achievement of a standardized operating system for millions of personal computers in an environment where many different computer makers and users existed offered a sort of interface that not only permitted computers to communicate with one another, but also permitted standardized application software to be utilized on a vast scale.

BACKGROUND

The electronic computer, as an embodied device that can be programmed and thereby controlled and manipulated to produce an output, can be considered to have originated around 1940. Though a computer initially was the size of a large room, a modern computer that uses solid-state transistors and integrated circuits can occupy only a tiny fraction of such space and is able to perform programmed operations faster, more reliably, and with little power.

The so-called personal computer (that is, generally a computer whose cost, size, and capabilities allow it to be used by individuals) dates from Kilby's 1958 invention of the integrated circuit with subsequent advances (see the chapter on Kilby and Wozniak), and is now in worldwide usage, but the so-called embedded computer (that is, a small, usually relatively simple computer that is associated with a device and that provides some control or operating function for the device, such as those in children's toys, fighter aircraft, digital cameras, machine tools, etc.) is even much more commonly employed. The complexity and capability of a computer can vary greatly. Almost all modern computers use at least one form of program or set of operating instructions, typically in a form called software. A program can be in various forms or of various types.

Software is an ordered set of prepared instructions (usually in electronic digital form) for changing or controlling the state or operation of computer hardware (that is, the physical components that make up a computer). Software is a general term that is inclusive of one or more programs, or groups of programs, and can include procedures and even documentation that can perform or indicate some function (including display) on or in a computer or computer system. The term generally includes system software, including computer operating systems, middleware (component connectors), programming software, applications software, machine language, and object code. Software may be written by a programmer in a computer language, such as a high-level programming language, and may be compiled by a compiler device for use in a computer, or be interpreted into object code or assembly language and formed into machine language object code by an assembler device for use in a computer. In a computer, software is loaded into its random access memory (RAM) and is executed by its central processing unit, the latter today typically being a microprocessor (or complex integrated circuit). Since the 1980s, software for particular applications is sold either installed on a computer as purchased, or in separate prerecorded packages for customer installation ("purchased software" or "purchased programs").

Purchased software usually comes with a license for its use, though software can also be in the form of freeware or shareware (software that a user may pay for later). A computer user may comprehend things somewhat differently from a software programmer. A user of a modern, general-purpose computer typically experiences three different types of software:

(a) Programs that allow a user to interact with a computer and its so-called peripherals (i.e., associated equipment). Commonly and collectively called the "platform software," such programs usually include the computer operating system, device drivers, and firmware (memory written to the read-only memory, which holds its content without electric power), and, perhaps, a graphical user interface (a user interface that allows interaction between electronic devices);

(b) Applications programs, including programs useful for office or personal operation, such as word processing, e-mail, games, and other programs independent from the operating system; and

(c) User software, such as may be written by a user, that acts in conjunction with purchased or installed programs, and is intended to improve performance.

BILL GATES

William (Bill) Henry Gates III, born October 28, 1955, was the son of prominent, well-to-do lawyer in Seattle, Washington, William H. Gates Jr. (now Sr.), and Mary Maxwell Gates, whose father was president of a national bank and who served on the board of directors for First Interstate Bank and the United Way. Bill's father assumed the identification "Senior," permitting his son to be designated as "III." Bill Gates III has an older sister, Kristi, and a younger sister, Libby. His father is now retired and his mother is deceased.

In public elementary school, he displayed substantial proficiency in mathematics, science, computing, business, and competition. His parents, perceiving that he could use intellectual challenge, enrolled him at age 13 in the Lakeside School, a junior high school that was Seattle's most exclusive private preparatory school.

In the spring of 1968, the Lakeside School decided that its students should be exposed to computers. When Bill was in eighth grade, the mothers of Lakeside students held a rummage sale, raised several thousand dollars, and the school bought some computer time on a nearby General Electric computer, identified as a DEC PDP-10. The computer was connected to the school by a teletype, having an associated 110-baud modem ("baud" is a measure of data transmission speed, now usually termed bits per second). The time sharing system included only a keyboard, a printer, a paper tape puncher, and a reader, plus the simple programming language BASIC (derived from "beginner's all-purpose symbolic instruction code"). At that time, Lakeside was undoubtedly one of very few schools in the country to undertake a computer program for its young students. Computers were then still large in size and costly. It was too expensive for the school to purchase and maintain one.

Though the school had bought what was thought to be more than enough computer time for the whole year, a small group of Lakeside students, including Bill Gates, Paul Allen, and others (some of whom later became the first programmers hired by Microsoft), immediately became obsessed with computers and spent all their time and energy on computer technology and staying in the computer room. Their homework was neglected and classes were skipped. They began programming computers. Bill was one of the principal users, and Bill and Paul studied BASIC together. The group soon used up all the purchased computer time. However, in the fall of 1968, Computer Center Corporation opened in Seattle, offering low rates for computer time (apparently on a DEC PDP-10), and one of the Corporation's chief programmers had a child in Lakeside. As a consequence, the school continued to provide students with computer time.

Bill and his young friends called the new corporation C-cubed, soon became hackers, caused system crashes, broke the computer's security code, and altered computer records regarding the quantity of computer time they were using. Gates and Allen managed to log in as system operators to get free time on C-cubed's PDP-10. When this was discovered, their group was banned by C-cubed from system use for a few weeks. Gates, Allen, and two others formed the Lakeside Programmers Group (LPG) in late 1968. C-cubed hired the group to find bugs and system weaknesses in return for unlimited computer time. Gates later said, "It was when we got free time at C-cubed that we really got into computers. I mean, then I became hardcore. It was day and night."

Gates became an expert at BASIC. Allen became an expert in PDP-10 computer operations by using DEC manuals and instructions and learned to use the macro assembler and programs that control computer hardware and input/output. By eighth grade, Gates had written two programs, and in high school he began teaching others computer skills. He placed in the top ten nationally in a math aptitude test.

In this early period, Bill began using some English-language phrases that he continued to use thereafter, such as "That's the stupidest thing I've ever heard of!" Late in 1969, C-cubed got into financial trouble, and in March 1970 it went out of business. Though Allen and Bill then found a few computers they managed to use for a while without authorization at the nearby University of Washington, it was about a year before a company called Information Services Inc. (ISI) hired LPG to make a payroll program. From this employment, LPG then had income, free computer time, and the possibility of receiving royalties on their programs sold by ISI.

In 1969, Lakeside managed to get two small DEC computers on loan. Gates got a paper tape with an assembler plus a source code for BASIC for a PDP-8. He used this equipment to begin work on a BASIC interpreter.

Gates and Allen on their own then started another company, Traf-O-Data. In the fall of 1972, Allen and Gates tried to build a small computer using an Intel 8008 chip. Their intent was to make a computer would count vehicular traffic. Allen built a software simulator of the 8008 chip using a USW IBM 360 computer and aimed to write software for the 8008. Apparently they could not get this computer to work well. However, using a small computer they made, they measured traffic flow in Seattle and helped identify traffic patterns in Seattle. They grossed about $20,000 when Gates, at age 15, and Allen sold it to the city. Traf-O-Data operated until Gates left for college. Allen graduated before Gates and entered Washington State University at Pullman. Meanwhile, in Gates's junior year, the Lakeside school administration offered Gates the job of computerizing and programming the school's scheduling system. Gates got Allen to help, and the following summer they prepared the program.

In Gates's succeeding and senior year, Gates and Allen were offered, apparently in late December 1972, by TRW, a defense contractor, a job of debugging a computer that was similar to that at C-cubed and that was used in a real-time data analysis system for the Bonneville Power Administration. Now they not only found the bugs (as at C-cubed), but also fixed them. Gates apparently dropped out of high school for a year but earned $30,000 and managed to get this operation treated as a senior project. Gates developed as a programmer. Gates and Allen began talking in earnest about forming their own software company.

The next fall, 1973, after graduating from high school, Bill left for Harvard University. Having no definite idea what he wished to study, he selected prelaw. He considered himself to be computer literate and had little interest in computer science. He took the conventional freshman courses, plus a tough elective math course. He located the university's computer center and spent all his available time there. His course work was satisfactory, but he was not an enthusiastic student. After long nights at the university's computer, he would fall asleep in class.

Gates and Allen continued their close contact and continued to discuss projects and starting a business. The following summer, they both got jobs at Honeywell in Boston, Massachusetts. Paul continued his efforts to get Bill to form a software company with him, but Bill was reluctant.

In December 1974, evidently en route to visit Bill, Paul chanced to come across the current issue of the magazine *Popular Electronics* and found on the cover a photo of the Altair 8080 with the headline, "World's First Microcomputer Kit to Rival Commercial Models." The kit was offered by Micro Instrumentation and Telemetry Systems (MITS) and had a selling price of $350. Paul bought the issue and rushed to Bill's dorm room. They both realized that this was their big opportunity. Someone would need to make the needed software for the new computer devices that were just entering the new, exploding home computer market.

From their high school days, Bill and Allen had held the belief that the computer would be a valuable tool for every office desktop and for every home, and they had thought of developing software for individual computers. This belief may have been high school idealism, but it has been central to Gates's success.

The Altair 8080 used the Intel 8080 chip with 256 bytes of RAM. Existing programming associated with the Altair was extremely awkward. As first assembled, lights and switches were the only interface. Commands had to be entered through a series of switches. Programs, for example, had to be entered in steps, with each step involving flipping toggle switches on the front of the machine followed by using an entering switch to load code into the machine's memory. This required thousands of toggles without an error. Output was represented in the form of flashing lights. The Altair had no display and no true storage. Since the device did not come with a terminal, most purchasers connected the computer to a teletype to use the computer. One also needed an input/output (I/O) card and a plug-in card for the I/O card. The Altair 8080 was great for hobbyists, for whom assembly was really a feature, but it was not suitable for the general public.

Gates and Allen saw that if they wrote a program in BASIC for the Intel 8080 chip-equipped computer of MITS, they could make money selling the program not only to MITS and Altair purchasers but also to other Intel 8080-based computers that would appear from other companies and compete with the Altair. They reached an agreement with their associate on the Traf-O-Data project that enabled them to write software for unrelated projects. They decided to make a direct contact with Ed Roberts of MITS, the owner of MITS and the maker of the Altair 8080, located in Albuquerque, New Mexico. Evidently, it was Gates who made the key telephone call. They offered MITS a BASIC interpreter for the Altair. Roberts, on behalf of MITS, expressed much interest in seeing their BASIC. In fact, though, neither Bill nor Paul had yet written even a line of this code. Also, they had neither an Altair 8080 nor the operating chip.

The boys began working intensively to produce their BASIC program. The BASIC was pursued by Bill, who wrote the assembly language code. Paul worked on simulating the Altair 8080 on the Harvard University's PDP-10 (i.e., creating an Altair 8080 emulator) and worked on the assembler. Monte Davidoff, a freshman at Harvard whom they chanced to meet, joined them and wrote several floating point math packages. Working long hours, they had their program ready and operating on the Altair 8080 emulator in eight weeks (mid-February 1975), but it had not yet been run on an actual Altair 8080.

Allen arranged to fly to MITS to make a demonstration of their program. The night before he left, Allen stayed up all night checking code and then making a paper tape. However, in mid-flight to Albuquerque, Allen realized that he did not have the necessary loader (the software that allowed the Altair to "communicate" with a connected teletype) so that the paper tape with the BASIC program could be input into the computer. Allen immediately wrote the loader in machine code before his flight landed.

At MITS, Allen loaded the loader in binary code into the Altair by the procedure of flipping switches on the front panel of the Altair. Apparently, Allen had never previously seen either the Altair or its Intel 8080 processor. One mistake would have caused the software not to run. Allen next loaded the paper tape into the Altair. The first time it ran, it displayed "Altair Basic" and then crashed. The paper tape had broken. On the next day, with a new paper tape, the system booted (that is, the computer loaded and the operating software operated) and worked as planned. Their BASIC worked. MITS was impressed. Allen (a future billionaire), though, had to borrow $40 for his motel bill.

Gates (then 19 years old) and Allen, on July 22, 1975, signed a deal with MITS for $3,000 plus a royalty for each copy of BASIC sold. MITS obtained the exclusive right to sublicense BASIC, but agreed to use its "best efforts" to commercialize BASIC.

Gates and Allen agreed that Allen would become an employee and Vice President and Director of Software at MITS. Also effective on this July 1975 date, Gates and Allen formed their own partnership, Micro-Soft, with Gates having 60% and Allen 40%. Gates proposed the larger share for himself because Allen was a full-time employee of MITS. Their initial investments were $910 and $606, respectively. Before long, the hyphen was dropped and their company name became Microsoft.

In addition to MITS, Microsoft (Gates) sold its BASIC system to NCR and to Intel. It was cheaper for these companies to buy Microsoft software than to develop their own. Besides, the Microsoft software worked well. Later, Allen got Gates to work part-time at MITS.

In October 1975, MITS decided to bring out a new Altair model, the 680, based on the Motorola 6800 processor. Gates and Allen immediately saw a new opportunity and decided to develop their original BASIC product for use with the new Altair 680. Allen created a simulator for the 6800 processor. They hired Ric Weiland, one of their friends from Lakeside, and he revised Gates's original BASIC interpreter for use on the 6800 processor.

An early form of their BASIC program became public knowledge and was being copied without any payment to Microsoft. Gates was distressed with the increasing and widespread piracy of Microsoft's software during the 1975–1976 period. He challenged the tradition of amateur software development. In February 1976, he published in all major computer publications his "Open Letter to Hobbyists" regarding this "theft." He maintained that unless software creators could recover their cost, there was no incentive to provide high-quality software. He received many negative responses.

Gates and Allen decided to place their codes in read-only memory (ROM; that is, recorded in silicon) rather than on paper tape to deter piracy, and they then began selling BASIC directly to manufacturers nonexclusively for flat fees.

Late in 1976, Microsoft contracted with Commodore International to adapt and place the Microsoft BASIC in ROM on the new Commodore PET (personal electronic transactor). This is believed to be the first complete personal computer that was able to operate as soon as a customer unpacked it. Since MITS only had license rights to the 8080 version of BASIC, MITS had to take a license for the 6800 BASIC from Gates and Allen, which cost MITS a starting fee of $31,200.

In November 1976, Allen quit MITS, and in February 1977, Gates left Harvard. Both worked at Microsoft full-time, but in February 1977 they changed the Microsoft ownership deal between them so that Gates held 64% and Allen 36%.

In 1977, MITS, then being taken over by Pertec, refused to license 8080 BASIC code to customers. Microsoft reminded MITS by letter of their contractual "best efforts" to commercialize, but MITS in a lawsuit then obtained a restraining order preventing Microsoft from disclosing the code to any third party and forcing arbitration. The arbitrator ruled in September 1977 that MITS had violated its contract with Microsoft and terminated the exclusive license.

Microsoft, as a result, could sell BASIC to all customers, which produced various deals. In an important deal with Radio Shack, Gates left so-called hooks in the BASIC code for the new Radio Shack TRS-80 computer that enabled Microsoft to load, if desired, extra functions unavailable to other software providers. This practice was successful and became a standard practice at Microsoft for years.

In 1978, Gates and Marc McDonald, in producing a BASIC version for NCR, achieved the innovative File Allocation Table (FAT) for Microsoft operating systems, which is an important part of the subsequent Standalone BASIC. Also in 1978, work began on programming for the Intel 8086.

Gates was an active creator of software in the early years of Microsoft. He worked especially on Microsoft programming language products. Until about 1989, Gates was known to write code that was sold in company products, but he is reported not to have been on a company software development team since working on the TRS-80 Model 100 (about 1978).

Microsoft's BASIC could function as a simple operating system. By 1977, Microsoft had completed development of its first version of FORTRAN, but this software required an operating system to run. In 1977, Gary Kildall of Digital Research in California had developed and released his CP/M program (a control program for computers). Gates thought a merger with Digital Research would be useful, but nothing was worked out.

Gates had a reputation for working harder than anyone else in the company. Sometimes he would work late and sleep on the office floor. Later, he hired a housekeeper who took care of cleaning his rooms, buying groceries, paying utility bills, and the like. Evidently as an outlet, he drove fast autos dangerously and got speeding tickets. Gates was not only a key programmer for Microsoft in the early, formative years, but also the key deal maker.

Gates continued to make good business deals for Microsoft. In addition to creating innovative software, he had good business ability. He seemed to perceive, usually correctly, the way in which the computer market was developing, the potential customers, and the places where software could be made and sold by Microsoft, time after time, situation after situation. In addition, he knew how to arrange things, how to make deals, and how to get Microsoft software products sold. He had ability to compete with cutthroat instincts.

Gates and Allen hired Marc McDonald and Ric Weiland, both originally high school friends, to help. Effective January 1, 1979, Gates moved Microsoft, then with sixteen employees, to Bellevue, Washington. Gates thought it would be easier to hire programmers in Seattle. For new employees, Gates and Microsoft thought gifted college graduates with little experience would be considered desirable and have no bias.

Management within Microsoft in the early years was unstructured and not organized. Microsoft, in its early years, did not own a word processor and kept its records in ledgers. Gates and Allen had never before managed a business. Gates was perceived to be quick-tempered, even hostile, toward managers and others. He frequently shouted about how stupid an idea or approach was, or he shouted things like "Why don't you just give up your options and join the Peace Corps?" or "Do you want me to do it over the weekend?"

1979 was an important year. In March, Microsoft reportedly had forty-eight original equipment manufacturer (OEM) customers for 8080 BASIC, twenty-nine for FORTRAN, and twelve for COBOL. In June, Microsoft ran its important Standalone BASIC at the National Computer Conference in New York. In August, Ross Perot of EDS (Electronic Data Systems) offered to buy Microsoft, but Gates and Perot failed to make a deal.

In June 1980, Steve Ballmer was hired by Gates and started working at Microsoft as assistant to the president, reportedly for $50,000 per year plus between 5% and 10% of the company, the percentage being defined by revenue growth. Gates had met Ballmer in 1974 at Harvard. Ballmer commenced to run the company in a businesslike manner. Ballmer was more confrontational than Gates.

In 1980, in the midst of its secret personal computer (PC) development project, IBM (Jack Sams) met with Microsoft (Gates and Ballmer) in Redmond, Washington. Evidently IBM, the biggest company in the computer industry, had done its research about who might be best to help them with software development. In approaching Microsoft, IBM's main objective was to see whether Microsoft could provide good software needed by IBM. Microsoft satisfied IBM, and IBM gave Microsoft a contract to develop the operating system for the IBM PC. IBM agreed to buy Microsoft's product line of languages, including BASIC, FORTRAN, and COBOL, for $600,000. Unlike other PC manufacturers, IBM decided to use an open architecture in its PC to encourage outside developers to produce new and useful applications. However, IBM wanted to be the licensor for all clones, saying they would just pay Microsoft an additional fee whenever IBM licensed a clone, but Gates refused to accept this arrangement. Gates insisted that Microsoft would retain its copyrights would handle software licensing. IBM accepted this arrangement. Also included was certain software for Intel 8086 and 8088 chips.

IBM needed a software operating system for its PC. Gates suggested to IBM that they consider the CP/M of Digital Research, but IBM could not make a deal there. IBM asked Microsoft to provide an operating system. To gain its own operating system for use with the IBM PC, Microsoft contracted in September 1980 with Seattle Computer (Rod Brock) to use and distribute a disc operating system (DOS) that Seattle Computer had just developed. Microsoft extended and improved this system for PC usage. In November 1980, Microsoft contracted with IBM under which, among other things, Microsoft supplied to IBM this improved operating system (called MS-DOS), but under the contract IBM could not itself license Microsoft's software to third parties.

In July 1981, a new Microsoft Corporation was formed from the original partnership. Stock ownership percentages were approximately as follows: Gates, 53%; Allen, 31%; Ballmer, 8%; Raburn, 4%; and Simonyl and Letwin, each 1.5%. About this time, Microsoft seems to have become aware of the importance of DOS relative to its own business. Microsoft broadened its range of products and tried to defend and enhance products where it had a dominant position in the market. In late July 1981, Microsoft bought all DOS rights from Seattle Computer for $50,000. Microsoft envisioned that its role was to provide the standard operating system so that the software industry was hardware independent. As Gates said later, "You had to have pure hardware competition that was orthogonal from software innovation, so that people could go to new models of hardware and applications guys didn't have to rev [do an upgrade] for every new piece of hardware" (Gates in *U.S. News*, August 20, 2001).

In August 1981, IBM announced its Personal Computer. Microsoft got royalties from IBM for all Microsoft software utilized by IBM in its sales of the IBM PC. Because a number of other businesses copied the IBM machine and used its operating system, this arrangement gave Microsoft income plus a large and flexible market for its products, including MS-DOS.

Also in August 1981, Steve Jobs visited Microsoft regarding software for the Macintosh Computer, then being developed within Apple (Jef Raskin). Gates negotiated a deal with Jobs and Apple under which Microsoft was to supply software for the Macintosh, and a contract was signed in January 1982. Microsoft provided a spreadsheet, a business graphics program, and a database, but Microsoft agreed not to commercialize the mouse before January 1, 1984.

In 1981, reportedly, Apple sold 150,000 computers and had a profit of $39.4 million, IBM sold 200,000 computers, profit unknown, and Microsoft netted $1.5 million. Although market experts had expected that the CP/M operating system, when it entered the market in spring 1983, would overwhelm DOS, this did not occur. Independent software developers wrote programs for new versions of DOS and almost none for the CP/M system. One reason for this result apparently was that Gates cut the official Microsoft price of DOS software, sometimes in half to OEM customers, in order to get DOS established before CP/M-86 entered the market. Another reason was that Microsoft produced, trained, supported, and promoted standardized, high-quality software. A third reason was that Microsoft kept DOS equal between manufacturers. Thus, Microsoft provided DOS codes to a company such as Compaq, which built the first legal IBM-compatible computer by reverse-engineering IBM's ROM Basic Input/Output System (BIOS) so that Compaq could produce and offer a fully compatible computer, and IBM and Compaq received the same DOS codes.

By 1982, Gates had come to believe that the graphical user interface (GUI) type of operating system associated with the Macintosh would be the future standard in PCs. He realized that the market demand was moving away from computer languages and toward applications and graphical user interfaces.

Unlike DOS, BASIC was both on disk and in ROM. As Compaq advised Microsoft of changes needed, Gates had a team of BASIC programmers implement the changes and the changes were furnished to Compaq. The changes in BASIC were included in new additions of ROM BASIC supplied to IBM. Both IBM and Compaq received the same codes, so all was equal and legal.

When it was released by Microsoft in late 1982, DOS 2.0 was substantially upgraded relative to the original DOS 1.0. Paul Allen acquired

Hodgkin's disease and resigned from Microsoft just before DOS 2.0 was released. In 1978, Microsoft had had sales of $1,355,655; by 1982, it had grown to have sales of $32,000,000 and two hundred employees.

In 1982, Gates and his programmers were proceeding as fast as possible on software for the Macintosh computer. Reportedly, Microsoft had as many programmers working on Macintosh software as Apple itself did. Gates, Allen, and their Microsoft programmers believed that the Macintosh software was the way of the future. Another competitive stimulus to Microsoft occurred at the Comdex trade show in the fall of 1982, when VisiCorp displayed on a Macintosh its GUI-type software.

Gates started developing the Microsoft Windows operating system (then at first called Interface Manager) to compete with VisiCorp. In November 1983, although it was apparently not yet fully developed, Microsoft announced Windows in New York City, trying to meet the VisiCorp new software. However, IBM refused to accept Windows and operated to oppose Microsoft. Gates remained steadfast in his determination to make Windows the new standard. He thought Windows would be the key to applications development.

To pursue Gates's aims, Microsoft developed its own version of the mouse and released it in June 1983. Its engineers had been able to power the mouse through the serial port, and they patented the invention. Although initially there was no software for using the mouse, by November 1983 Microsoft was able to release Word, its word-processing software, which supported the mouse and had many features that used the mouse.

Apple and Jobs became concerned by this Microsoft advance. In January 1984, they formally rescinded their contract for applications development with Microsoft only a week before the Macintosh was released.

In September 1983, the VisiCorp software was released but did not achieve a market. It can be classified as an expensive failure. To Microsoft, this was inept competition.

Lotus (Mitch Kapor) and its 1-2-3 system, though, were effective competition. In 1983, Lotus reportedly made more money than Microsoft. Lotus Symphony was an integrated, all-inclusive product that was of great concern to Gates. To meet the competition from Lotus, Gates changed Microsoft's Excel development from being a DOS spreadsheet to a Macintosh program.

Further competition appeared in late 1983, when Borland released Turbo Pascal for just $49.95 each. This program included an integrated compiler. It soon cut deeply into Microsoft's software language market. In 1983, though, in spite of the competition, Microsoft's gross sales were reportedly $55 million.

Between 1981 and 2000, Microsoft evolved and improved its DOS almost annually. By 1990, Microsoft dominated the operating systems market. Microsoft was concurrently developing, acquiring, and expanding its applications software. Because of its market dominance, it also influenced the designs of applications in packages developed by other software vendors. Microsoft produced software for word processing, spreadsheets, games, and

Windows. The latter software permitted a PC user to use pictures and a mouse as a pointing device to eliminate complicated keyboard commands, much as Apple had done with its Macintosh computer.

By 1984, about 10,000,000 personal computers were being sold per year. Most incorporated DOS. Microsoft by then had switched to per-machine contracts with vendors. Computer manufacturers received a price break from Microsoft on DOS if they also bought BASIC for a large number of computers, but most manufacturers overestimated their anticipated market share. A per-machine deal made DOS very difficult for competitors to attack. This was because a vendor who was paying already a per-machine royalty for DOS would not offer a different operating system unless it were a high-priced option. Every sale of a DOS computer that was not an IBM computer gave Microsoft a bonus.

In May 1984, Phoenix Software Associates succeeded in reverse-engineering the IBM PC ROM BIOS and sold the resulting ROM BIOS to anyone. As a result, the computer market suddenly expanded.

Also in 1984, IBM released its AT model with the Intel 80286 chip and Microsoft DOS 3.0 and having a keyboard with a new key identified as "SysRq" (system request). This key is believed by many to be an attempt by IBM to separate itself from Microsoft by providing the ability to run other operating systems merely by pushing this key. IBM hoped that with Top View, which supposedly allowed multitasking, Windows and other products would be blocked.

By January 1985, commercialization of Windows was about 2 years overdue relative to its introduction in January 1983. Microsoft moved Neil Konzen, its Macintosh expert, in to accelerate development. Gates delayed the product by insisting that Windows have also a keyboard equivalent to the mouse. Finally, in July 1985, Microsoft produced a beta version and in November 1985 released and shipped version 1.0 of Windows.

Gates was unable to get IBM to seriously consider Windows, but a joint development agreement with IBM was reached in June 1985 that replaced the 1980 agreement and released IBM from liability if DOS was disclosed to customers. Gates also urged Apple to license its Macintosh technology to allow Macintosh clones to be made by other companies, which would greatly expand the Macintosh market. Gates said he would help convince Compaq, Sony, and TI, among others, to make Macs. Apple, though, did not license its Macintosh technology, but it did enter into an agreement with Microsoft that allowed Microsoft to develop Windows. In March 1986, Microsoft stock went public. With the initial public offering, Gates was worth $311,000,000 on the first day of issue.

Seattle Computer sued Gates and Microsoft for attempting to deny Seattle Computer the right to sell its DOS license. In December 1986, Gates purchased Seattle Computer's DOS license for $925,000. Gates also bought Tim Paterson's DOS license for $1,000,000 and bought into his business. In addition, Gates employed Paterson under a large employment contract. Later, Paterson sold out and made $20 million.

IBM, with the initial aim of shutting out Microsoft, negotiated with Microsoft regarding a new operating system, the OS/2, intended to be a common interface for all IBM computers from PCs to mainframes. Gates and Ballmer tried to convince IBM to use and include Windows as an interface. A compromise was reached, and OS/2 changed into Presentation Manager. Between 1986 and 1988, Microsoft continued to try to protect its interests relative to IBM during the effort to develop OS/2. Ballmer and Gates sought to keep the association with IBM alive for fear that otherwise Microsoft could be squelched by IBM.

Internally, Microsoft code writers had little respect for IBM writers. The Microsoft writers felt the IBM writers spent too much attention on declining code languages and in preparation of code that included unnecessary lines of documentation, allegedly to make IBM software easy to service. Finally, in late December 1987, OS/2 1.0 was released without Presentation Manager. In October 1988, OS/2 1.1 with Presentation Manager was released, but only a small percentage of computer users with Intel 80286 and Intel 80386 chips purchased OS/2. Meanwhile, Gates and Microsoft sold Windows at a rate of about 50,000 per month.

In March 1987, Gates, at age 31, became a billionaire. Microsoft passed Lotus by the third quarter of 1987 as largest PC software vendor, with sales of $345,900,000.

Windows 2.0 was released in 1987. Compaq and Microsoft developed Windows to run on Compaq's machines with Intel 80386. Microsoft competitors complained about Microsoft's efforts to gain an advantage over competitors using DOS and Windows.

In March 1988 Apple filed its "look-and-feel" lawsuit against Microsoft. The suit was not decided until April 1992, but the decision was generally favorable to Microsoft.

In 1989, at the November Comdex in Las Vegas, IBM and Microsoft had an "open break." Neither would endorse the other's operating system.

In May 1990, Windows 3.0 was released. In the first year, 4,000,000 copies were sold. Microsoft control of the operating systems market became recognized. In the fall, IBM and Microsoft parted, with IBM retaining control of OS/2 1.X and 2.0 development and with Microsoft in control of OS/2 3.0. The latter was mainly Microsoft's Windows NT project.

In January 1991, Microsoft revealed WIN32, which it indicated would be the foundation for all its future operating systems. Microsoft incorporated into WIN32 features that IBM had "reserved" for its "high-end" operating system.

In 1992, Microsoft had a capitalization greater than Boeing or General Motors. Gates at age 36 was the richest man in the United States. In 1993, Windows NT 3.1 was released and was favorably received. Windows/DOS and Windows NT eventually were to be merged into one product running on all personal computers.

In 1995, Microsoft introduced Windows 95, which again revolutionized the PC market and became an industry standard. Microsoft continued to grow.

By the end of the century, Gates was a billionaire many times over. He became the world's richest person for a few years.

In 1994, Gates married Melinda French, and they had three children. In 1999, they established the Bill and Melinda Gates Foundation, with initial assets of about $5.4 billion. Initially, the foundation focused on promoting computer and Internet access, children's issues, global health, and projects concerning the Pacific Northwest.

In 1998, the Justice Department brought an antitrust suit against Microsoft, charging that Microsoft was using its control of the operating systems market to promote its own Internet Web browser and to prevent other companies from entering the market. In June 2000, federal District Judge Thomas Penfield Jackson ruled that Microsoft had violated the antitrust laws and should be divided into two companies.

Microsoft appealed. The federal appeals court threw out the decision by the lower court Judge Jackson to break the company into two and removed Judge Jackson, who had made public comments on the case that the appeals court judges found inappropriate. The appeals court did find that Microsoft had illegally maintained its monopoly on Windows through a series of acts that thwarted the Netscape Communications Web browser and the Sun Microsystems programming language Java. The court said that together they could have evolved into a threat to Microsoft's operating system because software developers could write applications to run them as a platform, much in the same way that they do for Windows.

Microsoft made an antitrust settlement with the Justice Department and nine states, but a coalition of state attorneys general from other states (California, Connecticut, Florida, Iowa, Kansas, Massachusetts, Minnesota, West Virginia, and Utah, plus the District of Columbia) asked the lower court on remand for stiffer sanctions. In the lower court, the case was assigned to Judge Colleen Kollar-Kotelly in the Federal District Court in Washington to carry out the strict instructions laid out in the appeals court decision. The judge was to fashion a remedy that would prevent Microsoft from repeating the violations, deprive it of the fruits of its misdeeds, and foster competition in the market for personal computer operating systems.

In her final opinion on November 1, 2002, the judge rejected nearly all of the stiffer sanctions sought by the coalition, but the judge did impose some new restrictions that would slow Microsoft's aggressive push into new markets. The judge also criticized the coalition states for suggesting that Microsoft should be punished for actions beyond those the federal appeals court found it liable. The Judge said that the suit was "not the vehicle through which plaintiffs can resolve all existing allegations of anticompetitive conduct which have not been proved." Microsoft was prohibited from retaliating or threatening retaliation against those who used non-Microsoft products.

The district court judge in effect agreed with Microsoft's argument that most of the states' proposals went beyond the Microsoft misdeeds and that there was very little relation between Microsoft's anticompetitive acts and their dominant position. Bill Gates had earlier taken the position that the penalties sought by the coalition of state prosecutors would cripple Microsoft, harm consumers, sow confusion, and drive up prices. "Instead of being able to buy a machine with Windows and knowing that all your applications would run, you wouldn't have that assurance," said Gates. The case was widely viewed as the most important antitrust action since the government's attack on Standard Oil at the turn of the last century.

In 2004, Microsoft reported fiscal year revenue of $36.8 billion, and the company employed 57,000 people in eighty-three countries. In 2006, billionaire businessman Warren Buffet pledged to make a large annual gift of stock to the Bill and Melinda Gates Foundation for as long as either Bill or Melinda remained involved in directing the foundation's activities. Buffet also announced that he was rewriting his will so that the stock donations would continue after his death. His first year's gift of 500,000 shares was estimated to be worth more than $1.5 billion.

Gates has received various awards and recognition. Various institutions have awarded him honorary degrees: Nyenrode Business Universiteit, Breukelen, the Netherlands (in 2000); Royal Institute of Technology, Stockholm, Sweden (in 2002); Waseda University, Tokyo, Japan (in 2005); Harvard University (in June 2007); and Karolinski Instituetet, Stockholm, Sweden (in January 2008). From Queen Elizabeth II of the United Kingdom, he received an honorary knighthood (in 2005). Bill and Melinda have received various awards for their philanthropic gifts. Whether or not Microsoft will remain the dominant supplier of software remains to be seen.

FURTHER RESOURCES

Michael Fairman. Bill Gates, the New Revolutionary Creator. Oxford, OH: Miami University. http://www.users.muohio.edu/shermalw/honors_2001_fall/honors_papers_2000/fairman_gates.htmlx.

William Gates. Business @ the Speed of Thought. *Business Plus*, 1999.

William Gates, Nathan Myhrvold, and Peter Rinearson. *The Road Ahead.* Upper Saddle River, NJ: Pearson Longman ESL, 1996.

Daniel Ichbiah and Susan Knepper. *The Making of Microsoft.* Rocklin, CA: Prima, 1991.

Stephen Manes. *Gates: How Microsoft's Mogul Reinvented an Industry and Made Himself The Richest Man in America.* New York: Touchstone, 1993.

Randall E. Stross. *The Microsoft Way.* New York: Basic Books, 1996.

James Wallace. *Hard Drive: Bill Gates and the Making of the Microsoft Empire.* New York: John Wiley & Sons, 1992.

AP Photo

The Internet and the World Wide Web: Kleinrock, Baran, Kahn, Cerf, and Berners-Lee

The work that resulted in the Internet began in 1958 with federal government funding and involved two inventions: packet switching and the protocol suite (transmission control protocol/Internet protocol [TCP/IP]). Packet switching (probably invented by Leonard Kleinrock and Paul Baran) enabled a computerized network of data communications. TCP/IP permitted information exchange between computers having different operating systems (probably invented by Robert Kahn and Vinton Cerf). The World Wide Web (WWW) (invented by Tim Berners-Lee) is an information retrieval system in which hypertext documents are interlinked and accessible through the Internet. About 1.24 billion people worldwide use the Internet and WWW as of October 2007, and the number of users is rapidly growing.

INTERNET AND WORLD WIDE WEB

The Internet is a worldwide network of computers that communicate using TCP/IP, various physical means (such as copper wires and fiber-optic cables), various wireless electromagnetic connections, and routing policies. The network includes individual computer networks that number in the millions, that are worldwide in distribution, and that each can hold and operate digitalized information. Communications in digital form between individual networks are accomplished using packet switching.

The Internet utilizes complex connections and operations that are facilitated and defined by protocols, specifications, and agreements, including commercial contracts, that together specify how data are exchanged over the Internet. Networking via the Internet enables social interactions.

The World Wide Web is a user-friendly interface in the Internet that is typically used for searching. "WWW" and "the Web" both commonly refer to either the Internet or the World Wide Web, or even both. The WWW is a system for finding and reviewing documents that are stored on computers worldwide. The documents are made up of Web pages formatted with hypertext using hypertext markup language (HTML) and linked by hypertext transfer protocol (HTTP) to other similarly formatted documents. The individual documents are located and retrieved at a particular computer using a program called a Web browser operated as a part of TCP/IP. Though of very great importance for purposes of document location and retrieval, the Web (that is, the World Wide Web) does not stand alone relative to the Internet as a system. Rather, it is a highly significant search tool that is part of the Internet and is useful for accessing and utilizing information.

THE INTERNET

This account is intended to consider iconic inventions and their inventors that are involved in the Internet, the World Wide Web (which is a facet of the Internet), and their present development. Mainly to identify and explain these inventions, the emphasis here is both on the incorporated technology and also on providing a description that is clear to those who may use the technology but are not themselves equipped with the knowledge of the originating technologists.

This account has little regard for matters relating to the utilization of the technology, such as the schemes used to commercially develop and exploit the technology. More information about these fields, if desired, is provided by much material that currently exists, as a visit to almost any bookstore or library will confirm. For ease of comprehension, the technological description has been somewhat simplified.

History

Two iconic innovations seem to have made possible the Internet in its present structure: *packet switching* and what is known as the *protocol suite.* The latter includes both transmission control protocol (TCP) and Internet protocol (IP). These two innovations seem to be best understood and appreciated from the context of Internet history.

The Internet grew and evolved from the work contributions of many gifted people. The earliest work on the Internet is usually dated back to the late 1950s and early 1960s. Concurrent developments in computer and computer network technology, and in the alacrity and enthusiasm for applying this technology, provided particularly by scientists and university teachers, without doubt promoted and hastened development of the Internet and the World Wide Web. Availability of funding, particularly from the federal government both directly and indirectly, was also of fundamental importance.

Those whose works culminated in the Internet, its components, and its applications have also created a specialized associated vocabulary. For convenient understanding by the reader, this text defines most of the specialized terms when they are first used.

ARPA

In 1958, the federal government created the Advanced Research Projects Agency (ARPA) to facilitate U.S. competition after the USSR launch of the *Sputnik* satellite. ARPA created the Information Processing Technology Office (IPTO) to extend networking research originally developed for national radar systems in the Semi-Automatic Ground Environment program. (Later, the word "Defense" was added to the agency title by Congress, so that ARPA became DARPA.)

J. C. R. Licklider from Harvard University and the Massachusetts Institute of Technology (MIT) was selected in 1962 to head IPTO. He considered computer networking to be important. He thought a globally interconnected group of computers was desirable, by which a user of one of these computers could rapidly access information from other computers in the group. A *network* is any connection between two or more computers (or combination of computers) that enables them to communicate. A combination of computers in a separate arrangement can itself constitute a network and may include various computers, connections, and devices; it can also be part of a large or main network. The term *network* is thus conventionally used to refer to (a) a combination of computers in a single system, and/or to (b) a group or collection of computers that is so interconnected that data can flow among them.

A piece of basic factual information: a computer works because it is following a program. A *program* is a sequence of instructions or rules that the computer interprets and executes. Operation of a computer network is controlled by at least one protocol. A *protocol* is a necessary set of rules, instructions, and conventions (usually written in a computer language) that specify and govern how computers in a network interact and exchange digital information.

Packet Switching: Kleinrock and Baran Were Probably Independent Inventors

In July 1961, Leonard Kleinrock, then a Ph.D. candidate at MIT, published a paper involving packet-switching theory. He developed his ideas and theories further in his 1963 Ph.D. thesis and in 1964 published a book including this subject matter. Kleinrock convinced Lawrence Roberts, an MIT classmate, that communications using packet switching were better than those using conventional circuit switching.

Meanwhile Paul Baran, working at RAND in California in the early 1960s, independently suggested (reportedly in a 1964 report to the Air Force) using packet switching, as distinct from circuit switching, for networking.

In *circuit switching*, a single or dedicated circuit or channel is established for a transmission, as in a telephone system. In *packet switching*, individual messages are broken up into discrete packets or bundles before being transmitted individually. An individual packet can follow a different route from other packets as it proceeds to its destination. When all packets that make up a message arrive at their destination, they are recompiled to form the original message. Baran evidently thought packet switching would inherently make a network robust and reliable, for example, in the event of extensive destruction from the use of atomic bombs.

Networked computers process and exchange information that is in a digital form. In their digital operation, computers use the mathematical two values, or "binary" language, of 0s and 1s. For communication and digital signaling purposes, such language of bits can be considered to be placed into substantially continuous and sequential electrical or optical waveforms before transmission. This waveform language after transmission is accurately recovered, and the binary language regenerated, so that computers and computer-compatible devices that can be controlled by digital information can be operated using transmitted signals.

In a telephone network, two separated telephones were originally manually interconnected through a "panel" having multiple connecting sites using at least one site-interconnecting patch cord that in effect acted as a switch. The result was called *circuit switching*, when an electrical circuit was established between the called telephone and the calling telephone. For a computer network, circuit switching was found not only to be too slow for making connections but also prone to generating errors in delivering digital data.

Licklider hired Lawrence Roberts in 1966 to implement a computer network. Roberts, evidently with knowledge of Baran's or Kleinrock's work, developed and published in 1967 a paper describing an early model of the ARPANET network. The ARPANET was created and developed under ARPA and comprised a computerized, packet-switched network of data communications for military and scientific operations. This network operated mainly in the late 1960s and early 1970s. With Thomas Merrill, Roberts connected the TX-2 computer in Massachusetts with the Q-32 computer in California using a slow-speed, dial-up telephone line (circuit switching), thereby producing the first wide-area computer network. This network successfully demonstrated that time-shared computers could work together running programs and transporting data, but it also showed that the circuit-switched telephone system was inadequate. The need for packet switching was thus confirmed.

Roberts' paper was published at a 1967 conference. At the conference, a paper by Donald Davies with Roger Scantlebury of the British National Physical Laboratory (NPL) was given on the packet network concept. It then developed that the MIT work, the RAND work, and the NPL work on packet switching networks had each been independently and separately carried out in the 1960s, though the NPL work came well after that of RAND and MIT. Owing to the usual importance given to the date of an invention, Kleinrock and Baran were prior inventors relative to Davies. Thereafter, though, the language "packet switching" was utilized from Davies' work to designate the packet concept, and the ARPANET line speed was upgraded from 2.4 to 50 kilobits per second.

For a computer network, packet switching was found to be a better choice not only because of the characteristic output communication associated with computers (a brief but intense "burst" followed by a period of silence before the next burst), but also because of the necessity of forwarding such a communication to a particular destination. Communication links and routers for handling such brief computer bursts became needed for computer networking and information transfer. A *router* acts as a switching device and directs and controls the flow of digital data through it. It routes data between interconnected computers (or networks). It receives data packets from computers (or networks) and forwards data packets only to computers (or networks) using preaddressed packets of data. Each receiving computer has a predetermined address and can receive particular data packets addressed to it. Thus, the address procedure is distinct from sending all data information to all computers (or networks) in a network. A network may contain more than one router. Packet switching, as developed, became an important and basic part of computer networking and the Internet.

From available evidence, it is not clear who first invented and implemented practical packet switching. Subject to reevaluation from further evidence, it now appears to be most probable that Kleinrock may have first developed the concept and theory of packet switching, while Baran independently proposed the concept and first proposed some implementations, but neither inventor seems to have achieved or produced actual embodiments of packet switches. The work of Donald Davies, though independently accomplished, seems to have occurred perhaps 2 or 3 years after that of Kleinrock and Baran. From available publications, it is presently unknown who the actual inventors were who first embodied and utilized packet switching and who first conceived and produced actual equipment embodiments that incorporated the packet switching concept and that achieved the needs of computer networking for packet-switching.

It is reported that in August 1968, responsive to request for quotation regarding a procurement proposal released by DARPA, Frank Heart at the Bolt, Beranek and Newman engineering firm (BBN), received a contract to build units of the important packet-switching router component for the ARPANET. This component was evidently a specialized, interconnecting, custom-configured router locatable between two computers (or separate networks) in a packet-switched computer network. Who proposed and embodied new software and/or hardware for this router is not now clear.

This specialized router was produced by BBN in September 1969 and was placed at the University of California, Los Angeles (UCLA), where Kleinrock, in view of his early work involving packet switching, operated the Network Measuring Center. When the Stanford Research Institute (SRI) and UCLA were connected together about one month later to form a primitive ARPANET, the first host-to-host message was sent from UCLA to SRI. Apparently, hypertext was tested.

The term *hypertext* refers to a word or words in a document that are retrievable from the Web (that is, for example, a Web page, which individually may involve more than one actual page or document). The words can be "read" variously through electronic links as well as by direct reading of text on a screen. Hypertext was first used about 1965 for a collection of documents containing cross-references (*links*) that permit a user, with a browser program, to move by computer from one document to another. Documents (Web pages) are stored and retrieved when hypertext is present.

Two Americans are credited with the invention of hypertext in computer documents: (1) Ted Nelson at Brown University coined the word "hypertext" in 1965 and worked with Andries van Dam to develop, in 1968, the hypertext editing system. The term hypertext, used by Nelson, designated a collection of documents containing cross-references or links that, with the aid of an interactive browser program, allow a user to move from one document to another. (2) Douglas Engelbart, in 1962 at SRI, started work on his NLS (oN Line System), but principal features thereof were not completed until 1968 because of funding, equipment, and personnel delays. Hypertext is selected to appear in a document being computerized and comprises particular text that is marked so that a user, using, for example, a mouse, can link or associate that text with correspondingly marked text in other documents. Hypertext is thus a way of presenting information online with connections between different pieces of information, and these interconnections are called links. (A *browser*, or browser program, is a software program that allows a computer to identify the links and permits a user to view various resources.)

With further additions, by the end of 1969, four host computers were connected into the ARPANET network, which represented achievements in network design and in utilization. Over succeeding time, other computers were added to this network. An early operational protocol, the Network Control Protocol (NCP), was completed, but it required all computers in a network to have about the same software, with one computer regulating others. Time-sharing computers that could run more than one computer program at once became available in the 1960s, and various research organizations developed programs that could exchange data and messages among computers in a network.

Robert Kahn in 1972 was hired by ARPA's IPTO, and he became involved in packet switching implementation and utilization. By 1972, network applications could be developed. Kahn successfully achieved a large demonstration of ARPANET in October 1972.

E-Mail

Although there had been some incidental prior efforts to develop e-mail, it was apparently in March 1972 that Ray Tomlinson at BBN, working in a group that was developing an early operating system called TENEX, wrote and practiced what became the historic and basic e-mail program, to send and read text messages; it was apparently in response to the need of ARPANET developers for an easy coordination system. Tomlinson used the @ character to mark a specified receiving computer address. Each computer that is connected to the network had (and still has) a numeric address, called its IP address, which is a thirty-two-bit numeric address written as four sequences of digits separated by dots. Apparently, by July 1972, Roberts had expanded e-mail practicality by writing a utility program that allowed users to list, selectively read, file, forward, and respond to messages. From this beginning, e-mail became the largest ARPANET network application for at least 10 years. File transfer and capability for remote login (TELNET) were important early applications for the Net and for developing the Internet, but e-mail achieved a practical model for communication and altered the nature of collaboration.

E-mail predated the Internet and was an important factor in its creation. Various protocols were developed to deliver e-mail among computer groups. Arrangements to distribute news (called newsgroups) appeared, along with various discussion groups and mailing lists.

After the successful October 1972 demonstration, ARPANET, the British Post Office, TELENET, DATAPAC, and TRANSPAC cooperated to produce the first international packet switching network. An early packet switching protocol standard, X.25, was developed about 1974 and approved in 1976 by the International Telecommunications Union. Later, X-25 became JANET. These protocols were independent of TCP/IP that were developed in about this same period by DARPA (which had by 1972 been renamed from ARPA) from its experimental work on ARPANET, the Packet Radio Net and the Packet Satellite Net.

Open Architecture: TCP/IP—Kahn and Cerf

The need for an open-architecture network that would, among other things, permit information exchange between computers having different operating systems became evident at about this time. Robert Kahn at DARPA decided to develop a suitable communications protocol. He asked Vinton Cerf (then at Stanford; he had worked on development of the NCP) to join him at DARPA designing a new protocol. They developed the first TCP in 1973 and published a paper about this network protocol in 1974. Later that year, Vinton Cerf, along with Yogen Dalal and Carl Sunshine, published the first complete specification for TCP. DARPA provided developmental contracts to Stanford (Cerf), BBN, and University College London (UCL) to work on TCP/IP.

History: Vinton Cerf and Robert Kahn

Vinton Cerf

Cerf's childhood was spent in Los Angeles. He was a good student and displayed aptitude in math. Even as a child, he tended to wear formal clothes (suits and ties): most days he was found wearing a tie and a jacket, a pattern he continued as an adult.

His early childhood interest in computers increased until, while attending Stanford and majoring in mathematics, he changed his major to computer science. After a brief period at IBM as a systems engineer, he enrolled in UCLA's computer science department as a Ph.D. candidate. There, with a group of students, he became involved with an ARPA project at UCLA and with solving problems involved in the construction of ARPANET. Subsequently, he there received a Ph.D. degree.

Robert Kahn

Kahn earned a B.E.E. from City College of New York and received a M.A. and then a Ph.D. in 1964 from Princeton University. After joining ARPA, he successfully demonstrated ARPANET in October 1972 by interconnecting forty different computers. He became Director of IPTO and started the federal government's billion-dollar Strategic Computing Program. While working on a satellite packet network project, he conceived of some ideas that later were used in the TCP.

Creation of TCP

One major problem with the ARPANET network was making two host computers (or networks) compatible so that they could communicate with one another. Computers and networks needed to be able to communicate regardless of the particular hardware or software involved in each host. It was determined that two computers needed to be connected through a router (or "gateway") computer, but while a router could accept and move packets, a router maker did not provide methods (or programs) for use in packet communication. The two computers had to communicate uniformly with each other without realization or identification of things such as something going on between them.

In 1969, the selected router manufacturer, BBN, delivered a first router to UCLA and a second to SRI. At about this time, Cerf met Kahn, and they worked well together, particularly beginning in spring 1973. Kahn (of DARPA and ARPANET) recruited Cerf (of Stanford University) to work with him on network and Internet protocols. Cerf was influenced by the French CYCLADES network (conceived by Louis Pouzin), Hubert Zimmerman, and Gerard LeLann.

The UCLA student group (which called itself the Network Working Group [NWG]) liked the idea of using "layers" or "building blocks" for protocols that could be joined to network communication protocols. In 1970, the NWG released a protocol for basic host-to-host communication that was called the NCP. It also created several additional protocols that operated on top of the NCP. For example, the Telnet protocol permitted remote logins. A group of protocols or protocol modules makes up a "suite," and each protocol module usually communicates with two others (one above, one below) in an imagined layered or modular relationship. The modularization model is considered to make protocol design and evaluation easier.

Kahn and Cerf thought a router (or gateway computer) located between two host computers, each perhaps an independent, self-contained network, should not care about what was going on in either computer network; it should only control and direct the passing packets using its own protocols. In a computer network, a router would thus be a device that "reads" packet destination addresses, selects the best route to that destination, and forwards packets to the next device along a pathway route.

By 1973, Kahn and Cerf created TCP as a common Internet protocol, which had various features:

- Each packet of information would have a destination address.
- Each packet of information would have a sequence number, thereby enabling assembly of the information in the correct order at its destination, and detection of the loss of any packet.
- A router would receive packets and forward same to destination.
- The destination computer would send information, in the form of a special packet (called an *acknowledgment*), back to the originating computer to show that the information was completely and successfully received.
- No portion of a network could either be a single location of failure or a location that controlled the whole network.
- Each packet sent would have an accompanying sum for checking, the sum being calculated by the originating computer, and being checked by the destination computer, thereby to show that the packet was not damaged in transit.
- If a packet sent was lost or damaged, the packet would be retransmitted after the loss was determined, as shown by a timeout that would show that the expected acknowledgment had not been received.

In September 1973, Kahn and Cerf presented to the International Networking Group a draft paper outlining their ideas, and in May 1974, they published their complete paper. Their main idea was to enclose the individual packets in standardized *datagrams* that were considered to be comparable to addressed envelopes. The paper presented a new transmission control protocol. It allowed only host receiving computers to "open" the envelope and "read" the enclosed packet. This TCP was found to work, and it made the Internet possible.

Cerf continued to develop TCP. In 1976 he obtained a job as program manager for what was then called the "ARPA Internet." In 1978, with several coworkers, Cerf split TCP into two components. One component became a separate protocol called the Internet protocol, which contained basic functions regarding data packets and is responsible for the sending and routing of packets over the network. The other protocol retained the name TCP and, like the original TCP, continued to have responsibility for dividing messages into datagrams, reassembling messages, detecting errors, assembling the correct order of packets, and resending lost packets. The TCP is thus responsible for error-free connection between two computers. TCP enables and governs the connection and exchange of sequential data between two host computers. The IP works only with packets; it routes outgoing packets and recognizes incoming packets. The two protocol components together became a new protocol that was called TCP/IP, or sometimes the Internet protocol suite, and became the basic standard for all Internet communications. Together, TCP/IP was designed to make a network robust and with capacity to recover automatically from any failure that might occur in a device.

Internet Protocol

The IP is regarded as the central and unifying protocol of the TCP/IP suite. It provides the basic delivery mechanism for data packets between all systems on the Internet regardless of system location. Moving packets is a fundamental function, and all other protocols of the suite depend on IP. IP is a simple protocol. It does not guarantee to (1) actually deliver data to a destination, (2) deliver a packet undamaged, or (3) provide only one packet of data at the destination. The simplicity enables good implementation even on systems of small memory or small processing power and facilitates deployment on many different networking systems. IP can be used to carry data from one station to another provided that the packets are differentiated by some convention from other data possibly seen by a receiver. To control size, IP must be able to break down large data packets into a train of smaller packets that are then arranged to be transmitted (this process is called fragmentation). The destination is responsible for rebuilding an original IP packet from the fragments. IP can use the address resolution protocol to find the station address that corresponds to a particular IP address associated with a packet or packets.

Over development time, different versions of IP were employed. The newest version, IP Version 6 (IPv6) is apparently not yet fully or widely deployed. It uses addresses of 128 bits instead of the 32 bits in IPv4 and so can support many more devices on the network. It incorporates features, such as authentication and multicasting, that previously were only appended to IPv4. IPv5 never existed.

Transmission Control Protocol

TCP achieves a reliable transfer service for a byte stream between two host computers or the like. Although TCP depends on IP to move packets around, IP is inherently unreliable, so TCP protects against errors, such as data loss or corruption, packet rendering, or data duplication by so-called checksums and sequence numbers to transmitted packets and on received packets sends back acknowledgments of receipts. Initially, before sending data, TCP establishes a connection with the receiver destination using packet management exchanges. Using management packet exchanges, a connection is terminated (destroyed) when the transmitter that is using TCP indicates that no more data are transferred.

TCP has a mechanism regarding multistate flow control that continuously adjusts the sender's data rate in an effort to achieve maximum data rate throughput yet avoid congestion and packet losses. TCP seeks to make the best use of network resources available by loading as much data as possible into a single IP packet subject to being overridden by applications that demand immediate data transfer.

TCP is still undergoing development. Later requests for comments (RFCs) define developed TCPs. An October 1989 RFC summarized the requirements for a TCP. A new TCP must support older TCP implementations.

To avoid overloading a network, TCP includes several features, such as "slow start," "congestion avoidance," "fast retransmit," and "fast recovery" algorithms. Another algorithm seeks to avoid transfer of very small data packets.

The TCP/IP suite usually includes *User Datagram Protocol* (UDP), which is used when reliable delivery is not required. UDP is a connectionless datagram protocol and does not use a "handshake" to start; it only transmits packets. UDP is widely used for applications such as *voice over IP*, video conferencing, and streaming audio and video, because there is no time available to retransmit lost or injured packets though packet sequence. Error notification can be written into UDP. UDP does not warrant data delivery to a destination, but it does warrant that packets will be delivered to a destination in the order sent. In UDP, if desired, the checksum generation can be disabled, but then corruption or truncation of data is undetected.

A *port* is a destination or interface on a computer to which a connection can be made. In TCP and UDP, a port is a numbered end point for a logical connection. Both destination and source port numbers are incorporated into TCP and UDP headers (lines at the start of a document) by the sender before a packet group is given to IP. Each received frame (i.e., part of a browser screen displaying a particular content) is examined by TCP and determines which server gets the packets. Port number designation is arbitrary, but it is fixed by tradition and by the Internet Assigned Numbers Authority. Well-known port numbers include port 23, for use of a Telnet (a utility protocol that allows a computer to connect with another computer) with a server receiving packets on TCP; port 80, for HTTP traffic; and port 21, for receiving packets on TCP. Periodically, an updating RFC is published that contains all assigned numbers.

Development

For about the next 9 years after 1974, work proceeded to develop and improve the component protocols of the protocol suite and to implement them on a wide variety of operating systems. As set up, the original TCP/IP model comprised four layers, but some now view the developed TCP/IP taken with the WWW as a reference model having five layers. The body having technical responsibility for the TCP/IP suite, the Internet Engineering Task Force (IETF), does not accept the five-layer model, but this model can be regarded as a teaching aid. The IETF emphasizes "architectural principles" rather than layering, such as connecting packet edges and maintaining speed and robustness.

In the four-layer version of the TCP/IP, the layer names that were early assigned by the IETF (from the top down) are as follows:

1. Process layer (or *application* layer) (where so-called higher-level protocols exist, such as file transfer protocol [FTP], SMIP, and HTTP)
2. Host-to host layer (or *transport* layer) (where the TCP and similar protocols exist for connection and flow control)
3. Internet working layer (or *data link* layer) (which defines IP addresses with routing schemes for sending packets from one IP address to another)
4. Network access layer (or *physical* layer) (which includes both protocols used to mediate access to shared media and also protocols used for communication between individual hosts and a medium)

Layers near the top of the model are closer to a user application (distinct from the human user), and layers near the bottom are closer to the transmission of data (information). The lower layers lack details, or knowledge of details, regarding each application and its protocol. The upper layers can provide services that the lower layers typically do not. A layer provides service(s) to the layer above it and receives service(s) from the layer below it. Basically, IP is a best-effort delivery protocol, while TCP is a protocol that that provides data integrity and delivery warrants. For example, a packet is redelivered by TCP until the addressed receiver receives it.

The application layer is used by most programs for network communication. Data are passed in a specific format and then encapsulated into a transport layer protocol. Received data are passed into the application layer, where they are encapsulated into the application layer protocol. Then they are passed down to the lower layer protocol of the transport layer.

The transport layer is responsible for end-to-end message transfer, including error control, fragmentation control, and flow control, independent of the underlying network. It can be connection-based or connectionless. This layer uses ports. It is the first layer of the TCP/IP stack to provide reliability, and protocols above it can also provide reliability. TCP involves and deals with many issues to provide reliability, including packet order, packet correctness, discarding of duplicate packets, and resending of missing packets.

The network layer was originally intended to solve the difficulties of getting packets across a network. In development, added functionality was placed in this layer relating to routing packets across a network, sometimes called "internetworking."

IP, if desired, can have protocols that carry data for various upper-layer protocols. These protocols are actually routing protocols and each has its own protocol number. One is known as ICMP (Internet Control Message Protocol) and is used to transmit diagnostic information. Another is known as IGMP (Internet Group Management Protocol) and is used to manage IP multicast data. Such protocols are considered by some to illustrate an incompatibility between the IP and the TCP or the Open Systems Interconnection Basic Reference Model (OSI Model). However, all routing protocols are regarded as part of the network layer because their product packets are results of the management of the network layer, and their encapsulation is irrelevant for layering purposes.

The data link layer in effect moves packets from the network layer. Perhaps it is not actually part of the suite, because IP can run over different data link layers. Control of packet transmission can be done in a software device driver, in firmware, or in special chips that can link data. Different protocols are used in the respective cases of dial-up modems, broadband Internet access, cable modems, and local wired networks. Depending upon the local network arrangement, the data link layer can be considered an application layer.

The physical layer encodes and transmits data over a network. It operates with bits of data that are sent from the sending computer and are received at the physical layer in the receiving computer. Various conventional devices function at the physical layer, such as cables, connectors, hubs, repeaters, the Ethernet, and so forth. The physical layer is involved with hardware design issues, such as local area networks (LANs) that share information and information resources and wireless communication.

The OSI Model was started in about 1977 at the American National Standards Institute and involved a seven-layered protocol design. Internet protocols, such as a common version of the TCP/IP model, split the OSI Model into about four layers, as indicated above. It happens that the TCP/IP model was developed before the OSI reference model, and the IETF has never become compliant with the OSI Model. The OSI Model has been widely used in teaching, and while it involves seven layers, it does not appear to be in accord with real-world protocol structure used in the common Internet composition.

Various particular programs and hardware can be involved with different layers, but implementation hardware or software is not stated in protocols. High-performance routers are usually nonprogrammable and involve digital electronics.

In July 1975, after ARPANET had been running for several years, DARPA turned the network over to the Defense Communications Agency, which was part of the Department of Defense. After the years of developmental work on TCP/IP (sometimes called the Internet protocol suite), these protocols were adapted for use with a large range of operating systems. Various additions, alterations, and complexities were adopted to improve utilization and functionality. Various other core protocols became included in the suite, such as those designated UDP and ICMP. The suite provides a networking framework for many different application protocols. The suite is also used widely in private networks that may not be connected to the Internet. Following development and planning, effective January 1, 1983, the first wide-area TCP/IP network became operational, and all hosts on ARPANET began using it (replacing the earlier NCP). About 3 years earlier, TCP/IP had been adopted as a defense standard. ARPANET subsequently developed into the Internet.

Gradually, use of the TCP/IP suite became substantially universal while OSI development became minimal or abandoned. As the Internet expanded and evolved and computer technology developed, problems occurred, and network developments were needed for solutions. When smaller, desktop computers first appeared, TCP implementations were made. The large development of LANs, personal computers, peripheral associated equipment, Ethernet technology (a popular protocol for linking computers in a LAN evolved by Robert Metcalfe at Xerox PARC in 1973), and the like actually encouraged the Internet and caused change and development in the original model of TCP/IP from usage in a few networks with few time-sharing hosts to usage in many networks with many hosts. Three network classes appeared: class A (large, national-scale networks), class B (regional-scale networks), and class C (many local area networks, few hosts). The network includes backbone networks, mid-level networks, and stub networks (which include commercial, university, and research networks).

All protocols of the TCP/IP suite are defined by documents called RFCs that are freely available. Anyone can compose and submit for approval an RFC. The RFC approval process is managed by the Internet Engineering Steering Group using recommendations from the IETF. There are more than 2,500 RFCs, each identified by a distinct number. An RFC index provides basic information on all RFCs. A significant culture developed on the idea that the Internet is not owned or controlled by any one entity, though it is recognized that some standardization and control are necessary for the system to function.

In 1984, the National Science Foundation (NSF) created the CSNET network, and in 1985, NSF began funding five new supercomputer centers at universities. The NSFNet network went online in 1986 and interconnected these centers without cost to users; it used a TCP/IP that was compatible with ARPANET. NSFNet experienced exponential growth. From about 1987 to

1995, the history of the Internet seems unclear, but IBM and MCI were apparently given a competitive advantage over other private companies. Federal funding in this period for the Internet seems to have been provided by NSF and by National Aeronautics and Space Administration (NASA), which established the NASA Science Internet (NSI). NSI was a high-speed, international network connecting more than 20,000 scientists on seven continents.

More developing countries provided access to the Internet. In 1990, the ARPANET project closed, and in 1994 NSFNet (renamed ANSNET) lost its position as backbone of the Internet. Various governmental and private groups created their own arrangements. The initial commercial restrictions ended. Regional network access points became the main interconnections between networks.

Kahn and Cerf: Post TCP/IP

Robert Kahn left DARPA after about 13 years and founded the not-for-profit Corporation for National Research Initiatives in 1986. He became chairman, chief executive officer, and president of this organization, which has the objective of providing leadership, funding for research, and development of national information infrastructure. He has received various honors and awards. He remains active in the field of computer networking.

Vinton Cerf is a cofounder of the Internet Society (ISOC), which is intended to promote the views of common users of the Internet and to serve as a body representing groups developing the Internet, including the Internet Engineering Task Force. Cerf served as the first president of ISOC from 1992 to 1995. He continues to pursue developments in the Internet. He has received various honors and awards. He has a hearing problem and serves on the board of Gallaudet University, the first school of higher learning for the hard-of-hearing and deaf. Kahn and Cerf received the Presidential Medal of Freedom from President George W. Bush in 2005.

Domain Names

To avoid numeric addresses, which were generally difficult to remember, assigned names for network hosts were used. The Domain Name System was created by Paul Mockapetris of the University of Southern California Information Sciences Institute (USC/ISI) in 1983, which achieved a mechanism for cataloging hierarchical host names relative to an Internet address. Almost all Web pages and e-mail addresses now use a domain name. A *Web page* is a computerized document that contains text and/or graphics that can be accessed through a Web browser.

The Internet Corporation for Assigned Names and Numbers (ICANN) is the authority that coordinates assignment of unique identifiers, including domain names, IP addresses, and protocol port and parameter numbers, on the Internet. ICANN is headquartered in Marina del Rey, California, but is under an international board of directors. The U.S. government has a primary role in approving changes to the root zone file in the domain name system. ICANN is probably the only central coordinating body of the global Internet.

Even the size of the original addressing tables used in routers stressed capacity of routers. With expansion of the Internet, routers and their algorithms had to be changed. The original single distributed algorithm for routing for all routers was replaced by a hierarchical model of routing having an interior gateway protocol (IGP) plus an exterior gateway protocol for tying regions together. Different regions could use different IGPs, thereby allowing for different requirements involving cost, rapid reconfiguration, robustness, and scale. New forms of address aggregation, such as classless interdomain routing, help to control router table size.

One problem was how to disseminate and incorporate software changes in existing host software. Experts feel that the strategy of incorporating Internet protocols into a supported operating system used in research was an important element in the successful adoption of the Internet.

Growth of the Internet

In 1985, the NSF in the United States funded construction of a 56-kilobit per second network backbone for university use, and in the following year a 1.5 megabit per second network backbone, which became the NSF/NET. The NSF/NET interconnected the five supercomputers (Princeton University, University of California at San Diego, University of Illinois, Cornell University, and the Pittsburgh Supercomputing Center) at no cost. The key decision to use here the previously developed TCP/IP is attributed to Dennis Jennings (from Ireland, who spent a year at NSF). His successor, Steve Wolff, took over the program and supported a wide-area networking infrastructure in the general academic and research community, with features for being independent of federal funding.

Federal agencies implemented various policy decisions that effectively advanced the Internet system. Federal money continued to flow into Internet development.

It was not until January 1, 1983, as indicated above, that the TCP/IP became practical and operational. At that time, all hosts of the ARPANET were changed over from the older NCPs to the TCP/IP. The new operation was successful. DARPA sponsored, encouraged, and funded development of TCP/IP implements for many operating systems and worked out scheduling for hosts to adapt TCP/IP. By mid-1980, the agencies of NASA, NSF, and Department of Energy had operating wide-area networks based on TCP/IP and the first high-speed multiprotocol wide-area network. Although many different networks were by then operating, the practicality of TCP/IP gradually caused the networks themselves to have responsibility instead of having a common network, as in the ARPANET.

The use of the Internet for commercial purposes became a topic for debate, but in 1988 the Internet was opened to commercial interests. Internet service providers (ISPs) appeared. Transmission facilities began spreading into Asia, Australia, Latin America, Africa, and elsewhere. Space does not permit a full review here of the growth of TCP/IP and the Internet. The role of a common network was reduced to a minimum, and almost any network could be interconnected with another. Interest in and new applications for networking grew and spread throughout the world, although there was considerable resistance in Europe to widespread use of TCP/IP. Although the philosophy that the Internet is not owned or controlled by any one is deep, as indicated above, some standardization and control is necessary for its functioning.

In 1988, a series of NSF-initiated conferences at Harvard's School of Government initiated a process for enabling privately financed commercial uses. A National Research Council committee chaired by Kleinrock and with Kahn and Clark as members produced a report commissioned by NSF in 1988, which was influential with then-senator Al Gore and which promoted high-speed networks. The network was opened to commercial interests in 1988, when the U.S. Federal Networking Council approved interconnection of the NSF/NET and the MCI Mail system, a link that was made operational in the summer of 1989. Soon, other email services were connected. Commercial Internet service providers appeared, and various separate networks were merged with the Internet. Commercial routers became available from various companies. Local area networking equipment and implementation of TCP/IP on the UNIX operating system further enhanced rapid growth. In 1990, the ARPANET was finally decommissioned, and it appears that the term "Internet" then became the name for the surviving network. In 1994, another report was issued by the same people (Kleinrock, Kahn, and Clark). In April 1995, NSF's so-called privatization policy resulted in the defunding of the NSFNET Backbone program. By 1990, TCP/IP was becoming the network protocol used worldwide on the Internet.

Augmenting usage of the Internet was the development of various books and search arrangements. For example, the Harley Hahn and Rick Stout *The Internet Yellow Pages*, published by Osborne McGraw-Hill, had hundreds of listings of various sites around the world.

Another Internet-promoting tool is FTP, which allows a computer user to access remote computers and retrieve files from them. Such retrieved files can be free and cover almost any imaginable subject. ISPs maintain *clients* (programs that enable a user to talk to and get downloads from distant computers). A client can access a server, which is a program (or sometimes computer having a program) that stores files (documents) of many sources and users, that can be accessed (shared), or that allows communication with a browser. FTP can involve files that have been compressed (to save space) through the use of compression software, but to decompress these files, one must know the compression method used. A search program called Archie became available for searching a collection of sites (file locations) to identify the location of a file of interest.

As the Internet grew, various ways to organize it appeared. A search program called Gopher became available for menu-driven applications. A Gopher client can have a site that is subsequently accessible through the use of a bookmark. To provide rapid access to Gopher, the Veronica program became available, by which a keyword entry is used to search through thousands of Gopher servers with entered individual items totaling in the many millions. Located files are retrieved and placed in a temporary browsable file.

The first hypertext interface to enter common usage was the search engine Gopher, which had menu items that were examples of hypertext, but which were not so perceived by users. Another popular Web browser was Viola WWW, modeled after Apple's HyperCard. Another was the Mosaic Web browser, which used a graphical interface and, after its introduction, soon became more popular than Gopher, but was overcome later in 1994 by Andreessen's Netscape Navigator. In turn, this browser was completely displaced by the later Internet Explorer and other browsers.

A search engine is associated with a Web site. A *Web site* is regarded as an organized and structured collection of one or more Web pages that can, for example, comprise the base of an e-commerce operation. A Web site can be located by a browser or by a search engine. A *search engine* is like a directory or searchable database and offers a computer user the ability to search for keywords or content (information, products, subjects, etc.) in Web pages and other Web sites. Perhaps the most popular search engine today is Google, but others include Yahoo, Excite, AltaVista, and Infoseek.

The American Standard Code for Information Interchange (*ASCII code*) became a standard format for transmitting textual data. Any computer can read an ASCII text file. Any ASCII text file will appear the same on any computer regardless of computer maker or operating system. Data (non-text) files must be saved in binary (0s and 1s) but can only be run with certain computers or programs.

One reason for the relatively rapid growth of the Internet was and is the free and open access to all of the basic documents, including the specifications of the protocols. As early as 1969, S. Crocker (then at UCLA) established and published the procedure of RFCs. As the FTP came into use, the RFCs were prepared as online files and accessed via FTP. Later, with the WWW, RFCs were accessed at sites around the world. RFCs can be made virtually by anyone. The effect of RFCs was and is to promote Internet development. Over time, RFCs have tended to focus on protocol standards and are important documents to those interested in Internet engineering and standards. E-mail, including specialized mailing lists, is used in the development of protocol specifications.

THE WORLD WIDE WEB

As the Internet was growing, there appeared various browsers, search engines, and the like, such as those indicated above, that attempted to organize the Internet. Projects to organize files and information and various Web browsers and the like appeared.

Before and after the advent of the World Wide Web, various search engines appeared that attempted to organize the Internet, such as, among others, Archie from McGill University in 1990, WAIS and Gopher in 1991, and Mosaic from the University of Illinois in 1993. Other search engines and Web directories appeared, such as Lycos in 1993, WebCrawler in 1994, and Yahoo! and AltaVista in 1995. The directory approach began to give way to search engines.

By 1996, public interest in what had previously been in the academic/technical world of the Internet was growing. Google arrived in 1998 and developed new approaches involving algorithms ranking relevancy, an approach that apparently now is used by all Web developers. During the 1990s, the Internet reportedly grew by about 100% per year. By 2001, the directory model was beginning to be replaced by search engines. Database size became replaced by relevancy ranking as it became evident that it was not practical to review whole lists of results. Rankings of search engines became important, and law has developed about matters that affect such rankings, such as the use of trademarks in meta tags.

Business plans based upon the idea that by using the Internet, existing distribution channels could be circumvented seem to have been the root cause for the famous burst of the "dot-com" bubble on March 10, 2000, after which more than half of the dot-coms ceased to exist.

None of these efforts were like the WWW. But even with such aids, it was still time-consuming and relatively complicated for a user to access information available through the Internet. Unfortunately, for many Internet users, access to the Internet continued to seem difficult and to involve specialized codes and time consumption, and not infrequently the results of searching were not adequate. All such conditions limited and discouraged use of the Internet.

Tim Berners-Lee

Tim Berners-Lee was born June 8, 1955 in London, England, to Conway Berners-Lee and Mary Lee Woods, both mathematicians, who were employed together on the team that fabricated the Manchester Mark I, an early computer. His parents taught Tim math and encouraged him to use it everywhere, including at the dinner table.

He attended Mount Primary School in Wandsworth, studied for his O-levels and A-levels at Emanuel School, and went to Queen's College,

University of Oxford. He built a computer there, but with a friend was caught hacking and banned from using the university's computer. He received a degree in physics in 1976.

Subsequently, he was employed in a succession of software and programming jobs at Poole and soon married his first wife, Jane, whom he had met at Oxford.

The Internet had developed by 1980 to a place where a user had various search tools available, as reviewed above. However, it was apparent to Tim, and generally to computer software creators, developers, and users, that improvements of various sorts were possible and needed. He appreciated the limitations and problems of using the Internet. In about 1980, working for a limited six-month period as a software consultant at CERN (the European Organization for Nuclear Research [in French, Organisation Européenne pour la Recherche Nucléaire]), the great particle physics laboratory on the border of Switzerland and France, Tim had written for his own personal use an unpublished notebook prototype program for storing information using random associations, which he named "Enquire" (this name being drawn from an old encyclopedia: "Enquire Within upon Everything"). The Enquire system provided links between essentially arbitrary nodes. Each node had a title and a class. He provided a list of bidirectional links (based upon classes). The program ran on Norsk Data computers under SINTRAN-III.

Later in the 1980s, Tim had promoted at various conferences the idea of a network-based implementation of hypertext as an avenue to permit computer users to readily access the Internet, but no one would implement this idea.

Origin of the World Wide Web

After working for a software developer in England from 1981 to 1984, Tim returned to CERN on a fellowship and worked in the area of scientific data acquisition and system control. While there, in March 1989, he proposed a developmental project. CERN was a significant user of TCP/IP on intranets, but this protocol use tended to be isolated (relative to Europe generally) until about 1989. Tim felt that "in an exciting place like CERN ... some place" was needed where the whole organization could access the work of others. His 1989 proposal related to information management based on hypertext. Tim was seeking a way to store, retrieve, and share CERN research information, and he proposed to use mainly known software to implement his plan.

In September 1990, the circulated proposal was approved for implementation by his manager, Mike Sendall, who also authorized purchase of a NeXT cube. Tim named the project the World Wide Web. In October, with encouragement from management, the original proposal was reformulated with Robert Cailliau. Tim undertook a network-based implementation of his hypertext idea. Development continued, and by November an experimental browser/editor named "What You See Is What You Get" and the first Web page appeared, along with a Web server.

The project looked interesting to management. Nicola Pellow (then a student) and Bernd Pollermann became associated with the project and pursued project details. In December, a line-mode browser and WWW browser/editor were demonstrated, and access to some hypertext files was achieved using HTML (created by Tim) and HTTP (created by Tim) using a uniform resource locator (URL; created by Tim). A server and path information used an HTML-coded source file or browser to locate another document.

WWW allowed interaction with the Internet through the use of icons (small images) and hypertext links. In WWW, Tim had used similar ideas to those that underlay his earlier Enquire idea. Using the NeXT cube, Tim created the first Web browser and editor and the first Web server. Tim used scalable, public domain software in the Web's basic structure, which enabled others to build inventions using this open architecture. His idea was to allow people to work together and to use their own knowledge of hypertext documents. One result was that others made millions of dollars on his WWW invention. He made his ideas freely available; no patent or other royalty fee was involved.

In 1991, development continued with work plans and internal presentations at CERN. In May, CERN released the WWW on central CERN computers. In June, CERN held a computer seminar on the WWW. Availability of files on the Internet by FTP was announced. Jean-François Groff joined the project in August. On August 6, 1991, Tim posted a brief summary of the project on the alt.hypertext newsgroup, which marks the date when the WWW became a publicly available Internet service. In October, auxiliary gateway programs were installed, along with establishment of WWW mailing lists and Telnet service. In December, a demonstration at Hypertext '91 in San Antonio, Texas, was given, and a CERN newsletter announced the WWW. The first Web server outside Europe was installed at the Stanford Linear Accelerator Center by Paul Kunz.

Thus, about 2 years after Tim Berners-Lee at CERN started using hypertext, HTML (as created and developed by him), HTTP (as also created by him), and URL (also created by him), he had created a few of the first Web pages and a format for them. CERN in August 1991 chose to publicize his WWW project. The timing was good, and by releasing this work for public use, Berners-Lee and CERN were certain this technology would be widely utilized. The technology could be used in cooperation with a search engine such as Google, was very simple and quick for a user to utilize, and tended to yield comprehensive results.

After WWW was released by CERN to the public, the simplicity and ease of use of this system for organizing and displaying information on the Internet rapidly increased its usage, and it soon gained worldwide acceptance. On April 30, 1993, CERN announced that WWW would be free for anyone, with no fees due. In some subsequent years, the usage of WWW is believed to have doubled over the preceding year. As of about October 1, 2007, it is estimated by Internet World Statistics that about 1.24 billion people around the globe use the Internet and WWW.

Composition of the World Wide Web

The *Web* is a system in which hypertext documents are interlinked and accessible through the Internet. When connected with the Internet, a user may begin to locate a Web page by typing the desired URL into a Web browser. The Web browser then undertakes a series of communications (not visible to the user) to retrieve and display a Web page having hypertext corresponding to the URL.

The WWW (or the Web) has come to identify a very large number of documents (including images, printed matter, textual material, photographs, graphics, video, multimedia, interactive matter such as demonstrations, game, applications, advertising, and the like) that are linked together by hyperlinks and URLs. The originals and copies are stored (cached) and are retrievable by machines such as Web servers or search engines. The individual documents are electronically delivered to a site as ordered by a user using, for example, HTTP, which is one of the main communication protocols employed in the Internet. HTTP, for example, enables software systems to communicate and thus share and exchange information, such as text or business data.

With the WWW, Internet research is keyword-driven and uses a Web browser, such as Internet Explorer or Firefox, or preferably a search engine, such as Google. Publishing a document on the Web for search and retrieval by others provides the potential for use by an extremely large audience. Content management software is available to aid in fabricating and maintaining a Web site. File sharing is available, as are version control systems that allow teams to work on shared documents without altering or accidentally erasing or overwriting each other's work.

In normal use of WWW, a computer user making a search inputs, using a keyboard, into a device, such as a browser, a key word or phrase. The device searches for hyperlinks and, using HTML and URL, accesses Web pages. Using HTTP, the device transfers the files found to a computer user's site. HTTP allows Web browsers to submit information to Web servers and to retrieve Web pages from them.

HTTP operates between clients and servers. A client, such as a Web browser, spider, or end user tool, is a user agent. A *server* stores or creates resources and responds to a client as an origin server. Intermediate devices may be located between a user agent and an origin server. HTTP can operate on top of any Internet protocol that provides reliable transport. HTTP is stateless, so that hosts do not need to retain information about users between requests. Resources on a server to be accessed by HTTP are identified using URLs. HTTP defines eight "verbs" (commands) specifying the desired action to be performed on an identified resource. The most commonly used verb on the Web is "get" (a request for a copy of the specified resource). Some verbs are considered safe because they normally do not have side effects, such as changing the state of the server. Some verbs are "idempotent," because multiple identical requests have the same effect as a single request. A user agent, such as a browser, may permit any idempotent request to be retried without advising the user. Various "backward-compatible" protocol versions of HTTP have developed. Two verbs are usually available for securing an HTTP connection.

The most commonly used HTTP is HTTP/1.1, which is supported by most Web browsers, but not Internet Explorer. The file format for a Web page is usually HTML and is identified in the HTTP protocol.

URL (sometimes identified as uniform resource identifier [URI]) identifies a resource and locates the resource by specifying its network location (or access mechanism). A URL is usually placed in the address location bar of a Web browser. A URL can be either "standard" or "clean." A clean URL typically is not tied to technical details or to organizational structure yet is consistent with the hierarchy of other URLs at the same site.

A Web page is accessible through a Web browser through information, usually in HTML, that may provide information to navigate to other Web pages. A Web server may construct the HTML when so requested by a browser, or a Web page may consist of static text file(s) stored with a Web server's file system. Most browsers support a variety of formats in addition to HTML.

The WWW is considered to be a part of the Internet. In the WWW, TCP/IP is used and remains a suite of communications protocols that forms the basis for and defines the Internet. The TCP is a communication protocol, that is, a layer located above the IP, which is among low-level communication protocols that allow computers to send and receive data. TCP governs the exchange of sequential data and enables two host computers to establish a connection and exchange data streams. The IP effectively routes outgoing messages and recognizes incoming messages. The TCP/IP suite design was intended to, and does, make a network robust and enable automatic recovery from any device or line failure. Also, the design permits construction of very large networks with little or even no central management.

A *Web browser* is a software program that enables a user to access documents (Web pages). A Web page contains HTML codes that are interpreted by a Web browser so as to allow connections from one document (or Web page) to another. When a user selects a Web page through a browser, the browser passes the request along to a server. *Servers* are regarded as the backbone of the Internet. A server allows communication with a browser and processes received requests for HTML in Web pages. Thus, a server awaits and processes received HTML code regarding requested information, locates the requested information, and sends the information back to the browser (sometimes called the client in this context). The browser displays the requested information on the screen of the user's computer.

It is important for WWW purposes that each computer that is connected to the Internet have a clearly identified, unique numeric address (the IP address). This address has been defined as comprising four sequences of digits, separated from one another by dots, the whole comprising thirty-two bits. The address is translated by a domain name server into a domain name readable or usable by a user. To locate a file that is either a Web page or another Internet entity, a path and address information is used in an HTML-coded source file. A URL is the complete path and address information of a file or other resource on the World Wide Web and includes the domain, the name of the file, and the protocol.

To transfer Web pages from one computer location to another, a set of rules is provided that comprises a communications protocol called HTTP. It is this protocol that allows browsing on the WWW so that a user browsing a hypertext document can jump to another document that may be located on another host perhaps thousands of miles away and retrieve information from that document.

Internet Growth

From the early 1990s until 1998, one company, Network Solutions, Inc. (NSI), kept track of domain names such as .com, .net, and .org. The National Science Foundation (NSF) funded most of the research and connection costs of the early Internet. When the domain name system arose in 1983, the NSF put out a request for bids for a contract to do the work of assigning and registering domain names, and the usual federal advantage was offered to minority-owned businesses. An African Americam entrepreneur established a new corporation, Network Solutions, and won the contract. Soon afterward, a corporation based in Washington, D.C., Science International Corporation (SAIC), realizing that the Internet would be a big thing, bought out the original owner of NSI and made it a wholly owned subsidiary of SAIC. It happens that SAIC has stockholders, directors, and officers who are or were top people in the national security structure, and SAIC does contract work that the federal government wants done by people with good connections. Although the federal government previously paid NSI to register domain names, SAIC got NSF to change the arrangement so that NSI was authorized to charge $50 per domain name per year for registration. Of this amount, NSI paid $15 to the federal government but kept $35. The actual cost of registering was estimated to be about $1 per domain name. When SAIC and its investors sold out NSI to the present owner VeriSign, they gained $21 billion.

(continued)

Originally, Jon Postel was the chairman and in effect the sole member of the Internet Assigned Numbers Authority, the organization that coordinates almost all Internet addresses. He was a professor at the University of Southern California, and he was trusted by everybody, including Vint Cerf and Bob Kahn, to technically coordinate the Internet addressing and domain name system. He had a long beard that came nearly to his waist in front and a ponytail that came nearly to his waist in back. Postel before long realized that the domain name system and the trademark legal system were conflicting.

Actually, no one is in charge of the Internet, which is organized as a hierarchy of networks. There are thirteen computer servers (called the root servers) at the hierarchy's top, of which ten are in the United States, some run by the government, the remainder by private interests. Of the remaining three, one is in Japan, one in Sweden, and one is in Amsterdam. There is no controlling legal authority, but all must run in synchronicity. All technical protocols are set by the Internet Engineering Task Force (IETF).

Representatives of organizations including the Internet Society, the International Telecommunications Union (a United Nations international treaty organization), the World Intellectual Property Organization, the International Trademark Association, and the IETF formed the International Ad Hoc Committee to restructure the domain name system. A report and the "Generic Top Level Domain Memorandum of Understanding" resulted; these restructured the domain name system with new domains and set up a quasilegal system for dealing with cybersquatters (trademark infringers). It was signed in May 1997. Eventually, two hundred significant parties and companies signed.

Although members of the Ad Hoc Committee liked to joke about reporting only to God, before long the U.S. federal government stepped in. Thereafter, technical coordination came from the U.S. Department of Commerce, not Jon Postel, and the federal government created the Internet Corporation for Assigned Names and Numbers (ICANN).

Berners-Lee Post-WWW

In 1991 to 1993, Tim continued work on the Web design, particularly coordinating feedback from users around the world. He refined his original specifications for URI, HTTP, and HTML, which were examined and discussed in increasingly large settings and references as Web technology dispersed.

The first International WWW Conference, which was held in May 1994 in Geneva, Switzerland, was greatly oversubscribed. M. Bangemann there reported on the European Commission Information Superhighway plan, and in July he reported on the new MIT/CERN agreement regarding WWW organization.

Tim became a senior researcher and holder of the 3Com Founders Chair at MITs Computer Science and Artificial Intelligence Laboratory. He continued to steer Web evolution, and in 1994 he founded the World Wide Web Consortium (W3C), which in an open forum develops standards for the Web. He remains a director of the W3C. In 2001, he became a patron of the East Dorset Heritage Trust in East Dorset, England. He currently resides in Lexington, Massachusetts, with his second wife, Nancy, and two children.

Though brought up in the Church of England, he left the church after being confirmed because he said he could not "believe in all kinds of unbelievable things." With his family, he later founded a Unitarian Universalist church in Boston.

He has received much recognition and many awards, several with accompanying substantial monetary grants. In 2002, the British public named him to be among the 100 greatest Britons of all time, according to a BBC poll, and in 2004 Queen Elizabeth II gave him the rank of Knight Commander, which is the second-highest rank in the Order of the British Empire.

He is interested in developing the so-called semantic web, which is an extension of the World Wide Web and which is evolving. Berners-Lee envisions the Web as a universal medium for information, data, and knowledge exchange. The semantic web involves various enabling technologies and includes some prospective future possibilities that have not yet been realized. The semantic web would involve information that is understandable by computers, thereby enabling them to perform tedious work such as is involved in locating, sharing, and combining information on the Web.

FURTHER RESOURCES

Janet Abbate. *Inventing the Internet*. Cambridge, MA: MIT Press, 1999.

Paul Baran, Sharla P. Boehm, and J. W. Smith. On Distributed Communications series: Volumes I–XI. RAND Corporation Research Documents, 1964. http://www.rand.org/about/history/baran.list.html.

Tim Berners-Lee. *The Original Design and Ultimate Destiny of the World Wide Web*. New York: HarperCollins, 1999.

Leonard Kleinrock. *Information Flow in Large Communication Nets*. Cambridge, MA: MIT, 1961. (Proposal for a Ph.D. thesis.)

Ruth Maran. *Internet and World Wide Web Simplified*. 3rd ed. New York: John Wiley & Sons.

Melissa Stewart. *Tim Berners-Lee: Inventor of the World Wide Web*. New York: Ferguson Publishing Company, 2001.

Epilogue

From the preceding twenty-four accounts of iconic inventions and their inventors, one can make various tentative observations and conclusions, such as those suggested below.

(1) Some comments about the achievement of usage for an iconic invention:

Based on the factual information presented herein, some may find helpful the information regarding **timing**. For examples, the approximate elapsed time from commercialization to the time when iconic invention status was achieved, or the age of the inventor at the time when an invention was made that later became iconic.

Based on the case histories presented, it appears that **marketing** of an invention that is to become iconic can be critical. Most inventions are not developed and manufactured until or unless someone, perhaps the inventor, has some information (sometimes only an idea) indicating that the embodied inventive product will sell. However, until an embodiment of a new product embodying the invention is actually marketed and offered for sale, one never knows for certain whether the prior market analysis was even approximately correct, or whether the embodiment is adequate, acceptable, and sellable in the actual market. Marketing can be more difficult than inventing or that an inventor realizes. Before marketing, it is best for an inventor and his backers to have a business model or plan that considers reasonable possibilities.

An inventor and his or her supporters, enablers, and backers can sometimes conclude that the creation and production of a suitable invention are simple and even easy. Sometimes, for example, a key reduction to practice of an early embodiment from a conception may seemingly be simple and even easy to a technically inclined person. A remedy to overcome an existing problem can seem to be easy to propose. Some inventions become easier to achieve and manufacture because of technological advances. But marketing an invention may well depend upon other factors. Ease of creating, developing, and manufacture may not correlate with acceptance, marketability, and commercialization.

Apart from innovation and manufacturing, from a marketing standpoint, it may sometimes be more difficult today to achieve an iconic invention than it was a century ago when society, industry, economics, and environmental circumstances and conditions were perceived differently and fewer products were available.

From the standpoint of marketing an invention, and in addition to contemporary perception and careful analysis (and also apart from possible patent protection), perhaps a **test program** should be formulated and used preliminarily before extensive costs of large production and marketing are undertaken. Testing can indicate whether or not a new product will sell to consumers, but testing may not produce wholly accurate results. Will the patent protection on a new product deter immediate competition and copying? Can a test program be used that will avoid extensive costs?

Perhaps the inventor is not a good marketer. If not, then a marketer is needed, preferably one with intelligence and experience. Contemporary experience generally suggests that a team of suitable people is more likely to achieve an iconic invention than an individual.

To a marketer, for whom marketing and sales are normally simple, the marketing of a new product invention can be difficult. History suggests that iconic inventions are achieved infrequently relative to the number of inventions occurring. Moreover, it has been suggested that less than 1% of the population at any given time are inventors/innovators. Perhaps the more complex or more original the innovation, the fewer the possible or likely buyers. Typically, few innovations become iconic in any given field at a given time period. Yet, for those with insight, or for those who are gifted, creation of an invention may seemingly be easy. Evaluating a given new product innovation, in spite of local appeal and tests, requires **judgment.**

Inventions that become iconic typically seem practical, simple to use, and save time. From the reviewed case histories, if such a new item seems to meet a need, even if that need has not previously been appreciated or recognized, then that item has a good chance of selling well if the item is essentially affordable and useful.

Most probably, the simpler the invention, and the easier it is to use, the more likely that it can be made and offered for sale at an acceptable price and actually sold in amounts deemed iconic. In addition, a relatively complex, expensive invention to make can be relatively harder to sell at a profitable price than a simple invention. However, if an invention offers advantages, is simple to use, and justifies its cost, then it can sell. **Circumstances** are important and can be controlling. Users tend to overlook objections, including even cost if a new device seems important and easy to use.

Commercial success of an invention can involve interlocked factors, such as unit price, availability, and cost of alternatives, desirability, size, utility, and usage potential. Relative to a new invention, where the cost to customer is more than trivial, a customer can make a "buy" analysis that involves a group of factors that may be lumped together and not individually or separately considered. It seems that peripheral factors sometimes may only be rarely considered separately. However, individual or special circumstances can change a particular situation.

Responses by potential consumers to a new invention can be uncertain and can be controlled by **individual circumstances** perhaps apart from finances or economics. For example, the age of a potential buyer can be significant. Judging from iconic inventions, sometimes the younger generations are more apt to like and buy an invention embodiment sooner than the older ones. Perhaps older customers are more likely to have experience, reasons, prejudices, circumstances, fear of the future, etc., that can overwhelm or suppress an inclination to buy a new invention. A decision to buy that is made by a committee may be correct but can be hard to evaluate.

(2) Some general comments about inventions that became iconic:

From the case studies presented, with a few exceptions, the respective inventors both before and during the period when their invention that became iconic was made lived in modest circumstances. In the main, these inventors seem to have come from homes where their respective parents lived in modest circumstances. However, these circumstances seem to have improved as civilization advanced. The sociological consequences may be interesting. Does this mean that a potential inventor is more likely to achieve an invention if he or she is a little "hungry"?

Although it is commonly said that technology is advancing at an ever increasing rate, the case studies presented may indicate that iconic inventions—which can be considered to be inventions of historic importance—occur at a different rate. The reasons are not known. All previous iconic inventions have not been identified in the present work.

From the case studies presented, it may be that most iconic inventions were made by inventors roughly between the ages of 30 to 45, though there are exceptions.

Iconic inventions here considered were taken mainly from considerations of occurrence and usage, not from inventorship. It so happens that of the iconic inventions here considered, only male inventors are involved. Perhaps as civilization advances, and women have more opportunities, female inventors will make iconic inventions.

Since about 1880, most new inventions seem to have been produced as identical embodiment reproductions (mass-produced) in factories. However, sewing machines (Singer) and agricultural harvesting machines (McCormick) were factory-produced by the 1850s. In addition, in about 1880, the appearance of a single inventive entity that is also responsible for manufacture and sale seems to decline and fade as technology advances.

From the case studies presented, it appears that without exception, between the time of first reduction to practice and successful commercialization, improvements in each invention were made.

(3) Some comments about particular iconic inventions:

It seems that inventors of inventions of universal significance may sometimes have more competitors initially than other inventions. Questions about what actually occurred at the time when an invention was made can sometimes seem to be very difficult to answer based on now available information. Contemporary investigation of original sources of information is needed and would be desirable.

It is possible to distinguish iconic inventions (as herein defined) from truly pioneering inventions. On the interesting and provocative question of the frequency of origin of iconic inventions, further research may be needed to establish type and the origin date. The dates involved in the present case studies may not entirely suitable or accurate in spite of effort to the contrary. For example, although Ford's Model A was made and sold out completely in about 1903, this was not the first automobile with an internal combustion engine, differential, transmission, and brakes. Such a vehicle apparently goes back to historic efforts by Daimler and Benz in the 1880s in Germany and to others in the United States. These prior efforts were not iconic inventions since the involved cars were not made and sold in significant quantities compared, for example, to Ford. However, this example could be considered to conflict with the invention of the aircraft by the Wright brothers. They flew their historic plane in 1903, but arguably iconic commercial aviation transport of large numbers of the public did not arrive until at least about 20 years later. Breakthrough and pioneering technology perhaps is distinguishable from iconic large-scale manufacture, sale, and use. Further investigation of this matter is needed.

In the Twentieth and early Twenty-first centuries, the inventors of iconic inventions seem to become more diversified and specialized as respects their fields of innovation than previously. Previous inventions seem to lie mainly in the mechanical and electrical fields, as illustrated by case histories presented. Relatively contemporary inventors tend to work in other fields, such as the inventors Carothers, Fermi, Kilby, and Pierce. Pierce recognized the potential to use artificial unmanned satellites for communication purposes and used his corporate position to achieve this purpose, though he is not apparently a classic inventor of any applicable patent rights. Kilby recognized the need for better electronic circuitry and invented the integrated circuit. Fermi recognized the need to demonstrate the chain reaction and achieved the atomic reactor. Carothers recognized the desirability of producing polymers from monomers and succeeded in producing useful polymers such as polyamides (nylon). These recent innovators can be regarded as achieving new innovations in new fields.

Creation and development of the Internet and the associated World Wide Web are treated here for convenience as a single iconic invention since both involve essentially an all-computer communication technology that has achieved enormous growth. Alternatively, each could be found to incorporate several distinct inventions that might individually be regarded as iconic, but at this time in history, the use of these inventions lies in computer communication technology. Accordingly, all are placed under the banner of electronic computer communication technology.

(4) Some comments about iconic inventions:

It is difficult, if not impossible, to generalize simply about the nature or character of an iconic invention. Iconic inventions come in many different forms and circumstances. Most iconic inventions, before becoming so, undergo development and are subject to improvements. Some iconic inventions can be simple in conception but complex as achieved. Determining when an iconic invention becomes part of a larger entity can be difficult to specify. Sometimes a long time is needed before an original embodiment of an invention becomes developed into a better or practical species and then perhaps becomes iconic. New and different inventors, and advances in technology, may be required for such a development and occurrence. It is not common for many different embodiments of relatively complex inventions with multiple components to be manufactured and such inventions are not commonly achieved by individual inventors. Rarely does a single inventor of a pioneering invention become a significant industrialist.

Bibliography

Janet Abbate. *Inventing the Internet*. Cambridge, MA: MIT Press, 1999.

Roger Adams. *A Biography, in High Polymers: A Series of Monographs on the Chemistry, Physics and Technology of High Polymeric Substances*. Vol. 1. Collected Papers of W. H. Carothers on High Polymeric Substances. New York: Interscience Publishers, Inc., XVIII, 1940.

Hugh G. J. Aitken. *The Continuous Wave: Technology and American Radio, 1900–1932*. Princeton, NJ: Princeton University Press, 1985.

Arthur Allen. *Vaccine: The Controversial Story of Medicine's Greatest Lifesaver*. W. W. Norton, 2007.

Richard T. Ammon. *The Rolls Royce of Reception: Super Heterodynes 1918–1930*.

Nicolas Appert Biography. In *Asimov's Biographical Encyclopedia of Science and Technology*. 2nd ed., p. 359. New York: Doubleday and Company, Inc., 1982.

E. H. Armstrong. A Method of Reducing Disturbances in Radio Signaling by a System of Frequency Modulation. *Proceedings of the IRE* 24:689–740, 1936.

Neil Baldwin. *Edison: Inventing the Century*. Chicago: University of Chicago Press, 2001.

John Bankston. *Enrico Fermi and the Nuclear Reactor*. Mitchell Lane Publishers, 2003.

Paul Baran, Sharla P. Boehm, and J. W. Smith. On Distributed Communications series: Volumes I–XI. RAND Corporation Research Documents, 1964. http://www.rand.org/about/history/baran.list.html.

Barsanti e Matteucci Web site. http://www.barsantiematteucci.it/inglese/motori_eng.htm.

Basic Antiques Web site. http://www.basic-antiques.com/prints-samuel-morse.htm.

Ellis Beers and Soule. *Wheeler & Wilson's Manufacturing Company Bridgeport, Conn*. Philadelphia: Worley and Bracher, 1867.

James W. Behrens and Allen D. Carlson. *50 Years with Nuclear Fission*. American Nuclear Society, 1989.

Mary Bellis. The History of the Telephone—Antonio Meucci. http://inventors.about.com/library/inventors/bl_Antonio-Meucci.htm.

Tim Berners-Lee. *The Original Design and Ultimate Destiny of the World Wide Web*. New York: Harper Collins Publisher, 1999.

Biography of Philo T. Farnsworth. University of Utah Marriott Library Special Collections. http://db3-sql.staff.library.utah.edu/lucene/Manuscripts/null/Ms0648.xml/Bioghist

Biography of Walter H. Brattain. Nobel Lectures, Physics, 1942–1962. Amsterdam, the Netherlands: Elsevier Publishing Company, 1964.

David Boothroyd. *Forgotten Hero, The Man Who Invented the Two-Stroke Engine*. http://the-vu.com./forgotten_hero.htm, 2001.

Ruth Brandon. *A Capitalist Romance: Singer and the Sewing Machine*. Philadelphia: J. B. Lippincott, 1977.

Anthony Burton. *Richard Trevithick: Giant of Steam*. London: Aurum Press, 2000.

Andrew J. Butrica, ed. *Beyond the Ionosphere: Fifty Years of Satellite Communication*. NASA SP 2417. 1997.

Andrew Carnegie. *James Watt*. Doubleday Page & Co., 1905; reprinted from the 1913 edition by University Press of the Pacific, 2001.

Herbert N. Casson. *Cyrus Hall McCormick: His Life and Work*. Chicago: A. C. McClurg & Co., 1909.

Basilio Catania. Basilio Catania's Work on Antonio Meucci. http://www.esanet.it/chez_basilio/meucci.htm.

Paul E. Ceruzzi. *A History of Modern Computing*. MIT Press, 1998.

Margaret Cheney. *Tesla: Man out of Time*. New York: Barnes & Noble, 1981.

Lewis Coe. *The Telegraph: A History of Morse's Invention and Its Predecessors in the United States*. Jefferson, NC: McFarland, 1993.

Editor of the *Commercial Motor. Compression Ignition Engines for Road Vehicles*. 2nd ed. London: Temple Press.

Margaret Connor. *Hans von Ohain: Elegance in Flight*. American Institute of Aeronautics, 2002.

Grace R. Cooper. *The Invention of the Sewing Machine*. Washington, DC: Smithsonian Institution, 1968.

Frank Crane. *George Westinghouse, His Life and Achievements*. Kessinger Publishing, 2003.

Tom D. Crouch. *The Bishop's Boys. A Life of Wilbur and Orville Wright*. New York: W. W. Norton, 1989.

C. Lyle Cummins, Jr. *Internal Fire*. Lake Oswego, OR: Carnot Press, 1976.

De Forest Sends Out the Opera from the Metropolitan. *New York Times*, January 1910.

P. Debré and E. Forster. *Louis Pasteur*. Baltimore, MD: Johns Hopkins University Press, 1998.

Degna Degna. *My Father, Marconi*. James Lorimer & Co., 1982.

H. W. Dickenson. *James Watt: Craftsman and Engineer*. Cambridge: Cambridge University Press, 1935.

H. W. Dickinson and Hugh Pembroke Vowles. *James Watt and the Industrial Revolution*. 1943. (New edition published in 1948 and reprinted in 1949; also published in Spanish and Portuguese in 1944 by the British Council.)

Hans Peter Diedrich. *German Jet Aircraft 1939–1945*. Atgien, PA: P. A. Schiffer Publishing, 1998.

E. Diesel, G. Goldbeck, and F. Schilberger. *From Engines to Autos: Five Pioneers to Engine Development and their Contributions to the Automotive Industry.* Chicago: Henry Regnery Co., 1960.

E. Diesel, G. Goldbeck, and F. Schilberger. *Nickolaus August Otto.* Chicago: Henry Regnery Co., 1960.

Eugen Diesel, Gustav Goldbeck, and Friedrich Schildberger. *From Engines to Autos.* Chicago: Henry Regnery Company, 1960.

Andrew Dunn. *Alexander Graham Bell.* Pioneers of Science series. East Sussex, United Kingdom: Wayland Ltd., 1990.

Election Returns Flashed by Radio to 7,000 Amateurs. *Electrical Experimenter,* January 1917, p. 650.

Doris Faber. *Enrico Fermi, Atomic Pioneer.* Prentice Hall, 1966.

Michael Fairman. *Bill Gates, The New Revolutionary Creator.* Oxford, OH: Miami University. http://www.users.muohio.edu/shermalw/honors_2001_fall/honors_papers_2000/fairman_gates.

Elma Gardner Farnsworth. *Distant Vision: Romance and Discovery on an Invisible Frontier.* Salt Lake City, UT: Pemberley Kent Publishers, 1989.

Russell Farnsworth. *Philo T. Farnsworth: The Life of Television's Forgotten Inventor.* Hockessin, DE: Mitchell Lane Publishers, 2002.

Laura Fermi. *Atoms in the Family: My Life with Enrico Fermi.* Chicago: University of Chicago Press, 1954.

Helen M. Fessenden. *Fessenden: Builder of Tomorrows.* New York: Coward McCann, Inc., 1940.

Reginald A. Fessenden. The Inventions of Reginald A. Fessenden. *Radio News,* 11-part series beginning with the January 1925 issue.

William Fewers and H. W. Baylor. *Sincere's History of the Sewing Machine.* Sincere Press, 1970.

Henry Ford. *My Life and Work.* NuVision Publications.

Lee de Forest, 87, Radio Pioneer, Dies; Lee De Forest, Inventor, Is Dead at 87. *New York Times,* July 2, 1961.

Lee de Forest. *Father of Radio.* Chicago: Wilcox and Follett, 1950.

Yasu Furukawa. *Inventing Polymer Science: Staudinger, Carothers, and the Emergence of Macromolecular Chemistry.* University of Pennsylvania Press, 1998.

Ann Gaines. *Wallace Carothers and the Story of du Pont Nylon (Unlocking the Secrets of Science).* Hockessin, DE: Mitchell Lane Publishers, 2001.

William Gates. Business @ the Speed of Thought. *Business Plus* (Magazine), 1999.

William Gates, Nathan Myhrvold, and Peter Rinearson. *The Road Ahead.* New York: Penguin Books, 1996.

Gerald L. Geison. *The Private Science of Louis Pasteur.* Princeton, NJ: Princeton University Press, 1995.

John Golley. *Genesis of the Jet: Frank Whittle and the Invention of the Jet Engine.* Airlife, 1996.

Michael E. Gorman. *Bell's Path to the Invention of the Telephone.* National Science Foundation. http://www3.virginia.edu/albell/homepage.html, 1994.

John Griffiths. "William Murdock." In *Oxford Dictionary of National Biography.* Oxford: Oxford University Press, 2004.

Edwin S. Grovenor. Comments. http://www.alecbell.org/mueccimemo.html.

O. Hahn and F. Strassmann. Über den Nachweis und das Verhalten der bei der Bestrahlung des Urans mittels Neutronen entstehenden Erdalkalimetalle. (On the Detection and Characteristics of the Alkaline Earth Metals Formed by Irradiation of Uranium with Neutrons.) *Naturwissenschaften* 27:11–15, 1939.

Eva-Maria Hanebutt-Benz (Director, Gutenberg Museum). Gutenberg und Seine Zeit in Daten (Gutenberg and His Times: Timeline). http://www.mainz.de/gutenberg/zeitgutb.htm.

Jacqueline Harris. *Henry Ford (An Impact Biography)*. Franklin Watts, 1984.

Matthew Hermes. *Enough for One Lifetime, Wallace Carothers the Inventor of Nylon*. Chemical Heritage Foundation, 1996.

Lillian Hoddeson and Vicki Daitch. *True Genius: the Life and Science of John Bardeen*. National Academy Press, 2002.

James Hodge. *Richard Trevithick*. Lifelines 6. Risborough, United Kingdom: Shire Publications, 2003.

D. A. Hodges, H. G. Jackson, and R. Saleh. *Analysis and Design of Digital Integrated Circuits*. New York: McGraw-Hill, 2003.

Sungook Hong. *A History of the Regeneration Circuit: From Invention to Patent Litigation*. University, Seoul, Korea (pdf).

Paul Horowitz and Winfield Hill. *The Art of Electronics*. Cambridge: Cambridge University Press, 1989.

Fred Howard. *Wilbur and Orville: A Biography of the Wright Brothers*. New York: Alfred A. Knopf, 1987.

William T. Hutchinson. *Cyrus Hall McCormick Harvest, 1856–1884*. New York: D. Appleton-Century Company, 1935.

William T. Hutchinson. *Cyrus Hall McCormick Seed Time, 1809–1856*. New York: The Century Co., 1930.

Daniel Ichbiah and Susan Knepper. *The Making of Microsoft*. Rocklin, CA: Prima, 1991.

Peter L. Jakab. *Visions of a Flying Machine: The Wright Brothers and the Process of Invention*. Washington, DC: Smithsonian Institution Press, 1990.

Steve Jobs and J. S. Young. *The Journey Is the Reward*. Scott Foresman & Co., 1988.

John Stevens Collection. 1808–1881 Archives Center. Washington, DC: National Museum of American History, Smithsonian Institution.

Glynn Jones. *The Jet Pioneers*. London: Methuen, 1989.

Jill Jonnes. *Empires of Light: Edison, Tesla, Westinghouse, and the Race to Electrify the World*. New York: Random House, 2003.

Matthew Josephson. *Edison: A Biography*. New York: McGraw-Hill Book Company, Inc., 1959.

Joseph N. Kane. *Necessity's Child: The Story of Walter Hunt, America's Forgotten Inventor*. Jefferson, NC: McFarland & Co., 1997.

Albert Kapr. *Johannes Gutenberg: The Man and His Invention*. Aldershot, United Kingdom: Scolar Press, 1996.

Jack Kilby. IEEE Virtual Museum. 2008. http://www.ieeeghn.org/

Leonard Kleinrock. *Information Flow in Large Communication Nets*. Cambridge, MA: MIT, 1961.

Robert Lacey. *Ford: The Men and the Machine*. Boston: Little, Brown and Company, 1986.

Lawrence Lessing. *Man of High Fidelity: Edwin Howard Armstrong*. Philadelphia: J. B. Lippincott Company, 1956.

Francis E. Leupp. *George Westinghouse: His Life and Achievements*. Boston: Little, Brown and Company, 1918.

Tom Lewis. *Empire of the Air: The Men Who Made Radio*. New York: E. Burlingame Books, 1991.

Ernest Bainbridge Lipscomb III. *Lipscomb's Walker on Patents*. 3rd ed. Vol. 1. Rochester, NY: Lawyers Co-Operative Publishing Co., 1984. (See also similar text in *Walker on Patents*, Deller's Edition. Vol. 1, 1937.)

J. W. Lowe. *British Steam Locomotive Builders*. Guild Publishing, 1989.

C. Lyle, Jr. *History of the Automotive Internal Combustion Engine*. Cummins Society of Automotive Engineers, 1976.

Carleton Mabee. *The American Leonardo: A Life of Samuel F. B. Morse*. New York: Knopf, 1944.

Catherine Mackenzie. *Alexander Graham Bell, The Man Who Contracted Space*. New York: Grosset & Dunlop, 1928.

John Man. *Gutenberg: How One Man Remade the World with Word*. Wiley, 2002.

Stephen Manes. *Gates: How Microsoft's Mogul Reinvented an Industry and Made Himself the Richest Man in America*. Touchstone, 1993.

Ruth Maran. *Internet and World Wide Web Simplified*. 3rd ed. John Wiley & Sons.

Guglielmo Marconi. *Wireless Telegraphic Communication: Nobel Lecture, December 11, 1909*. Nobel Lectures. Physics 1901–1921. Amsterdam, the Netherlands: Elsevier Publishing Company, 1967, pp. 196–222.

Thomas H. Marshall. *James Watt*. Edinburgh, United Kingdom: Leonard Parsons Ltd., Morrison C. Gibb Ltd., 1925.

Cyrus McCormick. *The Century of the Reaper*. Boston: Houghton Mifflin Company, 1931.

M. W. McFarland, ed. *The Papers of Wilbur and Orville Wright, including the Chanute-Wright Letters and other papers of Octave Chanute*. New York: McGraw-Hill, 1953.

Marshall McLuhan. *The Gutenberg Galaxy: The Making of Typographic Man*. University of Toronto Press; reissued by Routledge & Kegan, 1962.

Lise Meitner and O. R. Frisch. Disintegration of Uranium by Neutrons: A New Type of Nuclear Reaction. *Nature* 143:239–240, 1939. (February 11, 1939, paper dated January 16, 1939.)

John Micklos, Jr. *Alexander Graham Bell: Inventor of the Telephone*. New York: Harper Collins Publishers Ltd., 2006.

Edward Lind Morse, ed. *Samuel Finley Breese Morse, his Letters and Journals*, Vols. 1 and 2. Boston: Houghton Mifflin Company, 1914.

Samuel F. B. Morse. His Letters and Journals. http://www.gutenberg.org/test/11017.

Jacob Neufeld, George M. Watson, Jr., and David Chenoweth. *Technology and the Air Force: A Retrospective Assessment*. Diane Publishing, 1997.

Allan Nevins and Frank E. Hill. *Ford: Decline and Rebirth, 1933–1962*. New York: Charles Scribner's Sons, 1962.

Allan Nevins and Frank E. Hill. *Ford: Expansion and Challenge, 1915–1933*. New York: Charles Scribner's Sons, 1957.

Allan Nevins and Frank E. Hill. *Ford: The Times, the Man, the Company*. New York: Charles Scribner's Sons, 1954.

News of the Week. *Chemical and Engineering News*, September 29, 2008, p. 7.

November 17–December 23, 1947: Invention of the First Transistor. *American Physical Society* 9, 2000.

Edward F. Obert. *Internal Combustion Engines: Analysis and Practice*. 2nd ed. Scranton, PA: International Textbook Company, 1964.

James E. O'Neal. Fessenden: World's First Broadcaster?—A Radio History Buff Finds That Evidence for the Famous Brant Rock Broadcast Is Lacking. *Radio World Online*, October 25, 2006.

G. D. Padfield and B. Lawrence. The Birth of Flight Control: An Engineering Analysis of the Wright Brothers' 1902 Glider. *The Aeronautical Journal*, 2003: 697–718.

James Parton. *History of the Sewing Machine*. Lockport, NY: The Howe Machine Company, 1872.

James Parton. *History of the Sewing Machine*. Michigan Historical Reprint Series. Booksurge Publishing.

Sterling Michael Pavelec. *The Jet Race and the Second World War*. Westport, CT: Praeger Security International, 2007.

John R. Pierce and Ed C. Posner. *Introduction to Communications Science and Systems*. Springer, 1980.

Michael Pollard. *Henry Ford (Giants of American Industry)*. Blackbirch Press, 2003.

Frank Prager, ed. *The Autobiography of John Fitch*. Philadelphia: American Philosophical Society, 1976.

William S. Pretzer, ed. *Working at Inventing: Thomas A. Edison and the Menlo Park Experience*. Dearborn, MI: Henry Ford Museum & Greenfield Village, 1989.

Radio Set-up Eliminates All Noise. *Ogden Standard Examiner* (United Press), June 18, 1936, p. 1.

John B. Rae. *The American Automobile: A Brief History*. Chicago: University of Chicago Press, 1965.

T. H. Reid. *The Chip*. New York: Random House, 2001.

Richard Rhodes. *The Making of the Atomic Bomb*. New York: Simon and Schuster, 1986.

M. Richaria. *Satellite Communications Systems: Design Principles*. 2nd ed. New York: McGraw-Hill.

Michael Riordan and Lillian Hoddeson. *Crystal Fire: The Invention of the Transistor and the Birth of the Information Age*. New York: W. W. Norton & Company Ltd., 1998.

Robert Fulton Biography. In *Asimov's Biographical Encyclopedia of Science and Technology*, 2nd ed., pp. 385–386. New York: Doubleday and Company, Inc., 1982.

Robinson and Musson. *James Watt and the Steam Revolution*. 1969.

Howard B. Rockman. Alexander Graham Bell. In *Intellectual Property Law for Engineers and Scientists*, pp. 103–111. IEEE Press and John Wiley & Sons, Inc., 2004.

L. T. C. Rolt. *George and Robert Stephenson: The Railway Revolution*. Penguin, 1960.

Royal Air Force History. Frank Whittle and the Jet Age. http:www.raf.mod.uk/history_old/e281.html.

Paul Schatzkin. *The Boy Who Invented Television*. Silver Spring, MD: Teamcom Books, 2002.

Evan I. Schwartz. *The Last Lone Inventor: A Tale of Genius, Deceit & the Birth of Television*. New York: HarperCollins, 2002.

Victor Scholderer. *The Invention of Printing*. London: British Museum, 1940.

Charles Scott, ed. *George Westinghouse Commemoration: A Forum Presenting the Career and Achievements of George Westinghouse on the 90th Anniversary of his Birth*, conducted by the American Society of Mechanical Engineers, December 1, 1936.

Emilio Segre. *Enrico Fermi, Physicist*. Chicago: University of Chicago Press, 1970.

Shenandoah Valley Agricultural Research & Extension Center. Biography of Cyrus Hall McCormick. http://www.vaes.vt.edu/steeles/mccormick/bio.html.

Joel N. Shurkin. *Broken Genius: The Rise and Fall of William Shockley, Creator of the Electronic Age*. New York: Palgrave Macmillan, 2006.

Jack Simmons and Gordon Biddle. *The Oxford Companion to British Railway History*. Oxford: Oxford University Press, 1997.

Samuel Smiles. *The Life of George Stephenson*. 1857.

O. Sneedon. *Introduction to Internal Combustion Engineering*. London: Longmans, Green & Co., 1933.

L. Sprague de Camp. Bell and the Telephone. In *The Heroic Age of American Invention*, pp. 156–167. Garden City, NY: Doubleday & Company, Inc., 1961.

L. Sprague de Camp. *Henry, Morse, and the Telegraph*. Garden City, NY: Doubleday & Company, Inc., 1961, pp. 59–75.

L. Sprague de Camp. *The Heroic Age of American Invention*. Garden City, NY: Doubleday & Company, Inc., 1961.

Paul J. Staiti. *Samuel F. B. Morse*. Cambridge, 1989.

Melissa Stewart. *Tim Berners Lee: Inventor of the World Wide Web*. Ferguson Publishing Company, 2001.

Alfred T. Story. *A Story of Wireless Telegraphy*. New York: D. Appleton Company, 1904.

Randall E. Stross. *The Microsoft Way*. Perseus Books Group, 1997.

Erica Stux. *Enrico Fermi: Trailblazer in Nuclear Physics*. Nobel Prize Winning Scientists. Enslow Publishers, 2004.

John Hudson Tiner. *Louis Pasteur: Founder of Modern Medicine*. Fenton, MI: Mott Media, 1990.

John Hudson Tiner. *Samuel F.B. Morse: Artist with a Message*. Milford, MI: Mott Media, 1985.

U.S. Supreme Court. *Marconi Wireless Telegraph Co. of America v. United States*. 320 U.S. 1. Nos. 369, 373. Argued April 9–12, 1943. Decided June 21, 1943.

James Wallace. *Hard Drive: Bill Gates and the Making of the Microsoft Empire*. New York: John Wiley & Sons, 1992.

Steven Watts. *The People's Tycoon: Henry Ford and the American Century.* New York: Vintage Books, 2005.

Thompson Wescott. *The Life of John Fitch.* Philadelphia: J. B. Lippincott, 1857.

Westinghouse Electric Co. *George Westinghouse 1846–1914.* Westinghouse Electric Co., 1946.

David J. Whalen. *The Origins of Satellite Communications, 1945–1965.* Washington, DC: Smithsonian Institution Press, 2002.

Jim Whiting. *John R. Pierce: Pioneer in Satellite Communication.* Mitchell Lane Publishers, 2003.

Sir Frank Whittle. *Jet: The Story of a Pioneer.* Frederick Muller Ltd., 1953.

Wisconsin Historical Society Library Archives. Cyrus McCormick IHC collection. http://www.wisconsinhistory.org/libraryarchives/ihc/cyrus.asp.

Steve Wozniak and G. Smith. *iWoz: From Computer Geek to Cult Icon: How I Invented the Personal Computer, Co-Founded Apple, and Had Fun Doing It.* New York: W. W. Norton & Company, 2006.

The Wright Brothers & The Invention of the Aerial Age. Washington, DC: Smithsonian Institution, 1908.

Tsividis Yannis. *Edwin Armstrong, Pioneer of the Airwaves.* New York: Columbia University Web site, 2002.

Jeffrey Zygmont. *Microchip: An Idea, Its Genesis, and the Revolution It Created.* Scranton, PA: Perseus Books Group, 2003.

Index

Abbott, Charles Greely, 388
A. B. Dick Company, 257
Abelson, Phillip, 507
Ackroyd-Stuart, Herbert, 240–41; hot bulb engines, 240
Adair, Clement, 374
Adams, Milton, 253
Advanced Research Projects Agency (ARPA), 587–88
Aerial Experiment Association, 386
Aerocar Company, 353, 354
Aiken, Howard, 546
ailerons, 377
aircraft/airplane, 364–65, 387
Alexanderson, Ernst F. W., 399
Allen, Horatio, 94
Allen, Paul, 572, 575; and Bill Gates, 572–74; Lakeside Programmers Group (LPG), 572; resignation from Microsoft, 579–80; Traf-O-Data, 573
"all-glass globe", 308
Allston, Washington, 149
alpha rays, 487
Altair models, 549, 574, 575–76
alternating current (AC) electrical system, 295, 303–14; commercialization and war of currents, 307–11; Edison's position, 266
Ambroise, Pare, 326
Amelio, Gil, 553
American Association of Licensed Automobile Manufacturers (ALAM), 356–57
American automobile industry, development of, 333–34
American Bell Telephone Company, 26
American de Forest Wireless Telegraph Company, 403
American Institute of Electrical Engineers, 303, 313
American Marconi Company, 165, 404
American railroads, 92–97
American Speaking Telephone Company, 205
American Standard Code for Information Interchange (ASCII code), 603
American steam engine, and railroad development: disappointed early American inventors, 88; English influences, 88
American Telephone and Telegraph Company (AT&T), 517
American type of locomotive, 95
American Viscose Company, 465
Anderson, Herbert, 496
Anderson, John W., 32, 350, 351
ANIK, 567
anthrax, 324, 325
antibiotics, 430, 431
Apollo, 566
Apperson, Edgar, 333
Appert, Nicolas François, 100; death, 104; early years, 101; family, 101; food preservation techniques, 101;

Appert, Nicolas François (*continued*) House of Appert, 102, 103; *L'art de Conserver, Pendant Plusiers Annés, Toutes les Substances Animales et Végetales,* 102–3; Napoleon's prize to, 102
Apple Computer Inc., 552, 579; Jobs, influence of, 552; Jobs after, 1980, 553
Apple I, 544, 551
Apple II, 544, 552, 554–56
Appleton, Edward V., 166
Arbor, Ann, 511
Archimedean screw, 57
Arkwright, Richard, 50–51
Armat, Thomas, 286
Armstrong, Edwin Howard, 405, 406, 516; academic education, 407; birth, 406; childhood, 407; commercialization of radio after World War I, 411; and David Sarnoff, 412; deal with RCA, 411–12, 415, 416, 417; early life, 406–7; family, 406–7; FCC license, 415–16; fight with de Forest over regeneration, 412–14; fight with Levy over superheterodyne, 414; frequency modulation invention, 414–18; and H. J. Round, 409; investigation of audion tube, 407–8; marriage with Marion MacInnis, 412; patents, 408, 410; as Pupin's assistant, 407, 408; regenerative circuit invention, 407–8; suicide, 417; super-regenerative circuit invention, 411–12; superheterodyne circuit invention, 409–11; in World War I, 409–11
ARPANET, 589, 591–92, 599, 602
arsphenamine, 424
artificial silk, 464
Ashley, J. L., 255
Aston, Francis, 504
Astrotone, 472
AT&T, 216, 414, 562, 567; petition to Federal Communications Commission, 562–63
Atanasoff, J. V., 546
Atlantic and Pacific Telegraph Company, 258
atom, 485
atom bomb: building of, 503–10; centrifuging, 506; delivery, 509; developments at Los Alamos, 507–9; electromagnetic separation, 504–5; fissionable isotope separation, 504; gaseous diffusion, 505; history of, 489–90; liquid thermal diffusion, 507; perspective, 484–85; Pu-239, 506–7; terminology, 485–89
Atomic Energy Act of, 1954, 512
atomic energy, history of, 490–92
Atomic Energy Commission (AEC), 512
atomic mass, 486, 487
atomic number, 486–87
atomic weight, 486
audacious press campaign, 268
Augsburg Machine Works, 234
Aunt Sally, 57
"automatic brake", 299
Automatic Telegraph Company, 257, 258
automatic vote recorder, 254
automobile, 331; development of, 333–34; history of, 332–36
Aylsworth, J. W., 289

Babbage, Charles, 360, 545–46
bacterial antibiotic resistance, 430
bacteriophages, 422
Baekeland, Leo, 465–66
Baird, John Logie, 434
Bakelite, 465, 466
Balard, Antoine, 316
Baldwin, Frederick W., 386
Ballmer, Steve, 578, 582
balloons, 372–73
Baltimore and Ohio railroad, 94
Bangemann, M., 610
Baran, Paul, 588, 590
Bardeen, John , 515, 516, 517–18, 519, 524–25; birth, 524; death, 525; education, 524; family, 524; with Gulf Research Laboratories, 524;

marriage with Jane Maxwell, 524; with Naval Ordnance Laboratory, 525; Nobel Prize, 522; at University of Illinois, 525; withdrawal from Shockley, 522
Barker, George F., 264, 265
Barnett, William, 221, 222
Barrow, Bonnie, 366
Barrow, Clyde, 366
Barsanti, Eugenio, 221
Barthel, Oliver, 341, 342, 345
BASIC, 574–75
Batchelor, Charles, 257, 260, 262, 273, 275, 277, 284, 303
Battle of Britain, 427
BCS theory, 525
Bechamp, Antoine, 328
Becquerel, Antoine H., 490
Beighton, Henry, 28
Beijerinck, Martinus, 325
Bel, Joseph Le, 318
Belfield, Reginald, 305
Bell, Alexander Graham, 188–92, 254, 290, 261, 385; Bell Associates, 197; birth, 189; in Boston, 190, 194; British patents, 198; childhood, 189; commercialization, 196–200; Corporation, 197–98; death, 201; demonstration, 196, 197; ear phonoautograph, 191; in Edinburgh, 189; education, 189; and Elisha Gray, 202; experiments, 190, 194; federal courts, 209; and Gardiner Hubbard, 191–92, 197; and George Brown, 195; and George Sanders, 190; and Gray's caveat, 202–4; and Greene Hubbard, 190, 195, 196; harmonic telegraph, 192, 194; and Hermann von Helmholtz, 190; honeymoon, 197–98; honors, 201; and Joseph Henry, 194; in London, 189; and Mabel Hubbard, 190, 197–98; and Meucci's work, 188–89; musical telegraph, 192; observations, 191; other research and achievements, 200–201; patent application, 195–96; as professor, 190; *Science*, 201; speaking telegraph, 193; students, 190; as teacher, 189–90; telephone invention, 193–95; telephone inventions, challenges to, 201–2; and Thomas A. Watson, 194–95; and Thomas Sanders, 191–92, 197, 204; U.S. patent, 200; Western Electric, 197, 204–6
Bell, Alexander Melville, 189, 190
Bell Associates, 197
Bell Company, 204
Bell Laboratories, 416, 494, 516, 517, 518, 519, 520, 521, 522, 523, 525; patent application, 518; Project Echo, 561–62
Bell, David M., 343
Benger, Ernest, 471
Benz, Karl, 223, 230, 332
Benz & Cie, 230
Berchet, Gerard J., 474; 6–6 polyamide, 474
Bergmann, Sigmund, 257, 273, 277
Berliner, 292
Bernard, Claude, 322
Berners-Lee, Tim, 533, 604–5; after WWW, 610–11; birth, 604; at CERN, 605–6; education, 604–5; family, 604; HTML, 606; HTTP, 606; uniform resource locator, 606; Unitarian Universalist church, 611; World Wide Web Consortium (W3C), 611; WWW, origin of, 605–7
Bernoulli, Daniel, 54
Berry, Clifford, 546
Bertolino, Angelo, 212
beryllium, 494
Berzelius, Jons, 320
Best Friend, 94
beta rays, 487
Bethe, Hans, 508
Bicycle, 335
Bigot, M., 319
Binney, Charles Richard, 240
Biograph Company, 287

Biot, Jean Baptiste, 317, 318, 373; Rumford Medal, 318
bipolar dynamo, 271
bipolar junction transistor (BJT), 519, 521
Birdseye, Clarence, 106–7
Bishop, James W., 342
Bishop, Jim, 343
Black, Joseph, 31–33
Blackett, Christopher, 79
Blucher, 85
Bohr, Niels, 493, 494
Bolt, Beranek and Newman engineering firm (BBN), 590, 593
Bolton, Elmer K., 469, 471, 474
Boone, Gary, 543
Booth, Eugene, 505, 506
Borden, Gail, 108–9
Borland, 580
Bose, Sir Jagadischandra, 159
Boston Western Union, 254
Botanic Research Company, 294
Botox injection, 106
botulism, 105–6
Boulton, Matthew, 35, 36, 44–45, 220; sun and planet gearing, 39
Bourseul, Charles, 210
Branca, Giovanni, 27
Brandenberger, Jacques, 464–65
Brattain, Walter, 515, 516, 517–18, 519, 525–26; with Bell Labs, 525, 526; birth, 525; death, 526; education, 525; family, 525; marriage with Emma Jane Kirsch Miller, 526; marriage with Karen Gilmore, 526; Nobel Prize, 522; withdrawal from Shockley, 522
Braun, Karl Ferdinand, 167, 434, 516
Brayton, George, 222; gas-fueled engine, 333; petroleum-fueled engine, 333
Brayton engine, 358
breast wheel, 26
breeder reactor, 488–89, 506
Bresson, Jacob, 27
Briggs, Lyman, 495, 498, 499
British Bell Company, 261
British Edison Electric Lighting Company, Ltd., 275
British Swan United Electric Light Company, 275
Brock, Rod, 578; awards and recognitions, 611
Brooklyn, 310
Brown, George, 195, 367
Brown, Samuel, 221
Brunel, Isambard Kingdom, 62
Brunel, Marc Isambard, 62
Brush, Charles F., 265
Buffalo Gasoline Motor Company, 356, 357
Buffet, Warren, 584
Bush, Vannevar, 498, 499, 500, 501
Bushnell, David, 67
Buz, Carl, 234, 235
Byron, Ada, 545

Cailliau, Robert, 605
Cadillac Company, 346, 348
Calico Printers Association, 479
Calley, John, 28
Camden and Amboy railroad, 94
Campbell, Henry, 95
Campbell, Marion, 30
Campbell-Swinton, Alan Archibald, 435
canard, 378, 386
canning, 100; container technology, 107–8; development of, 104–5; history, 100–101; improved container fabrication, 109–11; microorganisms control, 105–7
capture cross-section, 487
carbon "burners", 268, 269
carbon filament suit, 280
carding machine, 50
Carhard, J. M., steam carriage, 333
Carnegie Institution, 501
Carnot, Nicolas Leonard Sadi, 221, 234
Carothers, Wallace H., 462; affair with Sylvia Moore, 472; Astrotone, 472; birth, 468; as chemistry tutor, 468; depression, 473–74; deterioration of, 474–75; and Donald Coffman, 473; early years, 468; education, 468;

elected to the National Academy of Sciences, 474; and Ernest Benger, 471; family, 468; and Julian Hill, 470–71; marriage with Helen Sweetman, 474; mental health, 238; neoprene, 469–70; nylon, 472–74; personal life problems, 472; polyesters, 470–71; psychiatric intervention, 473, 474–75; research (1931–1934), 471–72; suicide, 475
Carrell, Alexis, 387
Carroll, Bill, 212
Cartwright, Edmund, 51
Castaldi, Pamfilo, 18–19, 20
Catch-Me-Who-Can, 80
Caterpillar Company, 142
Cato, George, 342
Cavendish, Henry, 373
Caxton, William, 13; Black Letter type, 13, 21
Cayley, George, 373
Celanese Company, 464
cellophane, 464
cellular telephone, 416–17
Celluloid, 463
cellulose, 463–64
Central Air Data Computer (CADC), 541
centrifuging, 506
Cerf, Vinton, 592–93, 594–95; Internet Society (ISOC), 600
CERN, 605–6
Chadwick, James, 491
Chafe, Chris, 568
Chain, Ernst, 422, 426, 427; and Howard Florey, 426; Nobel Prize, in medicine, 429
chain reaction, 487–88
chain stitch, 173
chalk relay, 257
Chamberlain, Owen, 511
Chanute, Octave, 376, 378, 379
Chapin, Roy D., 335
Chardonnet, Comte de Louis Marie Hilaire Bernigaud, 464
Charles Brown and Company, 70
Charleston & Hamburg Railroad, 94
chemical reaction, 485
Chesnay, Cummings C., 313
Chicago Pile Number, 1 (CP-1), 501–3. *See also* nuclear reactor
chicken cholera, 325
China, 7
chiral molecules, 317
Christy, George H., 309
Cingular Wireless, 554
cipro, 431
circuit switching, 588, 589
CK703 transistor, 519
CK722 transistor, 520–21
Clark, Candi, 554
Clark, Sir Dougald, 222
Clark, Edward, 180
Clark, Rocky, 554
Clermont, development of, 68–69, 70
Clostridium botulinum, 105
Cochran, Michael J., 543
Coffin, Charles A., 281
Coffman, Donald, 473
Colossus computer, 546
Columbia and Electric Vehicle Company. *See* Electric Vehicle Company
Columbian high-pressure steam engine, 89
Commodore International Commodore PET, 552
Communications Satellite Corporation (COMSTAT), 558, 565, 567; Early Bird, 566
Compaq, 579
compass, 3–4
Compton, Arthur H., 498, 499, 500, 501
Computer Center Corporation (C-cubed), 572–73
Computer Terminals Corporation (CTC), 542
computers, 527
Conant, James Bryant, 468, 498, 499, 500, 501
concentrated liquids, 108–9
constant-voltage dynamo, 270, 271
Constitution. See *Oliver Evans*

container technology, 107–8; improved fabrication, 109–11
Cooper, Martin, 416
Cooper, Peter, 94; Baltimore and Ohio railroad, 94; *Tom Thumb,* 94
Cooper, Tom, 346
Cornell, Ezra, 156
Corporation for National Research Initiatives, 600
Coster, Laurens Janszoon, 19–20, 21
cottage industry, 49
Count Zeppelin, 372
Cour, Poul la, 210
Courier satellite project, 562
Couzens, James, 347, 348, 349, 350, 351, 353, 354, 357, 358, 361
Cowles, E. P., steam carriage, 333
Cox, Rachel, 62
CP/M program, 577, 579
Cramer Electronics, 551
Crawshay, Richard, 78
Crichton, Michael, 510
Cripps, Sir Stafford, 455
critical mass, 488
Crocker, S., 603
Crompton, Samuel, 50, 51
cross-section, 487
C. R. Wilson Carriage Company, 350
Cugnot, Nicolas-Joseph, 78, 220
curie, 487
Curie, Marie, 491
Curtiss, Glenn H., 386, 387
Curtiss-Wright company, 388
Curzon, Robert, 18
Cushing, Jack, 286

d'Abbans, Claude Jouffroy, 54
Daguerre, 155
Daimler, Gottlieb, 226, 230, 332
Dallery, Thomas-Charles-Auguste, 65
Dally, Clarence, 287
DARPA, 591, 592, 601
datagrams, 594
Datapoint. *See* Computer Terminals Corporation
d'Auxiron, Joseph, 54
Davidoff, Monte, 575
Davies, Donald, 589
Davis, Ari, 175–76
Davy, Sir Humphrey, 85–86
Day, Thomas, 36
de Caus, Salomon, 27
Deering, William, 141
Defense Special Weapons Agency, 512
De Forest, Lee, 293, 401–2, 408, 516; and Abraham White, 403; academic education, 402; American de Forest Wireless Telegraph Company, 403; and American Marconi Company, 404; and American Telephone & Telegraph Company (AT&T), 404, 405; audion, 403; audion development, 404–5; birth, 402; broadcasting efforts, 403–4; commercialization of radio broadcasting, 404; conflict with Reginald Fessenden, 398–99; contract with Bell Telephone Laboratory, 406; De Forest Radio Telephone Company, 403–4; death, 406; early years, 402; experiments involving diode, 403; with Federal Telegraph Company, 404; glow lamp, 406; J. Willard Gibbs, 403; job at Western Electric Company, 402–3; later life, 405–6; marriages, 406; military service, 402; motion picture, 405–6; parents, 402; patents, 403, 405, 406; triode tube invention, 403
De Forest Radio Telephone Company, 403–4
Degener, August, 348
Delamare-Debouteville, Edouard, 223
de Martinville, Edouard-Leon Scott, 263
de Rochas, Alphonse Beau, 222, 229
Detroit Auto Vehicle Company, 354
Detroit Automobile Company, 344–45
Detroit Dry Dock Company, 338
Detroit Free Library, 251
Detroit Free Press, 252
Detroit Young Men's Christian Association (YMCA), 339
D'Herelle, Felix H., 422
Dickinson, Robert, 80
Dickson, J. T., 479

Dickson, W. K. L., 285, 286
Diesel, Rudolf Christian Karl, 231; academic education, 232–33; accident due to cylinder explosion, 235; and Adolphus Busch, 236; as apprentice in refrigeration and steam engines, 233–34; association with Sulzer, Ltd., 243–44; in Berlin, 234; birth, 232; Carl Buz, 234, 235; commercialization of his engine, 236; compression-ignition engine, 235; cultural and political interests, 234; decline of, 237–38; diesel engine, 231, 234–37, 240–41; diesel engine, development, 241–46; early years, 232; as an engineer, 234; family, 232; influence of Carl von Linde, 233; influence of Carnot's theory, 234; leaving France, 232; and Martha Flasche, 234; in Munich, 233; nervous exhaustion, 237; patents, 235, 237; patent dispute with Herbert Ackroyd-Stuart, 237
diesel locomotives, 243–46
diesel-powered ships, 242
diffusion, 505; gaseous, 505; liquid thermal, 507
digital IC, 537
Digital Research, 577
Dillinger, John, 366
Dodge, Horace E., 349–50, 351, 361, 362
Dodge, John F., 349–50, 351, 361, 362
Doge, John, 173
Dolbear, Amos E., 206–7
Donkin, Bryan, 103
DOS, 578, 579
double-acting engine, 40–42
Dow, Alexander, 339, 344, 349
Dowd case, 204, 205
Drake, Alfred, 222
Drake, Frank, 510
Dreyfus, Camille and Henry, 464
du Pont, 462, 476, 477, 479; memorandum, 467–68
Duchesne Ernest, 422
Dudgeon, Richard, 221
Dudley-Williams, Rolf, 452, 453, 455
Dumas, Jean Baptiste, 316, 321
Dummer, Geoffrey W. A., 530
Duncan, John, 173
Dunning, John, 505
duplex telegraphy, 254, 258
Durand, Peter. *See* Pierre Duran
Durand, Pierre, 103, 107
Durant, William C., 336
Duryea, Charles, 333
Duryea horseless carriage, 342
Duryea Motor Wagon Company, 333
Dyer, Frank L., 287

E. A. Callahan stock ticker, 254, 256
Eagle, 364
Early Bird, 566
ear phonoautograph, 191
Eastman, George, 285
Eckert, J. Presper, 547
Edison, Charles, 252
Edison, Nancy, 250
Edison, Thomas Alva, 344, 516; appliance manufacturing business, 273–74; birth, 250; camera mechanism, 285; capacities of first-class operator, 253; childhood, 251; children, 252, 262, 279, 294; death, 295; Detroit Free Library, 251; developed an automatic vote recorder, 254; developed plastic composition for disks, 292; developments, 272–81; devised duplex system, 254; E. A. Callahan stock ticker, 254; "Edison effect" in 1883, 264; Electric Light Company, 269; electric train, 282–83; employment, 251; established workshops, 256; "etheric force" in 1875, 264; family, 250; family business, 292; fluoroscopic materials, 287; formal education, 250; General Electric Company, 279–81; hand-printing press, 252; hearing loss, 199; and Henry Ford, 367–68; high-resistance lamp, 270; improvements in Bell's telephone, 198–200; improvements in phonograph, 290–92; innovator, 250; inventions, 254, 257, 264–72; Laws indicator

Edison, Thomas Alva (*continued*)
system, 255; life, 281; light extinguished, 294–95; manufacturing companies, 277; marriage with Mary Stillwell, 259; marriage with Mina Miller, 278–79; mechanical and electrical apparatuses, 259; Menlo Park, 260–61; millionaire, 277; motion pictures, 284–87; multiplex telegraphy, 257, 258; Newark years, 256; operations, 277–78; organization, 277; partnership with Joseph T. Murray, 257; peep-show mechanism, 286; phenomenal inventor, 250; phonautograph, 263; phonograph, 261–64; Portland cement, 288; position regarding alternating current, 266; procedure for handling money, 256; producing concentrated iron ore, 283–84; public promotion, 269; research on "speaking telegraph" technology, 261; rubber, 294; small mechanical doll, 262; sound recording, 263; Speaking Phonograph Company, 262; stock printers, 257; storage battery, 288–90; supported preparedness movement, 293; supported Woodrow Wilson's presidential candidacy, 293; Tasimeter, 264; telegrapher, 252, 253; telephone technology, 261; Tesla, Nikola, 303; travelling abroad for remunerative work, 253; Universal Stock Printer, 257; West Orange Laboratory, 281–82; witnessed Black Friday, 255; Wizard of Menlo Park, 263; work on electric lights and power, 250, 264; World War I and manufacturing chemicals, 292–94
Edison & Swan United Electric Company, 275
Edison Company for Isolated lighting, 276
"Edison effect", 264
Edison Illuminating Company (EIC), 339, 344
Edison Machine Works, 274, 278
EDVAC, 547
Ehrlich, Paul, 424
Einstein, Albert, 495
electrical distributing system, 265
electrically ignited two-stroke engine, 228
"electric candles", of Jablochkoff, 264
Electric Illuminating Company, 274
Electric Light and Manufacturing, 303
electric light and power: Edison General Electric Company, 279–81; European endeavors, 275–76; financing, 268–69; growth and development, of Electric Industry, 276–78; install an independent lighting system, 274–75; manufacturing, 273–74; new industry invention, 272–73; origin and inducement, 264–65; preliminary investigation and efforts, 265–68; research and development, 269–71; success, 271–72; U.S. development, 276
Electric Light Company, 269, 272, 274, 277, 280, 309
electric motor-driven car, 229
electric motors, 332
electric train, 282–83
Electric Vehicle Company (EVC), 356, 357
electrified pen, 257
electromagnetic separation, 504–5
electromagnetic telephone, 211
electron, 485
element, 485
e-mail, 591–92
embedded computer, 570
Emerson & Fisher Company, 341
Energy Reorganization Act of 1974, 512
Engelbart, Douglas, 591
engineering plastics, 479
ENIAC (Electronic Numerical Integrator and Calculator), 547
Ericsson, 416
Erie Canal, 92–97
Erie Railroad Ring, 255

"Erskine Park", 296
"etheric force", 264
Evans, Oliver, 88–92, 221; Columbian high-pressure steam engine, 89, 90–91; early years, 89; employment, 89–90; family, 89; flour mill machines, 90; and George Washington, 90; litigation problems, 91–92; marriage, 90; Mars Works, 90, 91–92; *Oliver Evans* (renamed *Constitution*), 91; Orukter Amphibolus (Amphibious Digger), 91; patents, 90; Philadelphia Board of Health project, 91; wool cards machine, 90; *Young Mill Wright and Miller's Guide,* 90
Experiment, 86

fabric manufacture, in England, mechanization of, 48–52
Faggin, Federico, 542
Fairchild, 522, 524
Faraday, Michael, 516
Farmer, Moses, 265
Farnsworth, Philo, 434; achievement, 436–38; birth, 435; childhood, 435, 436; death, 444; early years, 435; employment with the Philco company, 439; and George Everson, 436–37; image dissector, 436, 437; interference, 439–41; and Leslie Gorell, 436; lost his young son Kenny, 441; marriage with Elma Gardner, 437, 444; nervous breakdown, 443; parents, 435; patents, 437, 439–40; and RCA, 438–40; Television and Radio Corporation, 443; and Vladimir Zworykin, 438–39; working with University of Pennsylvania, 441
Fawcett, E. W., 466
fermentation, 319
Fermi, Enrico, 483, 491; chain reaction on uranium, demonstration of, 495–97; Chicago Pile Number, 1 (CP-1), 501–3; early biography, 492–93; involvement with physicists, 493–95
Fessenden, Reginald Aubrey, 313, 393–94, 516; academic education, 394–95; alternating current (AC) generator, 398; amplitude modulation (AM) of radio waves, invention of, 393, 396–99; attitude, 396; barretter detector, 396; birth, 394; conflict with Lee de Forest, 398–99; contract with General Electric (GE), 399; contract with the United States Weather Bureau, 396; death, 401; development of, 398–401; dispute with his backers, 400; early years, 394–95; electrolytic detector, 396; employment with Thomas Edison, 395; Fessenden Wireless Company of Canada, 400; Fessenden Wireless Telegraph Company of Canada, 398; first radio broadcast of AM radio, 399; and Hay Walker, 397; heterodyne circuitry, 396; heterodyne principle to receiving modulated radio transmissions, 394; marriage with Helen Trott, 395; National Electric Signaling Company (NESCO), 397, 398; parents, 394; patents, 401; as professor, 395–96; rotary-gap (rotary-spark) transmitter, 397–98; submarine signaling and detection, 401; and Thomas H. Given, 397
field effect transistor (FET), 521
Field, Stagg, 501
finger tapping, 259
Firestone, Harvey, 294
Fischer, Emil, 469
Fish, Frederick P., 309
Fisher, George, 177
Fisk, James, 255, 494
fission, 486
fissionable isotope separation, 504
Fitch, John, 55, 57, 73; controversy with James Rumsey, 58; death, 59–60, 72; first steamboat design, 58; fourth steamboat design, 59; grants and backers, 59; second secondboat design, 58; third steamboat design, 58–59
flacherie, 321
Flasche, Martha, 234

Fleming, Alexander, 422; and Sir Almroth Wright, 423; article, to medical journal, 424, 426; bacterial antibiotic resistance, 430; birth, 423; in business, 423; as captain, 424; death, 429; discovery of, 426; education, 423; experiments of, 425; family, 423; and Howard Florey, 427; laboratory, culture plates in, 429; lysozyme, 425; marriage with Amalia Koutsouri-Vourekas, 424; marriage with Sarah M. McElory, 424; Nobel Prize, in medicine, 429; and Paul Ehrlich, 424; penicillin, 422, 425; penicillin, developments of, 431; penicillin, discovery and evaluation of, 425–26; private, 606, 424; as professor, 425

Fleming, John Ambrose, 516

Florey, Howard, 422, 426; and Ernst Chain, 427; Nobel Prize, in medicine, 429

fluorescent lighting, 313

fluoroscopic materials, 287

Flyer II, 384

Flyer III, 384

flying shuttle, 49

Fokker, Anthony, 389

Ford, Edsel, 362, 366, 368

Ford, Edward, 27

Ford, Henry, 291, 292, 293, 294, 295, 331, 332, 335; aircraft, 364–65; boyhood, 337; business failures, 344–45, 346; demise and achievements, 368; Detroit Automobile Company, 344–45; early years, 336; endeavors, 364–65, 366–67; first engine, 340–41; Greenfield Village, 365; hospital development, 365; marriage with Clara, 338–39; Model A cars, 348–52, 363, 366; Model B cars, 352, 366; Model C cars, 352; Model F cars, 352; Model K cars, 352; Model N cars, 352–53, 354; Model R cars, 353; Model S cars, 353; Model T cars, 354–56, 359, 361, 362–63; nature and character, 367–68; Otto-type engines, 332; quadricycle, 341–43; racing, 345–47; ships, 364; and Thomas Alva Edison, 367–68; tractors, 364; training, 337–38; V-8 Engine, 366; village industries and education, 365; and Warren G. Harding, 368

Ford, William, 337, 338

Ford & Malcomson Company, 348; growth of, 352–54

Ford Manufacturing Company, 353–54

Ford Motor Company, 353, 354; employee benefits, 360; mass production, 359–60; Seldon patent infringement, 356–59

Fordson tractors, 365

Fortescue, Charles Legeyt, 313

FORTRAN, 577

forty-two-line Bible project, 9–10

four-cycle gasoline engine, 220, 226–28

Fowler, Nancy (Nettie), 123–24, 135

Franco-Prussian war, 323

Franz, Anselm, 457–58

freezing, 106–7

frequency modulation invention, 414–18

Frisch, Otto, 497

Fry, Vernon C., 350

FTP, 602–3

Fulton, Robert, 57, 73, 74; "The Burning of Moscow", 65; canal boats, 64; *Clermont*, development of, 68–69; contract with the Britain, 67–68; contract with the French, 67; *Demologos,* 72; *Fulton the First,* 72; and John Stevens, 69, 70; and mission to destroy British warships, 65, 67; in Paris, 63–64; patents, 64; and Robert Livingston, 63–64, 65–68; steam-driven test boat, 66; submarine, 65, 67; "Treatise on the Improvement of Canal Navigation", 64

Fulton the First, 72

fusion, 486
Fust, Johann, 8–10, 19; partnership with Peter Schoeffer, 10

gadget, the, 503
gamma rays, 487
Gammon, Frank, 286
Gardiner Hubbard, 191–92, 197
Garfield, Ellery I., 344
Garrett AiResearch, 541
gaseous diffusion, 505
gas-fueled engine, 333
gas meter, 302
Gasmotoren-Fabrik Deutz AG, 226, 228
Gates, Bill, 570; awards and recognition, 584; Bill and Melinda Gates Foundation, 583; birth, 571; and Ed Roberts, 574–75; family, 571; key deal maker, 577; Lakeside Programmers Group (LPG), 572; and Marc McDonald, 576; marriage with Melinda French, 583; and Paul Allen, 572–74; ROM, 576; schooling, 571–72; and Steve Ballmer, 578; Traf-O-Data, 573
Gaulard, Lacien, 305
Gaulard-Gibbs "converters", 305
Gay-Lussac, Joseph, 373
Geison, Gerald L., 328
Geller, Steve, 541
Gell-Mann, Murray, 511
Gelthuss, Arnolt, 8
General Electric (GE), 279, 280, 281, 283, 293, 313, 520
General Electric Laboratory, 293
General Motors Company, 336, 362
Geneva mechanism, 285
germ theory, of disease, 321
germanium-based transistors, 520, 521
Gernsback, Hugo, 519
Gibbs, John D., 305
Gibney, Robert, 517, 518
Gibson, R. O., 466
Gilliland, Ezra, 279, 291
Giovanni Borelli, 373
Given, Thomas, 397, 400
Gladden, C. A., 416
"Glen Eyre", 296
Glenmont, 281
Globe Telephone Company, 208
Goddard, Robert Hutchings, 563–64
Gold & Stock Telegraph subsidiary, 256
"Golden Jubilee of Light", 295
goldenrod, 294
Gold Indicator Company, 254
"gold printer", 255
Gompers, Samuel, 311
Gopher, search engine, 603
Gould, Jay, 255, 258
Gould, Walter, 348
Grangers, 136
Grant, Thomas, 27
Graves, George, 474
Gray, Elisha, 193, 202, 261; caveat, 202–4; fraud allegations, 203; patent application, 203
Gray, John S., 350
Great Depression, 365
Great Eastern, 62
Great Train Robbery, The, 286
Greenfield Village, 295, 365
Gregory, Eugene, 362
Groff, Jean-Francois, 606
Grosvenor, Edwin S., 214
Groves, Leslie, 500–501, 503–4, 505; electromagnetic separation process, 504–5; gaseous diffusion separation process, 505
GTE, 567
guayule rubbers, 294
Guggenheim, Robert, 355
gun powder, 6–7
Gutenberg, Johannes, 2, 3, 20; and Archbishop Adolf von Nassau, 11; and Arnolt Gelthuss, 8; birth, 5; contemporaries, 18–21; death, 11; education, 5; equipment, 14–17; family, 5; fled to Eltville, 11; forty-line Bible, 10; forty-two-line Bible project, 9–10; as a goldsmith in Strasbourg militia, 5; ink, 16–17; invention, 13–21; and Johann Fust, 8–9, 10; lawsuit brought Georg and Klaus Dritzehn, 6; press, 17; printing, 17; printing plates,

Gutenberg, Johannes (*continued*) 22–23; sand castings, 22; type marker, 14–16; works, consequences of, 11–13; works, recent findings about, 21–23
G. Westinghouse & Company, 295

Haeckel, Ernst, 320
Hahn, Harley, 602
Hahn, Max, 456, 457
Hahn, Otto, 491, 492, 493
half-life, 487
Hall, Calder, 511
Hall, John, 103
Hammer, William, 275
hand sewing, 172
Harding, Warren G., 368
Hare, Ronald, 429
Hargreaves, James, 49–50
harmonic telegraph, 192
Hartford Rubber Works Company, 350, 351
Hata, Sahachiro, 424
Havilland, De, 389
Hawkins, Norval A., 354
Hayes, Elwood, 333
Hayes, Rutherford B., 263
Haynes-Apperson Company, 333
Heart, Frank, 590
Heatley, Norman, 422, 426, 428; penicillin, growth of production, 428; penicillin, testing on human, 428
Heaviside, Oliver, 166, 313
Heil, Oskar, 516
Heinkel, Ernst, 456
Heisenberg, Werner 491
helicoptere, 375
helium, 494
Henderson, James, 173
Henry, Joseph, 151, 152–53, 194, 304
Henry, William, 56
Henry Ford and Son Laboratories, 365
Henry Ford Company, 346
Henry II, 368
Hensen, William, 373
Hero (Greek engineer), 27, 81
Hertz, Heinrich Rudolph, 159, 516
Hideki Yukawa, 491
Hill, Julian, 470, 474
Hill, Murray, 517
Hinton, Christopher, 511
History of Modern Computing, A, 542
Hodgkin, Dorothy, 427
Hoemi, Jean, 533
Hoff, Jacobus Van't, 318
Hoff, Ted, 541, 542
Hollerith, Herman, 546
Holliday, Charles O., 480
Holonyak, Nick, 517, 525
Holt, Benjamin, 142; Caterpillar Company, 142
Holt, Ray, 541
Homfray, Samuel, 78, 79
Honda Motor Company, 228
Hooke, Robert, 463
Hoover, Herbert, 295
Hopper, Grace, 546
horizontal water wheel, 26
Hornblower, Jonathan, 48
Hornsby-Stuart Patent Oil Engine, 240
hot bulb engines, 240
Howe, Elias, 172; and Ari Davis, 175–76; birth, 176; childhood, 176; death, 179; as employee, 177; family, 176; and George Fisher, 177; and George W. Bliss, 178; and Isaac Merritt Singer, 178, 175–76; marriage with Elizabeth J. Ames, 176; in model shop; needles, 176–77; own company, 179; patents, 178; poverty, 177–78; and Quincy Hall Clothing Manufacturing Company, 177; royalty, 179; sewing machine, 176–79; and William Thomas, 177
Howe Machine Company, 178
Hoyt, Benjamin R., 344
HTTP, 607–8, 609
Hubbard, Greene, 190, 195, 196
Hubbard, Mabel, 190, 195
Huber, 354
Huff, Edward S. "Spider", 342, 344, 347
Huffman Prairie, 384

Hughes Aircraft, 566, 567
Hulls, Jonathan, 54
Humery, Conrad, 10
Hunt, Walter, 172; inventions, 174; in model shop, 174; in New York City, 174; patents, 175; sewing machine, 174–75
Huron Canal basin, 250
Hussey, Obed, 121
Hutchins, Robert Maynard, 501
Huygens, Christiaan, 220
Hyatt, John Wesley, 463
hydroelectric plant, 275
hypertext, 590–91

IBM, 520; AT model, 581; contract with Microsoft, 578; Personal Computer, 579
ignorameter, 293
Illuminating Company, 280
Imperial Chemical Industries, 479
IMSAI, 549
I. M. Singer & Co., 180, 184
incandescent lighting, 267
Indianapolis Chain Company, 342
India rubber, 294
induction motor, 303, 306, 313
Industrial Development Engineering Associates (IDEA), 520
Industrial Revolution, 26; roots of fabric mechanization in, 48–52
Information Processing Technology Office (IPTO), 587
Information Services Inc. (ISI), 573
integrated circuit (IC)528; background, 529–31; fabrication of, 534–36; invention of, 532–33; potential competition, 533–34; summary, 529; usage and growth, 536–38
integrated circuits, 522
Intel, 541–42, 548, 549
internal engine development, 220–23
International Ad Hoc Committee, 610
International Business Machines (IBM), 546
International Telecommunications Satellite Consortium (INTELSAT), 558, 565, 566
International Telephone and Telegraph Company (ITT), 443
Internet, 586; Advanced Research Projects Agency (ARPA), 587–88; American Standard Code for Information Interchange (ASCII code), 603; browsers, 603; commercialization, 602; domain names, 600–601; e-mail, 591–92; FTP, 602–3; Gopher, search engine, 603; growth of, 601–3, 609–10; history of, 587; interior gateway protocol (IGP), 601; IP, 595; open-architecture network, 592–93; packet switching, 588–91; requests for comments (RFCs), 596, 599, 603; search engines, 603; servers, 608–9; TCP creation, 593–96; TCP/IP, 586, 595, 597–600
Internet Assigned Numbers Authority, 610
Internet Corporation for Assigned Names and Numbers (ICANN), 600–601, 610
Internet Engineering Steering Group, 599
Internet Engineering Task Force (IETF), 610
Internet Protocol (IP), 595, 598
Internet Society (ISOC), 600
iPod media player, 554
iron ore, producing, 283–84
isolated lighting systems, 274
isotope, 487

Jackson, C. T., 151, 322
Jacquard, Joseph, 52, 545
Jacques Charles, 373
James Flower & Brothers' Machine Shop, 338
JANET protocol, 592
Jeannette, 274
Jehl, Francis, 275
Jenner, Edward, 324, 422
Jennings, Dennis, 601

Jervis, John B., 94
Jet engines, 448; and Frank Whittle. *See* Whittle, Frank; and Hans von Ohain. *See* von Ohain, Hans; invention of, 450–51; structure of, 449–50
Jet Propulsion Laboratory (JPL), 561
Jet wing, 459
Jobs, Steve, 549, 555, 579; early years, 549–50; development of, 551–52; small computer, building and marketing, 550–51
John Bull, 94
Johnson, Edward H., 257, 273, 275, 277, 278, 279
Johnson, Pat, 452
Joliot, Frederic, 491
Joy, Henry, 356
J. P. Morgan & Co., 216, 269, 274, 279, 281
juice, 427
Julius, Hadrianus, 19
"Jumbo" dynamo, 275, 276
junction transistor, 518

Kahn, Robert, 591, 592, 593, 594; Corporation for National Research Initiatives, 600
Kai Kaus, King, 372
Kapor, Mitch, 580
Karuesi, John, 262
Kay, John, 49
Kelley, John, 313
Kelly, Mervin, 517
Kennedy, Jeremiah J., 287
Kennelly, Arthur E., 166
Kensett, Thomas, 107
Kerner, Justinius, 105
Kettlesell, Dick, 348
Kilby, Jack, 522, 528, 529; awards and personal life, 538–39; early years, 531–32; integrated circuit, 528–38; Micromodule program, 531
Kildall, Gary, 577
Kill Devil Hills, 378
Kineto-phonograph, 285
Kinetoscope, 286
King, Charles B., 335; and quadricycle, 341–42
Kistiakowsky, George, 508
Kitty Hawk, 380, 381
Kleinrock, Leonard, 588, 590
Knight, Jonathan, 94
Knowles, John, 173
Knudsen, William, 366
Koch, Heinrich, 324
Kodak company, 285
Kompfner, Rudolf, 560
Korting, Berthold, 229
Korting, Ernst, 229
Krems, Balthasar, 173
Kreusi, John, 257, 260
Kullman, Ellen J., 480
Kunz, Paul, 606

Lady Morey, 57
Lakeside Programmers Group (LPG), 572
Lambert, John William, 333
lamp globes, exhausting air from, 271
Lancet, The, 424–428
Langen, Eugen, 225; and Gottlieb Daimler, 226; Otto, Nikolaus and Ludwig August Roosen-Runge, 226; and Wilhelm Maybach, 226
Langley, Samuel Pierpont, 374, 376
LANs, 599
latent heat, 33
Lawrence, Earnest, 491, 498, 499, 504
Lawrence Berkeley National Laboratory, 263
"Laws" gold price indicator system, 254
Lebon, Philippe, 221
Lee, Reverend William, 477
Lee, Yuan T., 511
Lefferts, Marshall, 256
Leibnitz, Gottfried, 545
Leland, Henry M., 346, 362
Lenoir, Jean Joseph Etienne, 222; engine, 224
Leonard, E. A., 344
Leonard, Mike, 286
Licklider, J. C. R., 587
Liebig, von Justus, 319

Life of an American Fireman, The, 286
lift balance equipment, 381
Lilienfeld, Julius Edgar, 516, 517
Lilienthal, Otto, 374, 376
Lincoln Motor Company, 362
Lindbergh, Charles: education, 387; family, 387; grant to Robert Goddard, 564
Lindemann, Frederick, 499
Lippincott, Jesse, 291
liquid thermal diffusion, 507
Lister, Joseph, 323
Little, Arthur D., 464
Little, George D., 257
Little Juliana, 63
Livingston, Robert, 57; and John Stevens, 60–61, 69–70; and Robert Fulton, 63–64, 65–68
Livingston/Fulton transportation monopoly, 71–72
Locomotion, 86
locomotives, American type, 95
Lodge, Sir Oliver, 160
London Crystal Palace Exposition, 275
London Steam Carriage, 78
Long-Waisted Mary Ann, 271
Lotus, 1–2–3 system, 580
Love, Harry, 348
low-cost personal computer, creation of, 548–49
Lowrey, Grosvenor P., 264, 265, 268, 280, 309
Lunar Society, 46
lysozyme, 425

Macintosh Computer, 579
Macintosh Portable, 553
Mackenzie, J. U., 252
Mac OS X system, 554
Madersperger, Josef, 173
Magnetic Telegraph Company, 157
Mainz Psalter, 10
Malcomson, Alexander T., 347, 352–54; Aerocar Company, 353; Model A, 348–52; Model B, 352; Model C, 352; Model F, 352; Model K, 352; Model N, 352–53, 354; Model R, 353; Model S, 353
Mallory, Walter, 284, 288, 289
Manhattan Project, 484, 489, 490, 491, 500–501, 512
Mannheim Gas Engine Company, 230
Manzetti, Innocenzo, 21
Marconi, Guglielmo, 159, 160, 313, 392, 516; advances in radio, 169; awards and prizes, 169–70; birth, 161; boyhood, 162; continuous-wave transmitter, 165; daylight effect, 166; developments and subsequent life, 169–70; family, 162; and Henry Jameson-Davis, 164; interest in science, 161–62; Italian Military Medal, 169; long-distance wireless telegraphy, achievement of, 163–64; Marconi International Marine Communication Company, Ltd., 165; Marconi's Wireless Telegraph Company Ltd, 164; marriage with Beatrice O'Brien, 169; marriage with Countess Bezzi-Scali, 170; Nobel Prize, 167; patents, 161, 165–66; and William Preece, 163–64; wireless communications, limitations of, 168; wireless telegraphy, development of, 164–68; yacht races, 165
Marcus, Siegfried, 223, 224
Maria, Black, 286
Mark, 1, 546
Markula, Mike, 552
Marsh, Charles W., 139
Marsh, William W., 139
Marshall, George C., 500
Marsh harvester, 139–40
Marvel, 354
Mary Ann dynamo, 282
mass defect, 486
Massie, William, 118
Matteucci, Felice, 221
Mauchly, John, 547
Maxim, Hiram P., 302, 333
Maxim, Hiram Stevens, 374
Maxwell, James Clerk, 313, 516

Maybach, Wilhelm, 226, 228, 230
Maybury, William C., 344
Mazor, Stanley, 542
McCormick, Cyrus, 114, 334; agreement with his brothers, 125–26, 134, 142–43; agreement with Orloff M. Dorman, 120; awards and prizes, 130; binders, 140–42; birth, 114; *Chicago Times,* 131; C. H. and L. J. McCormick, 136; company development after Civil War, 134–36, 137–38; company operations during Civil War, 131–33; defeat to "Long John" Wentworth, 134; early commercialization efforts, 117–19; early years, 114–15; effects of Panic of 1837, 118; and European marketing, 130; family, 114, 118, 119, 136–38, 139; farm machinery, developments in, 138–42; Gammon & Deering (competitor), 140; and Gorham, Marquis L., 141; and Grangers, 136; harvesters, 139–40; impact of Civil War, 130–31; invention of the reaper, 115–16; as inventor, 126–27; investment in real estate, 123; involving his brothers in operations, 120; knowledge on prior reaper attempts, 117–18; Leander (brother), 125; McCormick and Gray, 120; McCormick, Ogden, and Co., 120; McCormick's Patent Virginia Reaper, 119; as manager, 127–29; manufacture in Chicago, 119–21; manufacturing system, 121–22; market research, 118–19; marriage with Nancy (Nettie) Fowler, 123–24, 135; Marsh harvester patent, 140; money raising efforts, 117–18; mowers, 138; newspaper sold to Wilbur F. Storey, 131; non-company endeavors after the Civil War, 133–34; Obed Hussey (competitor), 121; participation in the Great Exhibition, 130; partnership with Gray and Warner, 120; patents, 119; payment to Gordon patent infringement claim, 141; personality, 144–45; positioning Cyrus Jr., 142, 145; program for the Presbyterian Church, 134; reaper, improvements, 117, 118; reaper, structure and operation, 116; rebuilding factory damaged due to Chicago Fire, 120–21, 135–36; recovery from yellow fever, 114; self-raking reapers, 138–39; seminary, 134; slave-holding, 131; success of, 122–23; traveling thresher, 122; Walter A. Wood & Company (competitor), 140; W. E. Jones, 120; and William (brother), 124–25; and William B. Ogden, 120; and William Massie, 118; Withington machines patent, 140–41
McCormick, Leander, 125, 134–35, 142–43
McCormick, Robert Hall, 142–43
McCormick, Williams, 132, 133
McCurdy, J. A. D., 386
mechanical power, 26–29
Meister, Joseph, 325, 326
Meitner, Lisa, 491, 492
Menlo Park, 260–61, 264
Merrill, Thomas, 589
metal containers, 107
Metcalfe, Robert, 599
Meucci, Antonio, 210–13, 214, 215; and Angelo Bertolino, 212; and Bill Carroll, 212; birth, 210; education, 210–11; electromagnetic telephone, 211; funding, 211; marriage with Ester Mochi, 211; patents, 213; sound telegraph, 212; "telettrofono", 211; in United States, 211; and William E. Ryder, 211
Meucci's telephone, 207–8
Micro Instrumentation and Telemetry Systems (MITS), 574, 575–76; taken over by Pertec, 576
Micromodule program, 531
microprocessors, 528; background, 540; history of, 541; in, 1980s, 543–44; summary, 539
Microsoft, 575; agreement with Macintosh, 581; antitrust suit against,

583–84; BASIC codes, 577; and Compaq, 579, 582; competition from Borland, 580; competition from Lotus, 580; contract with Commodore International, 576; contract with IBM, 578, 582; deal with Radio Shack, 576; FORTRAN, 577; management in early years, 577; Microsoft Windows operating system, 580; Microsoft Windows operating system, commercialization of, 581; MS-DOS, 578; per-machine deal, 581; ROM BASIC, 579–80; Seattle Computer lawsuit, 581; Steve Jobs's visit to, 579; WIN32 582; Windows NT3.1 583; Windows95 583; Word, 580
Mills, John R. Keim, 358
Mitchell, Charles, 348
Mitscherlich, Eilhardt, 317, 318
Mockapetris, Paul, 600
Model A cars, 348–52, 363, 366
Model B cars, 352, 366
Model C cars, 352
Model F cars, 352
model glider, 378
Model K cars, 352
Model N cars, 352–53, 354
Model T cars, 354–56, 359, 361, 362–63
moderator, 488
Montogolfier brothers, 372
Moore, Gordon, 522, 540, 541, 542
Moore, Hilbert, 517
Moore's Law, 540
Morey, Samuel, 57, 222
Morse, Samuel F. B., 148, 253; awards, 149; birth, 148; career, 149–50; childhood, 149; and C. T. Jackson, 151; and Daguerre, 155; death, 158; *The Dying of Hercules,* 149; family, 148; and Francis O. J. Smith, 154; *Gallery of the Louvre,* 151; implementation of telegraph, 156–58; interest in electricity and magnetism, 151; introduction to Joseph Henry, 152–53; invention, 155–56; and John Quincy Adams, 151; *The Judgment of Jupiter,* 149; *The Landing of the Pilgrims,* 149; *The Last Supper,* 150; and Leonard Gale, 152; Magnetic Telegraph Company, 157; marriage with Lucretia Walker, 149–50; marriage with Sarah Griswold, 158; as member of the Native American party, 151; *Mona Lisa,* 150; National Academy of the Arts of Design, 150; paintings, 149–50; patents, 158; as professor of literature of the arts and design, 152; as professor of sculpture and painting, 151–52; and Steven Vail, 154; system of telegraphs for the United States proposal, 154
Morse code, 191, 259, 393
Morton, Williams, 322
MOS, 6502 microprocessor, 549
MOS Technology, 543, 548, 549
Motion Picture Patents Corporation, 287
motion pictures, 284–87
Motorola, 543
Motorola MC6800 microprocessor, 543, 549
Motorola PowerPC, 7400 microprocessor, 554
mowers, 138
MRSA (Methicillin resistant Staphylococcus aureus), 431
multiplex telegraphy, 257, 258
Murdock, William, 47–48, 220
Murphy, William H., 344, 345, 346
Murray, Joseph T., 257, 259
Murray, Matthew, 73
musical telegraph, 192
Muybridge, Eadweard, 284

Napier, John, 545
Napoleon, 63–64
National Aeronautics and Space Administration (NASA), 600; Project Echo, 561–62; Science Internet (NSI), 600, 609
National Defense Research Council (NDRC), 496, 498

National Electric Signaling Company (NESCO), 397, 398, 400; for military use, 477
National Physical Laboratory (NPL), 589
National Science Foundation (NSF), 599–600, 601, 609
natural gas, 301
Nautilus, 65
Naval Research Laboratory, 293
Nelson, Ted, 591
neptunium, 497
network, 588
Network Control Protocol (NCP), 591, 593
Network Solutions, 609
Network Working Group (NWG), 593
neutrons, 485; thermal, 492
Nevin, Robert M., 385
Newark shops, 257
New Castle, 78
Newcomen, Thomas, 28
Newcomen steam engine, 28–29
The New England Telephone Company, 205
New Philadelphia, 72
Newton, Sir Isaac, 250, 545
New York Electrical Exposition, 287
NeXT Computer, 553
NeXT cube, 606
Niagara Falls, 310
Nicol prism, 317
Nicol, William, 317
Nieuwland, Julius, 469
Nipkow, Paul Gottlieb, 434
Nokia, 417
North American Phonograph Company, 291
Northumbrian, 87
Noyce, Bob, 541
Noyce, Robert, 522; awards and personal life, 539; planar IC, 533, 534
NSFNet, 599–600, 601
NTT, 416
nuclear physicists and U.S. government involvement, contact attempt between, 495
nuclear power, 512
nuclear reaction, 486, 494
nuclear reactor, 488. *See also* Chicago Pile Number, 1 (CP-1); aftermath, 510–12
Nuclear Regulatory Commission (NRC), 512
nucleus, 485
nylon, 462, 472–74; aftermath, 478–80; development of, 475–78; hosiery, 477, 478; for military use, 477, 478; problems, 475–77; stockings, 477–78; toothbrush bristles, 478

Official Gazette, 202
Ogden, William B., 120; *Sully* voyage, 150–51
Ohm, Georg Simon, 153; Ohm's law, 267; vehicle development, 230–31
Oldfield, Barney, 346–47
Olds, Ransom E.: gasoline-powered car, 335; single-cylinder engine, 335
Olds Company, 335
Oldsmobile Company, 336
Oliphant, Mark, 499
Oliver Evans (renamed *Constitution*), 91
Opdyke, George, 173
Open Systems Interconnection Basic Reference Model (OSI Model), 598
open-architecture network, 592
Oppenheimer, Robert, 503–4, 509, 511
Oregon Railway & Navigation Company, 274
Orson C. Phelps, 180
Orton, William, 258
OS/2 582
Osborne, Loyall A., 305
O'Sullivan, William J., 561
Ott, John, 256, 260
Otto, Nikolaus, 220; birth, 223; carburetor invention, 224–25; and C. Wigand, 229; early years, 223; employment, 223; and Eugen Langen, 225; family, 223; four-cycle internal combustion engine, 226–28; and Kortings, 229; loss of patent

protection, 229–30; and Ludwig August Roosen-Runge, 226; marriage with Anna Gossi, 223–24; Otto engine, 231–32
Otto-Langen atmospheric engine, 225–26
Otto-type gasoline engines, 332, 338
Oughtred, William, 545
overshot water wheel, 26
Oxford laboratory, 428

Packard, 354
packet switching, 588–91; router, 590
Panhard & Levassor, 334
Pantaleoni, Guido, 305, 306
Papin, Denis, 27, 54
parallel circuits, 267, 270
parallel motion device, 41
Parelman, M. A., 416
Paris Electrical Exposition, 275
Park Place Shop, 348
Parkes, Alexander, 463
Parsons, Charles, 311
Pascal, Blaise, 545
Pascaline, 545
Pasteur, Louis, 102, 103, 422; in Academy of Science, 321; achievements, 319; to Ales, 321; anthrax, 325; birth, 316; broths, 320; chicken cholera, 325; death, 326; discovery, 317; doctorate, 316; École Normale Supérieure, 316, 321; family, 317; first investigation, 317; in French Academy of Medicine, 323; germ theory, of disease, 321; graduation, 316; health, 323–326; and Heinrich Koch, 324; institut Pasteur, 326; and Jacobus Van't Hoff, 318; and Jean Baptiste Dumas, 316, 321; and Joseph Le Bel, 318; and Joseph Meister, 325, 327; Legion of Honor, 319; in Lille, 319; marriage with Laurent, Marie, 317; Nicol prism, 317; pasteurization, 320; private papers, 328; rabies, 325; Rumford Medal, 318; silkworm parasite, 321; spontaneous generation, 320; tartaric acid, 317, 318; tartrates, 317, 318; treatment, for dog bite, 326; at University of Lille, 319; at University of Strasbourg, 318; vaccination, 324; work, 318
Pasteur effect, 319
Pasteur's Deception, 328
pasteurization, 320, 328–29
PDP-10 575
Pearl Street system, 277
Pearson, Gerald, 517
pebrine, 321
Peddle, Chuck, 543
peep-show mechanism, 286
Pegram, Dean, 498
Pellow, Nicola, 606
Penicillium mold, 429
Penicillium notatum, 428
Perier, J. C., 54
Perry, Stuart, 222
personal computer, 570
Pertec, 576
petroleum-fueled engine, 333
Pfennig, E., 351
Pfister, Albrecht, 10
Philadelphia Company, 302
Phoenix, 70–71, 94
Phoenix Software Associates, 581
phonautograph, 263
phonograph, 261–64, 285; improvements in, 290–92
Pierce, John, 518, 558, 559, 566; at Bell Labs, 559–60, 561, 567; birth, 559; death, 568; early years, 559–60; education, 559; family, 559; IRE lecture, 560–61; at Jet Propulsion Laboratory, 568; J. J. Coupling (pseudonym), 560, 567; John Roberts (pseudonym), 567; later works, 567–68; "The Philosophy of PCM", 567; Pierce gun, 567; prizes and awards, 568; professor of music at CCRMA, 568; Project Echo, 561–62; science fiction story writer, 560, 567; *The Science of Musical Sound,* 568; Telstar, 562–63
Pierce-Arrow, 354

Pilcher, Percy, 376
Pi Sheng, 2
planar IC, 533, 534
plastic, history of, 462–67
platinum burners, 269
pleomorphic theory, 328
Plunkett, Roy, 466
Pohl, Robert Wichard, 456
point contact transistor, 520
poisonous lead solder, for sealing cans, 104, 107
Pollermann, Bernd, 606
polyester, commercialization, 479–80
polyolefins, 480
Polyphase motors, 305, 314
polyvinyl chloride, 466
polyvinylidene chloride, 466
Pope, A. A., 333
Pope, Franklin L., 254, 255, 306, 310
Pope-Hartford, 333
Pope-Robinson, 333
Pope-Toledo, 333
Pope-Tribute, 333
Popov, Alexander Stepanovich, 159–61
Pope-Waverly, 333
port, in TCP and UDP, 596
Porta, Giambattista della, 27
Portland cement, 288
positron, 487
Postel, Jon, 610
post mills, 26
Potter, Humphrey, 28
printing, 2–4; body, 14; equipment type, 14–16; ink, 16–17; ligature, 16; matrix, 14; patrices, 14; press, 17
private, 606, 424
Private Science of Louis Pasteur, The, 328
Project Echo, 561–62
protocol suite, 587, 588
protons, 485
Prudden Company, 350
Pseudomonas aeruginosa, 431
Pu-239, 506–7, 508, 509
Pu-240, 508, 509
Puffing Devil, 77
Pullman, George, 95, 97, 311
pumping water, using steam, 27–28
Pupin, M., 287
putting out system, 49

quadricycle, 341–43
quadruplex system, 258

rabies, 325
racing, 345–47
Rackham, Horace H., 350
radio, background, 392–93
radioactivity, 487
Radio Corporation of America (RCA), 400, 436, 438, 520; Philo Farnsworth, 438–40
Raff, Norman, 286
railroad air brake innovation, 297, 298–99
railroad signaling and switching, 297, 299–300
Ramelli, Agostino, 27
RAND project, 558
Raskin, Jef, 579
rayon, 464
Raytheon, 520
Read, Nathan, 56–57, 63
ready-made clothing, 173
Red Flag Act, 222
Regency TR-1 520
Reiff, Josiah, 258
Reis, Johann Phillip, 192
Reis transmitter, 192
Remington Arms, 334
remote temperature-sensing gadget, 264
Renaissance period, 26
Renault, Louis, 334
requests for comments (RFCs), TCP, 596, 599, 603
Rhodes, Richard, *The Making of the Atomic Bomb*, 508
Rickover, Hyman, 511
Righi, Augusto, 160
Rivault, Florence, 27
road carriages, 221–22
Roberts, Lawrence 589, 591–92; with Thomas Merrill, 589

Roberts, Ed, 574–75
Roberts, Richard, 52
Robison, John, 32
Rocket, 87
rockets, 563–65
Roebuck, John, 34–35; bankruptcy, 35–36
Roentgen, W. W., 287, 490
ROM BIOS, 581
Roosen-Runge, Ludwig August, 226
Roosevelt, Franklin Delano (FDR), 489, 495, 499, 500
Roosevelt, Theodore, 385, 386
rotary engine, 295
Round, H. J., 409; V24 vacuum tube, 409
router, 589–90
rubber, 294
Ruddimen, Edsel, 365
Rumsey, James, 57; and Robert Fulton, 57
Ruston & Hornsby Ltd., 240
Rutherford, Ernest, 490
Ryder, William E., 211

Samuel Courtalds and Co., Ltd., 465
Sanders, George, 190, 204
Sanders, Thomas, 191–92
sandwich transistor, 518
Sarnoff, David, 412, 438
satellite communications, 558; background, 558–59; developments, 565–67
Savery, Thomas, 27
Scantlebury, Roger, 589
Schickard, Wilhelm, 545
Schenectady laboratory, 293
Schlack, Paul, 467
Schmidt, Alfred, 305
Schoeffer, Peter, 8
Schutzenberger, Paul, 464
Schweizer, Eduard, 464
Science, 201
Science and Invention, 435
Science International Corporation (SAIC), 609
Scientific Revolution, 48–49
Scott, Charles F., 305
Sculley, John, 553
Seaborg, Glenn T., 498, 508
search engines, 603
Seattle Computer, 578
Seeman, Fred W., 348
Segre, Emilio, 497, 498
Selden, George, 356–59
self-binder, 140
self-propelled vehicles, 222, 339
self-raking reapers, 138–39
semiconductors, 516–17
Semmelweiss, Ignaz, 323
Semon, Waldo, 466
Sendall, Mike, 605
SERENDIP (Search for Extraterrestrial Radio Emissions from Nearby Developed Intelligent Populations), 510
servers, 608–9
SETI (Search for Extraterrestrial Intelligence) Institute, 510
sewing machine, 172–85
Sewing Machine Cartel, 178
Sewing Machine Combination, 178
Shallenberger, 308
Shallenger, Oliver B., 305
Shanck, H. K., 338
Shapiro, Aaron, 368
shells, 485
Shepardson Microsystems, 551
Shepherd, Mark, 533
Shima, Masatoshi, 542
shipbuilding, 56
ships, 364
Shockley, William, 402, 499, 515, 516, 517, 518, 519, 523–24; at Bell Labs, 523; with Beckman Instruments, 523; bipolar junction transistor, 519–20; birth, 523; childhood, 523; death, 524; during war period, 523; *Electrons and Holes in Semiconductors,* 518; family, 523; marriage with Emmy Lanning, 524; marriage with Jean, 523–24; sandwich transistor, 518; *Electrons and Holes in Semiconductors*, 518; Nobel Prize, 522

Sibley, Hiram, 158
Siemens & Halske, 280
Signal Communication by Orbiting Relay Equipment (Score), 562
signal system, for customer protection, 257
silicon-based transistors, 521
silkworm parasite, 321
Silver Lake, 289, 293
Singer, Isaac Merritt, 172, 334; advertisements, 181–82; birth, 182; in Boston, 182; childhood, 182; customer payments, 181; death, 184; demonstrations, 181; and Elias Howe, 178, 179, 181; in England, 184; in Europe, 184; factory in New York, 180; factory in Scotland, 182; and George Zieber, 180, 183; I. M. Singer & Co., 180, 184; machine sales, 181; marriages, 183–84; and Orson C. Phelps, 180; patents, 179, 180; Singer Sewing Machine Company, 185
Sisson, Jonathan, 32
Slaby, Adolph, 161
Small, William, 35
small computers: background, 544; history of, 545–48; summary, 544
Smeaton, John, 26, 29, 103
Smeaton coefficient, 381
Smith, F. L., 335
Smith, Francis O. J., 154
Smith, Frederick, 356
Smith, Willoughby, 434
sodium ammonium tartrate crystals, 317
software, background, 570–71
Soho factory, 35, 36, 38, 39, 42
Sony, 520
Sorensen, Charles, 366
sound telegraph, 212
Space Surveillance Network (SSN), 559
Spallanzani, Lazzaro, 102, 320
Spandex, 478
Speaking Phonograph Company, 262, 291
speaking telegraph, 193, 261
spinning jenny, 49–50
spinning mule, 50, 52
Spirit of St. Louis, The, 387
Sprague, Frank J., 279
Sprague Electric Company, 280
Sprague Electric Railway and Motor Company, 279
Sprengel pump, 271
S.S. *Columbia,* 274
standard American car, 95
Stanley, Morgan, 269
Stanley, Wendell, 325
Stanley, William Jr, 295, 312–13; alternating current (AC) electrical system, 305; birth, 300; early years, 300, 302; family, 300; schooling, 300, 302
Stanley Manufacturing Company, 313
stationary steam engine system, 78
Staudinger, Hermann, 469
steam, utilization of, 27–28
steamboat, 54–74; American patent laws, 58; commercialization, 74; development, 72–73; early American innovators, 56–57. *See also individual inventors*; early efforts, 54–55; in United States, 55–56
steam carriage, 333
steam-driven road carriage, 42
steam engines, 332; American steam engine (*see* American steam engine); development, 46–48 (*see also* Watt's steam engine)
steam locomotive, 74–97; in Britain, 93; in United States, 93, 94
steam turbine, 311–12, 337
Stearns's duplex system, 258
Steinmetz, Charles P., 313, 399
Stephenson, George, 74, 85; at the Killingworth Colliery, 85; *Blucher,* 85; death, 87; early years, 85–86; *Experiment,* 86; family, 85; improvement of Trevithick's device, 86; Liverpool and Manchester Railway project, 86; *Locomotion,* 86; miners' safety lamp, 85–86; Robert Stephenson and Company, establishment of, 86; Stockton and Darlington Railway project, 86

Stephenson, Robert, 83, 86–87; death, 87; railway projects, 87; *Rocket,* 87; *John Bull,* 94
stepped reckoner, 545
Stevens, John, 57, 73, 74, 92, 93, 94; break with Loyalist friends, 62; Camden and Amboy railroad, 94; ferryboats, 62; first federal patent statute draft, 60; hiring Marc Isambard Brunel, 62; *Little Juliana,* 63; marriage with Rachel Cox, 62; multitubular boiler, 63; patents, 63; *Phoenix,* 70–71; and Robert Fulton, 69, 70; and Robert Livingston, 60–61, 69, 70; and steamboat development, 61–63, 93; twin-screw stern propeller system, 63
Stevens, Robert, 71, 72, 92; cowcatcher invention, 96; as president of the Camden and Amboy Rail-Road and Transportation Company, 96; T-rail, 96
Stevenson, Alexander, 252
Stillwell, Lewis B., 305
Stine, Charles, 467, 471, 477
Stone, John, 161
Stone, Thomas, 173
stopper lamp, 309
storage battery, 288–90
Stout Metal Airplane Company, 364
Stout, Rick, 602
Strassmann, Fritz, 492, 493
Strelow, Albert, 349, 351, 353
Stringfellow, John, 373
Studebaker, 290
superbugs, 431
Swan, Joseph, 275, 464
Swan Electric Company, 302
SWEET, 16, 552
Switch Company, 305
swivelable trucks, 94–95
Symington, William, 54–55
Syncom, 563, 566
syphilis, 424
Szilard, Leo, 494, 495; chain reaction on uranium, demonstration of, 495–97
Tabulating Machine Company, 546
Taft, Howard, 385
Tainter, 290
talking machine, 262, 263
talking telegraph, 206–7
Tandy Corporation, 552
tartaric acid, 317, 318
Tasimeter, 264
Tate, Dan, 298
Taylor, Charlie, 382
Taylor, Frederick W., 360
Teflon, 466
Telegrapher, The, 255
telegraphy, prior history, 148, 159–61
telephone, 188, 261; commercialization, 196–200, 214, 216; contemporary research, 210; Edison's improvements in Bell's telephone, 198–200; invention, by Alexander Graham Bell, 193–95; telegraph message traffic, 191
"telettrofono", 211
television: history of, 434; and Philo Farnsworth. *See* Farnsworth, Philo
Teller, Edward, 503
TELNET, 592, 593
Telstar, 562–63
TENEX, 591
Terrell, Paul, 551
Terry, Charles A., 309
Tesla, 399
Tesla, Nikola, 160–61, 266, 295, 313–14; alternating current (AC) electrical system, 306, 307, 313; birth, 302; brushless motor, 303; childhood, 303; coil, 303, 313; death, 314; early years, 302–3; Edison, 303; education, 302; Electric Light and Manufacturing, 303; family, 302; fluorescent lighting, 313; induction motor, 303, 306, 313; inventions in radio, 313; polyphase system, 306, 314; three-phase electricity, 313; x-rays, 303
Texas Instruments (TI), 521, 522, 529, 533, 543
thermal reactor, 488
Thimonnier, Barthelemy, 173

Thomson, George, 498
Thomson, J. J., 490
Thomas, William, 177
Thomas A. Edison, Inc., 286, 292, 293
Thomlinson, John, 291
Thomson-Houston Company, 280, 281
Thor Delta vehicle, 566
thorium (Th-90), 493
thread, 172–73
three-phase electricity, 313
Tihanyi, Kalman, 435
Tillotson, Roy, 562
tin can, 103–4
Tomlinson, Ray, 591
Tom Thumb, 94
tower mills, 26
tractors, 364
T-rail, 96
transistor, 515, 516; bipolar junction, 519, 521; creation of, 517–20; development of, 520–22; field effect, 521; history of, 516–17; impact of, 526; point contact, 520; public demonstration of, 519; sandwich, 518
transmission control protocol (TCP), 594–96, 608; creation of, 593–95; requests for comments (RFCs), 596
transmission control protocol/Internet protocol (TCP/IP), 586, 595, 599, 601, 608; development, 597–600; four-layer version, 597; User Datagram Protocol (UDP), 596
transmutation, 486
trans-uranium, 492. *See also* uranium
Trevithick, Richard, 74, 75–76, 221; arrival to Cartagena, 83; bankruptcy, 81–82; *Catch-Me-Who-Can,* 80; closed-cycle steam engine, 84; column designing, 84; condenser-free engine, 81; Cornish engine, 80, 81; cylindrical boiler, 80, 81; death, 84; during War of Emancipation, 83; early years, 76–77; employed in J. & E. Hall Limited, 84; and Francisco Uville, 82; heater, 84; *London Steam Carriage,* 78; marriage with Jane Harvey, 76; *New Castle,* 78; patents, 78, 80; in Peru, 82–84; *Puffing Devil,* 77; return to England, 83–84; and Robert Dickinson, 80; and Robert Stephenson, 83; service in Simon Bolivar's army, 82; stationary steam engine system, 78; steam engine based on Hero device, 81; steam engine locomotives, 78–79, 80; steam engines, 77, 80–81; and Thames Archway Company, 79; tram-waggon, 78; vertical tubular boiler, 84; and William Murdock, 76
triode, 401–2, 403
TRS-80, 552
Tsai Lun, 3
Turbo Pascal, 580
Turbofan jet engine, 450
Turboprop engine, 450
Twombly, Hamilton M., 268
Twort, Frederick W., 422
two-stroke (two-cycle) diesel locomotives, 239
two-stroke engine, 228
Tyndall, John, 320
typography, 3

U-234, 493
U-235, 493, 494, 496, 497–500, 504–5, 507
U-238, 493, 494, 497, 504, 505
UCLA student group, 593
Union Signal, 305
United States, 389; Circuit Court Room, 309; Electric and Consolidated Electric, 280; Electric Company, 283, 302; of high-grade iron ore, 283; patent laws, 58; sewing machine, 185; shipbuilding, 56; steamboats in, 55–56; Western Electric, 261
Universal Stock Printer, 257
Upton, Francis R., 269, 273, 277
uranium, 484. *See also* trans-uranium; chain reaction on, 495–97; fission, 493; isotopes, 493, 494
uranyl nitrate hexahydrate (UNH), 498
URL, 607, 608, 609

U.S. Army involvement, in nuclear development, 500–501
U.S. Army Signal Corps, 531
U.S. Bell Telephone Company, 261
User Datagram Protocol (UDP), 596
U.S. House of Representatives, 2002 Resolution, 213–14, 215
utility gas transport system, 297, 301–2
Uville, Francisco, 82

V-8 Engine, 366
Vail, Steven, 154
van Dam, Andries, 591
Vanderbilt, W. H., 268
Velox, 465
variable electric current distribution, 270–71
Varley, Cromwell, 210
Vazie, Robert, 79
Verbiest, Ferdinand, 78
Verlinden, Ed, 345
VIC-20, 552
Victor Talking Machine, 292
Villard, Henry, 274, 279, 282
viscose, 434
Visible Speech, 189
VisiCorp, 580
Vitascope, 286
von Braun, Wernher, 564–65
von Helmholtz, Hermann, 190, 192–93
von Justinius, Emile, 105
von Linde, Carl, 233
von Nassau, Adolph, 11
von Neumann, John, 508
von Ohain, Hans, 451; in Air Force Aerospace Research Laboratories, 459; birth, 455; death, 460; development of, 457–58; early years, 455–56; education, 455–56; and Ernst Heinkel, 456; invention of jet engine, 456–57; jet wing, 459; and Max Hahn, 456; patent, 456; retirement, 459; and Robert Wichard Pohl, 456; with University of Dayton Research Institute, 459; in U.S. government service, 459; after the war, 458–60

Walcott, Charles D., 388
Waldvogel, 6, 7
Walker, Hay, 397, 400
Walker, Lucretia, 149–50
walking beam, 28; steam-driven road carriage, 42
Wallace, Henry A., 499
Wallace, William A., 265, 280
Wallace-Farmer, 8-horsepower dynamo, 270
Wandersee, John, 346, 348
Wang Chen, 2
Warsitz, Erich, 457
Washington, George, 90
water power, 26
waterwheels, 26
Watson, Thomas A., 188, 194
Watt, James, 26, 77, 220–21; after retirement, 45–46; as an instrument maker's apprentice, 31; in Birmingham, 35, 36; birth, 30; as a canal surveyor, 35; childhood, 30; condenser-equipped steam engine, 34, 35; customer agreement, 38–39; death, 46; document copier, 43; double-acting engine, 40–42; family, 30; Galton, Mary Anne, 31; honors, 46; improving Newcomen engine steam, 33–35; indicator device, 43; inventions in 1778 and 1780, 40; James Watt & Co., 43; and John Wilkinson, 37, 38, 40; and Joseph Black, 31–32, 34; Marion Campbell on, 30; marriage with Ann MacGregor, 39; marriage with Margaret Miller, 33; and Matthew Boulton, 35, 36–40; mental health, 30, 238; parallel motion device, 41; partnership with Dr. John Roebuck, 34–35; patents, 36, 39, 41, 42, 44; precision tools for machining, 39; retirement, 44; revolution counter, 43; steam engines, 29; steam engines, impacts of, 43–44; steam engines,

Watt, James (*continued*)
with the speed governor, 42; steam pressure gauge, 43; sun and planet gearing, 39; techniques for reducing the amount of fuel, 34; tilt hammer, 43; unit of power, 43; in University of Glasgow, 31–32; and William Murdock, 47–48
wax paper, 257
Wayne, Ronald, 551
Web browser, 608
Web pages, 600
Web sites, 603
Weekly Herald, 252
Weiland, Ric, 576
WESTAR, 567
Western Electric, 197, 258, 261, 517, 519, 520; lawsuit, 204–6
Western Union, 158, 255, 258, 261, 567; Callahan stock printer, 256
Westinghouse, George, 97, 295, 516; alternating current (AC) electrical system, 295, 304; career, 297; child, 296; death, 312; early years, 295–96; Electric & Manufacturing Company, 280, 303, 306; Electric Company, 307, 308; Electric Corporation, 306; parents, 295–97; G. Westinghouse & Company, 295; innovations, 297; marriage with Marguerite Erskine Walker, 296; New York National Guard, 295; railroad air brake innovation, 297, 298–99; railroad signaling and switching, 297, 299–300; reversible cast steel frog, 296; rotary engine, 295; steam turbine, 311–12; U.S.S. *Muscoota,* 296; utility gas transport system, 297, 301–2; Wilmerding, 311
Westinghouse Engine Company, 338
West Orange laboratory, 281–82, 285
West Point Foundry, New York, 94
Whinfield, J. R., 479
White, Abraham, 403
White, Albert E. F., 346
White, Samuel S., 203
Whitfield, John, 79
Whitney, Eli, 334
Whittle, Frank, 450; birth, 451; with British Overseas Airways Corporation (BOAC), 455; with British Thomson-Huston Company (BTH), 452, 453; childhood, 451; death, 460; development of, 454–55; early years, 451–52; education, 451–52; invention of jet engine, 452–54; *Jet: The Story of a Pioneer,* 455; motorjet, 452; negative impact on, 453; parents, 451; and Pat Johnson, 452; patent, 452; Power Jets, 454–55; and Rolf Dudley-Williams, 452, 453, 455; with Shell Oil, 455; after the war, 455
Wigginton, Randy, 551
Wiley, Ralph, 466
Wilkinson, John, 36–38, 40
Williams, Charles Jr., 254
Williston Academy, 302
Wills, C. Harold, 346, 347–48
Wilmerding, 311
windmills, 26–27
wing-wraping system, 377
Winton, Alexander, 333, 345–46
Winton Motor Carriage Company, 333, 356, 357
wire self-binder, 140
Withington, Charles B., 140
Wizard of Menlo Park, 263
Wolff, Steve, 601
Woolf, Arthur, 48
World Columbian Exposition, 308
World War I, 292, 361–62, 424
World War II (WWII), 546
World Wide Web (WWW), 586, 604; composition of, 607–9; origin of, 607
World Wide Web Consortium (W3C), 611
Wozniak, Steve, 528, 544; Apple II, 554–56; childhood, 548; low-cost personal computer, creation of, 548–49
Wright, W.L., 222
Wright, Lorin, 384
Wright Brothers, 374; achievement, 374; aeronautical experimentation, 377;

aftermath, 384; balances, 379; canard, 378; childhood, 375; commercialization, 386; company, 386; design change, 386; efforts of, 378; entering history of, 388; experiments and developments, 376; family, 375; flight demonstration, in France, 386; Flyer II, 384; Flyer III, 384; and Glenn Curtiss, 387; glider, 378, 380; glider, problems in, 382; involvement, 376; Kill Devil Hills, 378, 383; Kitty Hawk, 380, 381; learning, from bird, 377; lifestyles, 376; lift balance equipment, 381; 1900 glider, 378; 1901 glider, 378; 1902 glider, 380, 381; and Octave Chanute, 376, 378, 379, 381; Orville's death, 388; powered, controlled flights, 382; practical airplane, developing, 384; propellers, 382; sale, of airplanes, 385; Smeaton coefficient, 381; speech, in Chicago, 379; Success of, 385, 388; Wilbur's death, 386; wind tunnel experiments, 380; wing testing, 380; wing-wrapping system, testing, 377, 378; Wright company, 386; *Wright Flyer I*, 382, 383; Wright, Lorin, 384

Wright Flyer I, 382, 383

X-25 protocol, 592

x-rays, 287, 303, 490

xylography, 2, 3

Yang, Chen Ning, 511

Zieber, George, 180, 183

Zilog, 542

Zilog Z80, 542

Zuse, Konrad, 547

Zworykin, Vladimir, 438–39; to Philo Farnsworth's laboratory, 438; and RCA, 440

ABOUT THE AUTHOR

JOHN W. KLOOSTER is a retired attorney-at-law who specialized in intellectual property law.